Lecture Notes in Computer Science   **10497**

Commenced Publication in 1973
Founding and Former Series Editors:
Gerhard Goos, Juris Hartmanis, and Jan van Leeuwen

More information about this series at http://www.springer.com/series/7408

Philipp Reinecke · Antinisca Di Marco (Eds.)

Computer Performance Engineering

14th European Workshop, EPEW 2017
Berlin, Germany, September 7–8, 2017
Proceedings

 Springer

Editors
Philipp Reinecke
Bristol
UK

Antinisca Di Marco
University of L'Aquila
L'Aquila
Italy

ISSN 0302-9743   ISSN 1611-3349   (electronic)
Lecture Notes in Computer Science
ISBN 978-3-319-66582-5   ISBN 978-3-319-66583-2   (eBook)
DOI 10.1007/978-3-319-66583-2

Library of Congress Control Number: 2017949517

LNCS Sublibrary: SL2 – Programming and Software Engineering

Printed on acid-free paper

This Springer imprint is published by Springer Nature
The registered company is Springer International Publishing AG
The registered company address is: Gewerbestrasse 11, 6330 Cham, Switzerland

Preface

This volume of LNCS contains the proceedings of the 14th European Performance Engineering Workshop, held in Berlin, Germany, September 7–8, 2017. EPEW was part of the week-long umbrella conference QONFEST, which co-located QEST, CONCUR, FORMATS, and EPEW, along with several workshops. This gave researchers the opportunity to explore and engage with a broad range of topics and colleagues across the space of performance, dependability, and security modelling, verification, evaluation, and engineering. We wish to express our gratitude for the support QONFEST received from the Freie Universität Berlin, the Technische Universität Berlin, the Ernst-Reuter-Gesellschaft, the DFG, and the Max-Planck-Gesellschaft.

The goal of the annual EPEW workshop series is to gather academic and industrial researchers working on all aspects of performance engineering. The papers presented at the workshop reflect the diversity of modern performance engineering, with topics ranging from the analysis of hybrid Petri nets and Markov decision processes, even under uncertainty; to performance, security and energy analysis of computer systems and networks; to machine-learning techniques for predictive analysis and testing. The domains of the application studies are diverse and at the cutting edge of current developments, ranging from cloud computing environments to cyber-physical systems and to communication protocols.

EPEW 2017 received submissions from 14 countries all over the world. There were 30 submissions. Each paper was peer reviewed by an average of four reviewers from the Program Committee (PC) on the basis of its relevance, novelty, and technical quality. After the collection of reviews, the PC members discussed the quality of the submissions for one week before getting the final decision. Based on the reviews and discussions, 18 high-quality contributions were selected for publication in the proceedings and presentation at the workshop.

This year, we were honored to have two keynote speakers: Prof. William Knottenbelt, from Imperial College London (UK), who works in applied quantitative analysis; and Antonino Sabetta, a senior researcher at the Security Research department of SAP Research (Sophie Antipolis, France), who works in the analysis and management of vulnerabilities of open-source components when embedded in large-scale enterprise applications.

We thank our keynote speakers, as well as all PC members and external reviewers for their terrific work in the review process. We also express our thanks to the Organizing Committee, especially to the two General Chairs, Uwe Nestmann (TU Berlin) and Katinka Wolter (FU Berlin) for their continuous and valuable help, the EasyChair team for their conference system, and Springer for their continued editorial support.

Above all, we would like to thank the authors of the papers for their contribution to this volume. We are sure that these contributions will be as useful and inspiring to the readers as they were to us.

September 2017   Philipp Reinecke
                                                                  Antinisca Di Marco

Organization

EPEW Program Chairs

Antinisca Di Marco   University of L'Aquila, Italy
Philipp Reinecke   Bristol, UK

QONFEST General Chairs

Katinka Wolter   Freie Universität Berlin, Germany
Uwe Nestmann   Technische Universität Berlin, Germany

EPEW Program Committee

Davide Arcelli	Università de L'Aquila, Italy
Rena Bakhshi	Netherlands eScience Center, The Netherlands
Simonetta Balsamo	Università Ca' Foscari di Venezia, Italy
Marta Beltran	Universidad Rey Juan Carlos, Spain
Marco Bernardo	University of Urbino, Italy
Ana Busic	Inria and ENS, France
Laura Carnevali	University of Florence, Italy
Giuliano Casale	Imperial College London, UK
Dieter Fiems	Ghent University, Belgium
Jean-Michel Fourneau	Université de Versailles St Quentin, France
Stephen Gilmore	University of Edinburgh, UK
Boudewijn Haverkort	University of Twente, The Netherlands
András Horváth	University of Turin, Italy
Gábor Horváth	Budapest University of Technology and Economics, Hungary
Alain Jean-Marie	CNRS University of Montpellier, France
William Knottenbelt	Imperial College London, UK
Samuel Kounev	University of Würzburg, Germany
Vasilis Koutras	University of the Aegean, Greece
Lasse Leskelä	Aalto University, Finland
Catalina M. Lladó	Universitat Illes Balears, Spain
Andrea Marin	University Ca' Foscari Venice, Italy
Raffaela Mirandola	Politecnico di Milano, Italy
Marco Paolieri	University of Southern California, USA
Roberto Pietrantuono	University of Naples Federico II, Italy
Agapios Platis	University of the Aegean, Greece
Anne Remke	WWU Münster, Germany
Markus Siegle	Universität der Bundeswehr, München, Germany

Miklos Telek	Budapest University of Technology and Economics, Hungary
Nigel Thomas	Newcastle University, UK
Catia Trubiani	Gran Sasso Science Institute, Italy
Petr Tuma	Charles University, Czech Republic
Aad Van Moorsel	Newcastle University, UK
Maaike Verloop	CNRS Toulouse, France
Joris Walraevens	Ghent University, Belgium
Qiushi Wang	Nanyang Technological University, China
Huaming Wu	Tianjin University, China
Armin Zimmermann	Technische Universität Ilmenau, Germany

EPEW Additional Reviewers

Alnafessah, Ahmad	Iffländer, Lukas
Baltas, Ioannis	Meszaros, Andras
Herbst, Nikolas	Pekergin N., Nihal
Horvath, Illes	Pilch, Carina
Hüls, Jannik	von Kistowski, Jóakim

Abstracts of Invited Talks

Cryptocurrency and Blockchain Technology: Challenges and Opportunities

William J. Knottenbelt

Imperial College Centre for Cryptocurrency Research and Engineering,
Imperial College London, London, UK
wjk@imperial.ac.uk

The meteoric rise of blockchain-enabled cryptocurrencies, and Bitcoin [2] and Ethereum [1] in particular, has received global attention, not least from governments, entrepreneurs and researchers. Cryptocurrencies, of which there are now more than 800[1], provide an attractive alternative to traditional fiat currencies via a distributed, trustless and self-governing framework which not only enables low-friction financial transactions around the globe but also preserves the freedom and privacy of spending inherent in cash transactions.

Cryptocurrency and blockchain technology brings with it a host of new challenges from the quantitative modelling perspective. Indeed, a range of issues including performance, security, energy use, incentives and scalability are poorly understood, as are the inherent trade offs between them, despite these being critical barriers to mass adoption. What analyses are carried out often do not take into account problems posed by the lack of diversity that emerges from a natural tendency towards dominant concentrations of computational and other power. These can arise from something as simple as the majority of network participants flocking to deploy the most energy-efficient cryptocurrency mining hardware. Indeed it is estimated that up to 70% of the computational power assuring the integrity of the Bitcoin network is provided by a single model of a hardware device. This device was recently found to have a backdoor that could be used by the manufacturer to shut the device down[2].

This talk will cover some of the challenges and opportunities posed in this context, with a special emphasis on the performance evaluation and quantitative modelling perspectives. It turns out that classical performance evaluation techniques, especially Markovian analysis and queueing theory, are readily applicable to the study of cryptocurrencies and blockchains. Further, a judicious combination of analytical modelling, simulation and benchmarking techniques can be effectively applied to yield insights. Building on [3], we will illustrate this in the context of a study of a queue-based Ethereum mining pool [4] whose superficially fair reward scheme turns out not only to penalise more powerful miners, but also to incentivise a number of attacks which can

W.J. Knottenbelt—The content of the talk is the result of joint work with A. Zamyatin, K. Wolter, C. Mulligan, P. Harrison, S. Werner and I. Stewart, amongst others.

[1] Source: http://coinmarketcap.com. Accessed 5 July 2017.

[2] Source: http://antbleed.com. Accessed 5 July 2017.

increase rewards, including the donation of mining power to other participants in certain circumstances. Examples of such attacks observed in the real world will be presented.

The talk will conclude by outlining student-led spinout activity and ongoing directions of research in the Imperial College Centre for Cryptocurrency Research and Engineering. The former includes Gradbase[3], a qualification verification startup, Aventus[4], a blockchain-based ticketing company and Kotiva Technologies[5], who are seeking to use blockchain technology to increase the integrity of supply chains. The latter includes work being supported by industrial partners such as Blockchain.com and Outlier Ventures, as well as grants sponsored by government-related bodies such as Innovate UK.

Biography

William Knottenbelt is Professor of Applied Quantitative Analysis and Director of Industrial Liaison in the Department of Computing at Imperial College London, where he became a Lecturer in 2000. He is a founder of the Imperial Blockchain Forum, is co-Director of the Centre for Cryptocurrency Research and Engineering and is Director of the Data Economy Lab in Imperial's Data Science Institute. He serves on the editorial board of the cryptocurrency/blockchain journal Ledger, is an editor of Performance Evaluation Journal, and has served as general or program chair of numerous conferences and workshops related to quantitative modelling and analysis. A keen supporter of student-led innovation, he is the Innovation Fellow for the Department of Computing and serves on the Entrepreneur First Science Partners panel. In June 2017, he presented his Inaugural Lecture entitled "Memoirs of the Memoryless: A Markovian Meander from Disk Drives to Digital Money", which is available online[6].

References

1. Vitalik B.: Ethereum: A next-generation smart contract and decentralized application platform (2014). https://github.com/ethereum/wiki/wiki/White-Paper.Accessed 18 June 2017
2. Satoshi, N.: Bitcoin: A peer-to-peer electronic cash system (2008). https://bitcoin.org/bitcoin. pdf. Accessed 18 June 2017
3. Meni, R.: Analysis of bitcoin pooled mining reward systems. arXiv preprint arXiv:1112.4980 (2011)
4. Zamyatin, A., Wolter, K., Werner, S., Mulligan, C.E.A., Harrison, P.G., Knottenbelt, W.J.: Swimming with fishes and sharks: beneath the surface of queue-based Ethereum mining pools. In: Proceedings of the 25th IEEE International Symposium on Modeling, Analysis and Simulation of Computer and Telecommunications Systems (MASCOTS 2017), September 2017

---

[3] See http://gradba.se.

[4] See http://aventus.io.

[5] See http://kotiva.tech.

[6] See https://youtu.be/TTQOwyXXKHw.

Open-Source Libraries Included in Enterprise Applications: Workhorses or Trojan Horses?

Antonino Sabetta

SAP Labs, France
antonino.sabetta@sap.com

The adoption of open-source software (OSS) components in the software industry has grown at a spectacular pace over the last decade. By some estimates [3], the average commercial software product contains 100 distinct open source components whose code weights as much as 35% of the overall application size[1].

At the same time, new vulnerabilities affecting open-source software (OSS) are reported on a daily basis, sometimes hitting the headlines of mainstream media (as it happened, for example, with Heartbleed[2] and ShellShock[3]).

The relevance of this problem has been well documented by now [1, 2] and establishing effective vulnerability management practices for OSS is broadly understood as a priority in the software industry.

Despite the deceiving simplicity of the existing solutions (especially of the most obvious: *updating to a recent, non-vulnerable version*), OSS libraries with known vulnerabilities are found to be used for quite some time after a fixed version has been released [3].

As a matter of fact, updating a library to a more recent release is quite straightforward at *development* time. However, things become considerably more difficult when vulnerable OSS libraries are part of large enterprise systems that are already in *operation* and serve business-critical functions. Any change (including corrections) may cause costly system downtime and comes with the risk that new unforeseen issues could arise.

For this reason, it is extremely important to properly assess whether an application requires an urgent patch to update an OSS dependency, or whether the update could be scheduled for the next regular release cycle. Just the presence of a vulnerable dependency is not enough to justify a urgent update, with its high costs and even higher risks. The real question is whether a given vulnerability is indeed *exploitable* given the particular way the dependency is used.

Unfortunately, assessing the exploitability and the potential impact of a vulnerability found in a dependency is difficult, expensive, and error-prone. Vulnerabilities are

---

A. Sabetta—The content of the talk is the result of joint work with Serena E. Ponta and Henrik Plate, SAP Labs France.

[1] The same study reports that for applications developed for internal use, the proportion is as high as 75%.

[2] http://heartbleed.com/.

[3] https://shellshocker.net/.

documented in advisories that consist of short, high-level, textual descriptions expressed in natural language, whereas a reliable assessment demands much lower-level, detailed, technical information.

The consequences of a wrong assessment can be expensive. If an exploitable vulnerability is not identified as such, users remain exposed to attackers. When, on the contrary, a correction is produced for a non-exploitable vulnerability, the effort of developing, testing, and deploying the correction is spent in vain.

This talk summarizes the key elements of our research on how to make the assessment of OSS vulnerabilities more efficient and systematic [4]. Our approach aims to automatically produce concrete evidence (when it can be found) supporting the case for urgent patching. Such evidence consists of concrete call sequences (traces) that start from application methods and reach the vulnerable methods of a dependency. We complement *potential* traces obtained through static analysis with *actual* observations of runtime executions collected through dynamic instrumentation. Our approach relies on the availability of detailed (code-level) vulnerability information, which we extract by mining software repositories with the support of machine learning. The initial research prototype that we implemented to validate our approach evolved over time into an enterprise-grade OSS vulnerability analysis toolkit (internally known as *Vulas*), which is used regularly in hundreds of development (and maintenance) projects across our company.

Biography

Antonino Sabetta is a senior researcher at the Security Research department of SAP. The main focus of Antonino's recent work is the analysis and management of vulnerabilities of open-source components embedded in large-scale enterprise applications. In particular, Antonino is interested in the application of machine-learning to the mining of open-source software repositories and the automation of the vulnerability management workflow.

Before moving to SAP in 2010, Antonino was a researcher at CNR, Pisa, Italy. He earned his PhD in Computer Science and Automation Engineering from the University of Rome Tor Vergata, Italy in 2007. From the same university he had received in 2003 his "Laurea *cum Laude*" degree in Computer Engineering.

References

1. Arce, I., et al.: Avoiding the top-10 software security design flaws. Technical report, IEEE Center for Secure Design. IEEE Computer Society (2014)
2. OWASP Foundation. OWASP Top 10 – 2013 (2013). https://www.owasp.org/index.php/Top_10_2013-Top_10
3. Pittenger, M.: Open source security analysis: The state of open source security in commercial applications. Technical report, Black Duck Software (2016)
4. Plate, H., Ponta, S.E., Sabetta, A.: Impact assessment for vulnerabilities in open-source software libraries. In: Proceedings of the IEEE International Conference on Software Maintenance and Evolution (ICSME) (2015)

Contents

Performance, Energy and Security

Case Studies

Advances in Markov Models

Analysis of Markov Decision Processes Under Parameter Uncertainty

Peter Buchholz, Iryna Dohndorf[(✉)], and Dimitri Scheftelowitsch

Department of Computer Science, TU Dortmund, Dortmund, Germany
{peter.buchholz,iryna.dohndorf,
dimitri.scheftelowitsch}@cs.tu-dortmund.de

Abstract. Markov Decision Processes (MDPs) are a popular decision model for stochastic systems. Introducing uncertainty in the transition probability distribution by giving upper and lower bounds for the transition probabilities yields the model of Bounded Parameter MDPs (BMDPs) which captures many practical situations with limited knowledge about a system or its environment. In this paper the class of BMDPs is extended to Bounded Parameter Semi Markov Decision Processes (BSMDPs). The main focus of the paper is on the introduction and numerical comparison of different algorithms to compute optimal policies for BMDPs and BSMDPs; specifically, we introduce and compare variants of value and policy iteration.

The paper delivers an empirical comparison between different numerical algorithms for BMDPs and BSMDPs, with an emphasis on the required solution time.

Keywords: (Bounded Parameter) (Semi-)Markov Decision Process · Discounted reward · Average reward · Value iteration · Policy iteration

1 Introduction

Markov Decision Processes (MDPs) are a commonly used stochastic model in various areas like operations research, control theory, model checking or artificial intelligence [15,18]. Often the parameters of a Markovian model are not exactly known. Reasons might be that parameters result from measurements where each parameter is a point estimate whereas a confidence interval would be a much better choice, or the states are chosen in a way that the memoryless property is only approximately fulfilled and the future behavior depends slightly on the past behavior. In these cases, the parameter values of the resulting MDP are best described by an *uncertainty set* to which the parameter values belong. MDPs with parameter uncertainty have been defined in different variants in the past [10,16,20]. Most prominent became recently Bounded Parameter MDPs (BMDPs) [10] where parameters are defined by intervals rather than single values.

Computation of optimal policies for MDPs can be done with value iteration, policy iteration or linear programming [15]. Different variants of these algorithms

© Springer International Publishing AG 2017
P. Reinecke and A. Di Marco (Eds.): EPEW 2017, LNCS 10497, pp. 3–18, 2017.
DOI: 10.1007/978-3-319-66583-2_1

have been investigated in the past. For BMDPs the situation is different, the basic paper introducing BMDPs [10] proposes a value iteration algorithm for the discounted reward. Only very few additional papers on numerical approaches for BMDPs are available. This is surprising since the computation of optimal policies for BMDPs is much harder than the computation for MDPs and the approach is only useful in practice if optimal policies can be computed for BMDPs of a reasonable dimension. The main focus of this paper is on numerical techniques for computing optimal policies and corresponding reward vectors for BMDPs. We present and compare algorithms based on value iteration and on policy iteration. Furthermore we extend BMDPs to Bounded Parameter Semi-Markov Decision Process (BSMDPs) and extend the algorithms to this class of models.

The paper is structured as follows. Section 2 reviews related work. Then, in Sect. 3 MDPs, Semi-MDPs, BMDPs and BSMDPs are defined and it is shown that it is sufficient to consider the discrete time case. Section 4 introduces numerical algorithms for BMDPs and BSMDPs. The algorithms are evaluated experimentally, results are presented in Sect. 5. The papers ends with the conclusions.

2   Related Work

An enormous number of papers about MDPs and BMDPs exists. We review approaches that are related to the numerical computation of optimal policies for BMDPs which is the topic of the current paper. The basic results can be found in [10]. In the papers [7,9] discounted rewards are computed for a class of MDPs with imprecise transition rates. For the analysis, mathematical programming approaches are considered, which result in the problem to solve a bilevel or multilinear program. Due to the high computational effort only small instances can be solved. In [7] a specific algorithm for factored MDPs, based on approximated multilinear programming is presented. Factored MDPs with imprecise transition probabilities are also considered in [8]. There, a value iteration approach to compute the discounted reward is presented which exploits the factored structure using a BDD based implementation. A further extention of MDPs with imprecise parameters are *parameteric MDPs* where transition probabilities are given as functions over a set of parameters. The computational effort for solving these problems is high such that more sophisticated solution techniques are required in general [5,6].

The average reward case for BMDPs is handled in [19]. Based on the basic approach for BMDPs in [10], a value iteration approach is presented which consecutively increases a weight value that weights the reward accumulated in the next step in relation to the reward accumulated in the current step. In our experiments this algorithms shows a bad convergence behavior and numerical instabilities. In none of the mentioned papers, larger sets of experiments are performed for BMDPs to evaluate the algorithms empirically. Furthermore, the extension to semi-Markov processes, which is available for MDPs, is investigated for BMDPs in this paper. Such an extension is, to the best of our knowledge, not available in the literature.

3   Background and Definitions

In this section we introduce the basic definitions and notation. We begin with the definition of Markov and semi-Markov decision processes following the standard literature [2,15]. Afterwards, BMDPs are introduced and the new class of BSMDPs is defined. We consider MDPs with a finite state and action space.

Definition 1. *A (Discrete Time) Markov Decision Process is a 5-tuple* $\left(\mathcal{S}, \mathcal{A}, (\boldsymbol{P}^a)_{a \in \mathcal{A}}, (\boldsymbol{r}^a)_{a \in \mathcal{A}}, \boldsymbol{p}\right)$, *where* $\mathcal{S}$ *is a (finite) set of states of cardinality* n, $\mathcal{A}$ *is a (finite) set of actions,* $(\boldsymbol{P}^a)_{a \in \mathcal{A}}$ *is a set of* $n \times n$ *stochastic matrices,* $(\boldsymbol{r}^a)_{a \in \mathcal{A}}$ *is a set of non-negative reward vectors,* $\boldsymbol{p}$ *is the initial probability distribution.*

To simplify the notation we assume $\mathcal{S} = \{1, \dots, n\}$ such that states can be identified by their numbers. We assume here that the MDP is unichain [15]. Furthermore we assume that rewards are bounded.

A policy π assigns at each time $t \in \mathbb{N}$ to each state $i \in \mathcal{S}$ a probability distribution over the set of actions $\mathcal{A}$. A policy is deterministic if the distribution is a Dirac distribution, it is stationary if it does not depend on t and it is pure if it is deterministic and stationary. Let Π be the set of pure policies. A pure policy can be described by a vector π where $\pi(i) \in \mathcal{A}$ is the action chosen in state $i \in \mathcal{S}$. An MDP with a pure policy defines a Markov Reward Process $(\mathcal{S}, \boldsymbol{P}^\pi, \boldsymbol{r}^\pi, \boldsymbol{p})$, where $\boldsymbol{P}^\pi$ is a stochastic matrix of size $n \times n$ whose i-th row is the i-th row of $\boldsymbol{P}^{\pi(i)}$ and the reward vector $\boldsymbol{r}^\pi$ is of size $n \times 1$ with $\boldsymbol{r}^\pi(i) = \boldsymbol{r}^{\pi(i)}(i)$. Our results apply to unichain models. An MDP is called unichain, if for each strategy π the Markov chain induced by π is ergodic [13].

We consider optimization of MDPs over infinite horizons. For the discounted case we have to solve the optimization problem

$$\boldsymbol{g}^* = \max_{\pi \in \Pi} (\boldsymbol{r}^\pi + \gamma \boldsymbol{P}^\pi \boldsymbol{g}^*) \text{ and } \pi^* = \arg\max_{\pi \in \Pi} (\boldsymbol{r}^\pi + \gamma \boldsymbol{P}^\pi \boldsymbol{g}^*), \tag{1}$$

where $\gamma \in [0, 1)$ is the discount factor. It can be shown [15] that the optimum is reached by a pure policy. $\boldsymbol{g}^*$ is the optimal gain vector and π^* an optimal policy which is not necessarily unique. For the average reward we have

$$\bar{\boldsymbol{g}}^* = \max_{\pi \in \Pi} \left(\lim_{K \to \infty} \left(\tfrac{1}{K} \sum_{k=1}^{K} (\boldsymbol{P}^\pi)^k \, \boldsymbol{r}^\pi \right) \right), \bar{\pi}^* = \arg\max_{\pi \in \Pi} \left(\lim_{K \to \infty} \left(\tfrac{1}{K} \sum_{k=1}^{K} (\boldsymbol{P}^\pi)^k \, \boldsymbol{r}^\pi \right) \right). \tag{2}$$

Like in the discounted case, the maximum is reached by a pure policy. MDPs can be generalized by defining *Semi-Markov Decision Processes* (SMDPs).

Definition 2. *A Semi-Markov Decision Process is a 6-tuple* $\left(\mathcal{S}, \mathcal{A}, (\boldsymbol{P}^a)_{a \in \mathcal{A}}, (F^a(i, t))_{a \in \mathcal{A}, i \in \mathcal{S}}, (\boldsymbol{r}^a)_{a \in \mathcal{A}}, \boldsymbol{p}\right)$, *where* $\left(\mathcal{S}, \mathcal{A}, (\boldsymbol{P}^a)_{a \in \mathcal{A}}, (\boldsymbol{r}^a)_{a \in \mathcal{A}}, \boldsymbol{p}\right)$ *are defined as in Definition 1 and for all* $i \in \mathcal{S}$, $a \in \mathcal{A}$: $F^a(i, t)$ *is a distribution function with* $F^a(i, t) = 0$ *for* $t < 0$, *some* $\delta, \epsilon > 0$ *exist such that* $F^a(i, \delta) < 1 - \epsilon$.

Let $f^a(i,t) = (F^a(i,t))'$ be the density of the sojourn time in state i under decision a. For discrete distributions we can define probabilities rather than densities. The dynamics of a SMDP is described by a set of decision epochs which start and end at decision points. At a decision point, the process enters a state i and the decision maker selects an action a. Then the process stays in state i a time which is distributed according to $F^a(i,t)$ and afterwards jumps to state j with probability $\boldsymbol{P}^a(i,j)$. During each time unit the process stays in state i with decision a, it earns a gain of $r^a(i)$. The definition of $F^a(i,t)$ assures that the process can make in each finite interval only a finite number of jumps with probability 1. The probability distribution $\boldsymbol{p}$ defines the probability to start in a given state at time $t = 0$.

If all $F^a(i,t)$ are constant distributions with the same mean, then the SMDP is an MDP. If all $F^a(i,t)$ are exponential distributions, then the SMDP is a MDP in continuous time. In this case it can be transformed into a discrete time MDP using uniformization and the discounted and average reward remain the same [15,17]. Hence, we consider here the general case with general sojourn time distributions in the states.

For the discounted reward with discount rate $\beta > 0$, the process earns reward $e^{-\beta t}r^a(i)$ at time t in state i with decision a. The relationship between the discount parameters in discrete and continuous time is $\gamma = e^{-\beta T}$, if T is the time spent in a decision epoch. To compute the optimal policy and gain vectors, the SMDP is transformed into an equivalent discrete time MDP [13]. Define for $i \in \mathcal{S}$, $a \in \mathcal{A}^1$

$$s^a(i) = r^a(i) \int_0^\infty (1 - F^a(i,t))e^{-\beta t}\mathrm{d}t, \text{ and } \boldsymbol{Q}^a(i,j) = \boldsymbol{P}^a(i,j) \int_0^\infty f^a(i,t)e^{-\beta t}\mathrm{d}t.$$
(3)

Vector s^a includes the discounted rewards accumulated between two decision points and $\boldsymbol{Q}^a$ is a substochastic matrix that includes the effect of discounting. Let $F^a(i,t)$ be $(\boldsymbol{p}, \boldsymbol{D}_0)$ phase-type distributed (PHD) with parameters $\boldsymbol{p}, \boldsymbol{D}_0$, where $\boldsymbol{p}$ is an initial distribution and $\boldsymbol{D}_0$ is a subgenerator [4]. Then $\boldsymbol{d}_1 = -\boldsymbol{D}_0\mathbb{1}$ [4] and the integral evaluates to

$$\int_0^\infty \boldsymbol{p}e^{-\beta t}e^{\boldsymbol{D}_0 t}\boldsymbol{d}_1\mathrm{d}t = \int_0^\infty \left(\boldsymbol{p}\sum_{k=0}^\infty \frac{(\boldsymbol{D}_0 - \beta \boldsymbol{I})^k}{k!}\boldsymbol{d}_1 \right) \mathrm{d}t.$$
(4)

The integral can then be evaluated using uniformization [11]. The optimal discounted gain vector and policy can then be computed as solution of the following equations.

$$\boldsymbol{h}^* = \max_{\phi \in \Pi} \left(s^\phi + \boldsymbol{Q}^\phi \boldsymbol{h}^* \right) \text{ and } \phi^* = \arg \max_{\phi \in \Pi} \left(s^\phi + \boldsymbol{Q}^\phi \boldsymbol{h}^* \right)$$
(5)

---

[1] We consider in the following and subsequent equations continuous random variables where the integrals are well-defined for sojourn times in the states. For discrete random variables, the integrals have to be substituted by sums and the densities by probabilities, respectively.

Q^ϕ is a substochastic matrix constructed from matrices defined in (3) according to the policy ϕ. For the optimization of the average reward let $y^a(i) = \int_0^\infty t f^a(i,t) \mathrm{d}t$ be the average sojourn time in state i under action a. Then

$$\bar{h}^* = \max_{\phi \in \Pi} \left(s^\phi - H^* y^\phi + P^\phi h^* \right) \text{ and } \bar{\phi}^* = \arg\max_{\phi \in \Pi} \left(s^\phi - H^* y^\phi + P^\phi h^* \right)$$

(6)

are the equations to be solved for the average reward [15]. H^* is the long run average gain and vector $\bar{h}^*$ contains the short term deviations from the average. The equations have a unique solution if one value $\bar{h}^*(i_0)$ is fixed. Observe that (6) with $y^\phi = \mathbb{1}$ is the solution of (2). For further details about SMDPs and their analysis we refer to [15, Chap. 11] and [13, Chap. 9.5].

3.1   Bounded Parameter Markov Decision Processes

Most times the parameters of a stochastic model are only estimates resulting from measurements or expert guesses. Consequently, parameters are uncertain and are given by intervals rather than point estimates. This is the idea of *Bounded-Parameter MDPs* (BMDPs). In the following the comparison of vectors and matrices is pointwise.

Definition 3 (Bounded-parameter Markov decision process [10]).   *A Bounded-Parameter MDP is a 5-tuple $\left(\mathcal{S}, \mathcal{A}, (P_{\updownarrow}^a)_{a \in \mathcal{A}}, (r_{\updownarrow}^a)_{a \in \mathcal{A}}, p_{\updownarrow} \right)$, where $\mathcal{S}$ and $\mathcal{A}$ are defined as in Definition 1, $P_{\updownarrow}^a = \left(P_{\downarrow}^a, P_{\uparrow}^a \right)$ with $P_{\downarrow}^a, P_{\uparrow}^a \in \mathbb{R}_{\geq 0}^{n,n}$. For all $a \in \mathcal{A}$ it holds that $P_{\downarrow}^a \leq P_{\uparrow}^a, P_{\downarrow}^a \mathbb{1} \leq \mathbb{1} \leq P_{\uparrow}^a \mathbb{1}$ and $r_{\updownarrow}^a = \left(r_{\downarrow}^a, r_{\uparrow}^a \right)$ with $r_{\downarrow}^a, r_{\uparrow}^a \in \mathbb{R}_{\geq 0}^{n,1}, r_{\downarrow}^a \leq r_{\uparrow}^a. p_{\updownarrow} = (p_{\downarrow}, p_{\uparrow})$, with $p_{\downarrow}, p_{\uparrow} \in \mathbb{R}_{\geq 0}^{1,n}, p_{\downarrow} \leq p_{\uparrow}$ and $p_{\downarrow} \mathbb{1} \leq 1 \leq p_{\uparrow} \mathbb{1}. P_{\updownarrow}^a, r_{\updownarrow}^a$ and $p_{\updownarrow}$ define the following sets of matrices and vectors, respectively.*

$$P_{\updownarrow}^a = \left\{ P^a \mid P_{\downarrow}^a \leq P^a \leq P_{\uparrow}^a \wedge P^a \mathbb{1} = \mathbb{1} \right\}, \ r_{\updownarrow}^a = \left\{ r^a \mid r_{\downarrow}^a \leq r^a \leq r_{\uparrow}^a \right\},$$
$$p_{\updownarrow} = \left\{ p \mid p_{\downarrow} \leq p \leq p_{\uparrow} \wedge p \mathbb{1} = 1 \right\}.$$

(7)

Thus, each BMDP defines a set of MDPs and the *best* policy is no longer unique. Commonly considered are two cases, computation of the policy that behaves best in the worst case scenario or in the best case scenario, where the worst and best case scenarios are defined over the set of MDPs defined by the BMDP. Analysis of the worst case is often more important and can be interpreted as a form of robust optimization. Therefore we consider in the sequel only the worst case scenario. The best case scenario can be handled similarly after exchanging minimum and maximum. We denote a BMDP as unichain, if all MDPs described by the BMDP are unichain. In the sequel we will assume that this is the case.

For the discounted reward we then have the following equations to solve.

$$g_{\downarrow} = \max_{\pi \in \Pi} \min_{\boldsymbol{P}^{\pi} \in P_{\updownarrow}^{\pi}} \left(\boldsymbol{r}_{\updownarrow}^{\pi} + \gamma \boldsymbol{P}^{\pi} g_{\downarrow} \right) \quad \text{and} \quad \pi_{\downarrow} = \arg\max_{\pi \in \Pi} \min_{\boldsymbol{P}^{\pi} \in P_{\updownarrow}^{\pi}} \left(\boldsymbol{r}_{\updownarrow}^{\pi} + \gamma \boldsymbol{P}^{\pi} g_{\downarrow} \right)$$

(8)

Due to the nested max, min operations (8) is harder to solve than (1). The expressions for the expected average reward of BMDPs are

$$\bar{g}_{\downarrow} = \max_{\pi \in \Pi} \min_{\boldsymbol{P}^{\pi} \in P_{\updownarrow}^{\pi}} \left(\lim_{K \to \infty} \left(\frac{1}{K} \sum_{k=1}^{K} (\boldsymbol{P}^{\pi})^k \, \boldsymbol{r}_{\updownarrow}^{\pi} \right) \right)$$

$$\bar{\pi}_{\downarrow} = \arg\max_{\pi \in \Pi} \min_{\boldsymbol{P}^{\pi} \in P_{\updownarrow}^{\pi}} \left(\lim_{K \to \infty} \left(\frac{1}{K} \sum_{k=1}^{K} (\boldsymbol{P}^{\pi})^k \, \boldsymbol{r}_{\updownarrow}^{\pi} \right) \right).$$

(9)

3.2 Bounded Parameter Semi-Markov Decision Processes

To the best of our knowledge the concept of BMDPs has not been extended to SMDPs yet, although such an extension seems to be important from a practical point of view.

Definition 4. *A Bounded-Parameter Semi-Markov Decision Process (BSMDP) is a 6-tuple $\left(\mathcal{S}, \mathcal{A}, (\boldsymbol{P}_{\updownarrow}^a)_{a \in \mathcal{A}}, (F_{\updownarrow}^a(i,t))_{a \in \mathcal{A}, i \in \mathcal{S}}, (\boldsymbol{r}_{\updownarrow}^a)_{a \in \mathcal{A}}, \boldsymbol{p}_{\updownarrow} \right)$, where $\left(\mathcal{S}, \mathcal{A}, (\boldsymbol{P}_{\updownarrow}^a)_{a \in \mathcal{A}}, (\boldsymbol{r}_{\updownarrow}^a)_{a \in \mathcal{A}}, \boldsymbol{p}_{\updownarrow} \right)$ is defined as in Definition 3 and for all $i \in \mathcal{S}$, $a \in \mathcal{A}$ it holds that $F_{\updownarrow}^a(i,t) = \left(F_{\downarrow}^a(i,t), F_{\uparrow}^a(i,t) \right)$ with $F_{\downarrow}^a(i,t) \geq F_{\uparrow}^a(i,t)$ for all t is a pair of distribution functions with $F_{\downarrow}^a(i,t) = F_{\uparrow}^a(i,t) = 0$ for $t < 0$, and for all $1 > \epsilon > 0$ some $\delta > 0$ exists such that $F_{\uparrow}^a(i,\delta) < 1 - \epsilon$.*

Observe that $F_{\uparrow}^a(i,t)$ is stochastically larger than $F_{\downarrow}^a(i,t)$ which implies that for any non-decreasing function g, $\int_0^{\infty} g(t)\mathrm{d}F_{\downarrow}^a(i,t) \leq \int_0^{\infty} g(t)\mathrm{d}F_{\uparrow}^a(i,t)$ [14]. We denote by $f_{\downarrow}^a(i,t) = (F_{\downarrow}^a(i,t))'$ and $f_{\uparrow}^a(i,t) = (F_{\uparrow}^a(i,t))'$ the corresponding probability density functions. Again the approach can be easily extended for discrete time models by considering probabilities rather than densities. $F_{\updownarrow}^a(i,t)$ defines a set of distribution functions $F_{\updownarrow}^a(i,t) = \left\{ F^a(i,t) \mid F_{\downarrow}^a(i,t) \geq F^a(i,t) \geq F_{\uparrow}^a(i,t) \right\}$.

Thus, a BSMDP defines a set of SMDPs. Each SMDP is defined by choosing one element from each of the sets $(\boldsymbol{P}_{\updownarrow}^a)_{a \in \mathcal{A}}$, $(F_{\updownarrow}^a(i,t))_{a \in \mathcal{A}}$, $(\boldsymbol{r}_{\updownarrow}^a)$ and $\boldsymbol{p}_{\updownarrow}$. For the optimization of the discounted reward it is not possible to use (3), instead we have to compute state and action dependent discount factors and combine them with the matrices in $\boldsymbol{P}_{\updownarrow}^{\phi}$. For discount rate β we obtain

$$s_{\downarrow}^a(i) = r_{\downarrow}^a(i) \int_0^{\infty} (1 - F_{\downarrow}^a(i,t)) e^{-\beta t} \mathrm{d}t, \quad s_{\uparrow}^a(i) = r_{\uparrow}^a(i) \int_0^{\infty} (1 - F_{\uparrow}^a(i,t)) e^{-\beta t} \mathrm{d}t,$$

$$\gamma_{\downarrow}^a(i) = \int_0^{\infty} f_{\downarrow}^a(i,t) e^{-\beta t} \mathrm{d}t, \qquad \gamma_{\uparrow}^a(i) = \int_0^{\infty} f_{\uparrow}^a(i,t) e^{-\beta t} \mathrm{d}t.$$

(10)

For policy $\phi \in \Pi$ define diagonal matrices $\boldsymbol{\Gamma}_\downarrow^\phi, \boldsymbol{\Gamma}_\uparrow^\phi$ with $\boldsymbol{\Gamma}_\downarrow^\phi(i,i) = \boldsymbol{\gamma}_\downarrow^{\phi(i)}(i)$ and $\boldsymbol{\Gamma}_\uparrow^\phi(i,i) = \boldsymbol{\gamma}_\uparrow^{\phi(i)}(i)$. Then the following equations have to be solved for the discounted reward of BSMDPs.

$$h_\downarrow = \max_{\phi \in \Pi} \min_{\boldsymbol{P}^\phi \in \boldsymbol{P}_\downarrow^\phi} \left(\boldsymbol{s}_\downarrow^\phi + \boldsymbol{\Gamma}_\downarrow^\phi \boldsymbol{P}^\phi \boldsymbol{h}_\downarrow \right) \quad \text{and} \quad \phi_\downarrow = \arg \max_{\phi \in \Pi} \min_{\boldsymbol{P}^\phi \in \boldsymbol{P}_\downarrow^\phi} \left(\boldsymbol{s}_\downarrow^\phi + \boldsymbol{\Gamma}_\downarrow^\phi \boldsymbol{P}^\phi \boldsymbol{h}_\downarrow \right)$$
(11)

Observe that for $\boldsymbol{\Gamma}_\downarrow^\phi = \boldsymbol{\Gamma}_\uparrow^\phi$ and $\boldsymbol{P}_\downarrow^\phi = \boldsymbol{P}_\uparrow^\phi$ the relation $\boldsymbol{Q}^\phi = \boldsymbol{\Gamma}_\downarrow^\phi \boldsymbol{P}_\downarrow^\phi$ holds. For the average reward we first define $\boldsymbol{y}_\downarrow^a(i) = \int_0^\infty t f_\downarrow^a(i,t) \mathrm{d}t$, $\boldsymbol{y}_\uparrow^a(i) = \int_0^\infty t f_\uparrow^a(i,t) \mathrm{d}t$ and $\boldsymbol{y}_\updownarrow^a = (\boldsymbol{y}_\downarrow^a, \boldsymbol{y}_\uparrow^a)$. Then the average reward in the worst case and the policy can be computed, following [15]

$$\bar{h}_\downarrow = \max_{\phi \in \Pi} \min_{\boldsymbol{P}^\phi \in \boldsymbol{P}_\downarrow^\phi} \left(\boldsymbol{r}_\downarrow^\phi - H_\downarrow \boldsymbol{y}_\uparrow^\phi + \boldsymbol{P}^\phi \bar{\boldsymbol{h}}_\downarrow \right), \quad \bar{\phi}_\downarrow = \arg \max_{\phi \in \Pi} \min_{\boldsymbol{P}^\phi \in \boldsymbol{P}_\downarrow^\phi} \left(\boldsymbol{r}_\downarrow^\phi - H_\downarrow \boldsymbol{y}_\uparrow^\phi + \boldsymbol{P}^\phi \bar{\boldsymbol{h}}_\downarrow \right)$$
(12)

To obtain a unique solution one has to fix $\bar{\boldsymbol{h}}_\downarrow(i_0) = 0$ for some $i_0 \in \mathcal{S}$. Then $H_\downarrow$ is the minimal expected average reward and $\bar{\boldsymbol{h}}_\downarrow$ is the expected total deviation vector.

4 Numerical Analysis

Before we introduce numerical algorithms for BMDPs and BSMDPs, we briefly review the available methods for MDPs. The pseudocode of the presented algorithms is available online [1].

4.1 Numerical Methods for Markov and Semi-Markov Decision Processes

Optimal policies and gain vectors for MDPs can be computed with *value iteration*, *policy iteration* or *linear programming* (LP). LP formulations are useful to prove some results and they are currently the standard method to solve MDPs with additional constraints. However, for most MDPs, LP solvers are significantly slower than the other two approaches [15]. Additionally, it is not possible to analyze BMDPs or related models using LP [9]. Therefore we briefly introduce the basic realizations of value and policy iteration for discounted and average reward.

We start with value iteration for discounted rewards. Let $\boldsymbol{v}^{(0)} \geq \boldsymbol{0}$ be the initial vector and set $k = 0$. Then compute for all $i \in \mathcal{S}$ and $k \in \mathbb{N}$

$$\boldsymbol{v}^{(k+1)}(i) = \max_{a \in \mathcal{A}} \left(\boldsymbol{r}^a(i) + \gamma \sum_{j \in \mathcal{S}} \boldsymbol{P}^a(i,j) \boldsymbol{v}^{(k)}(j) \right)$$
(13)

until $\left\| v^{(k+1)} - v^{(k)} \right\| < \epsilon \frac{1-\gamma}{2\gamma}$. If the condition holds, then $\left\| v^{(k+1)} - g^* \right\| < \epsilon$ and

$$\pi^*(i) \in \arg\max_{a \in \mathcal{A}} \left(r^a(i) + \gamma \sum_{j \in \mathcal{S}} P^a(i,j) v^{(k)}(j) \right) \tag{14}$$

is an ϵ-optimal policy[2]. Policy iteration evaluates a policy before the policy is improved. Let $\pi^{(0)} \in \Pi$ some pure initial policy and set $k = 0$. Then, for $k \in \mathbb{N}$,

$$\text{Solve } r^{\pi^{(k)}} = \left(I - \gamma P^{\pi^{(k)}} \right) v^{(k)}, \quad \pi^{(k+1)} = \arg\max_{a \in \mathcal{A}} \left(r^a(i) + \gamma P^a v^{(k)} \right) \tag{15}$$

If $\pi^{(k+1)} = \pi^{(k)}$, then the optimal policy has been found.

To compute the optimal policy for the long term average reward, in principle (13) can be applied, then $\bar{g}^* = \lim_{\gamma \to 1} (1 - \gamma) \lim_{k \to \infty} v^{(k)}$. However, since the entries in $v^{(k)}$ are unbounded, it is preferable to use *relative* value iteration [15]. One state i_0 is chosen, vector $v^{(0)}$ is initialized and

$$\begin{aligned} w^{(k)} &= v^{(k)} - v^{(k)}(i_0)\mathbf{1} \\ v^{(k+1)}(i) &= \max_{a \in \mathcal{A}} \left(r^a(i) + \sum_{j \in \mathcal{S}} P^a(i,j) w^{(k)}(j) \right) \text{ for all } i \in \mathcal{S} \end{aligned} \tag{16}$$

is computed until

$$sp(v^{(k+1)} - v^{(k)}) = \max_{i \in \mathcal{S}} \left(v^{(k+1)}(i) - v^{(k)}(i) \right) - \min_{i \in \mathcal{S}} \left(v^{(k+1)}(i) - v^{(k)}(i) \right) < \epsilon. \tag{17}$$

(14) can then be applied to compute the optimal policy $\bar{\pi}^*$. The major advantage of relative value iteration is that values in the vectors $v^{(k)}$ remain small because they include the difference of the value of a state i and the value of state i_0. To obtain the optimal value vector $\bar{g}^*$ (6) with $y = \mathbf{1}$ is used. Thus, by solving

$$\bar{r}^{\bar{\pi}^*} = \left(I - P^{\bar{\pi}^*} \right) \bar{g}^* + H^* \mathbf{1} \tag{18}$$

with $\bar{g}^*(i_0) = 0$ the optimal average gain H^* and the deviation vector are computed. For unichain MDPs the equations have a unique solution. In policy iteration for the average reward case, one can also use the representation in (18). In this case policy $\bar{\pi}^{(k)}$ is evaluated using (18) and a new policy is computed with (15). The approach is iterated until the policy remains.

The approaches for SMDPs are very similar. For the discounted case reward vectors s^a rather than r^a and matrices Q^a rather than γP^a (computed with (3)) are used in (13) or (15). For the average case we apply uniformization to transform the SMDP into an MDP that is equivalent w.r.t. expected average reward [3]. Define $\eta = \min_{i \in \mathcal{S}} \min_{a \in \mathcal{A}} y^a(i)/(1 - P^a(i,i))$, vector $\bar{s}^a$ with $\bar{s}^a(i) = r^a(i)/y^a(i)$ and matrix $\bar{Q}^a$ with

$$\bar{Q}^a = I + \eta \text{diag}(y^a)^{-1}(P^a - I) \tag{19}$$

$(\bar{Q}^a)_{a \in \mathcal{A}}$ and $(\bar{s}^a)_{a \in \mathcal{A}}$ define an MDP which can then be analyzed using value or policy iteration for MDPs as described above.

---

[2] ϵ-optimality means that the optimal value is reached up to ϵ.

4.2 Discounted Rewards for Bounded Parameter (Semi-) Markov Decision Processes

The approach can be applied for BMDPs as well as for BSMDPs which can be transformed into BMDPs by computing expected rewards with (10 and 11). Value iteration for BMDPs has been introduced in [10]. For $P_{\updownarrow}^{\phi}$ and vector v define

$$f_{\downarrow}(P_{\updownarrow}^{a}(i\bullet), v) = \min_{P \in P_{\updownarrow}^{a}} (P(i\bullet)v) \text{ and } M_{\downarrow}(P_{\updownarrow}^{\phi}, v) = \arg\min_{P \in P_{\updownarrow}^{\phi}} (Pv), \qquad (20)$$

where $P(i\bullet)$ is the row in a matrix P corresponding to state i. The effort to evaluate $f_{\downarrow}$ is proportional to the number of non-zero elements in row $P_{\uparrow}^{\phi}(i\bullet)$ and the effort to evaluate the function $M_{\downarrow}$, resulting in a stochastic matrix, is proportional to the number of non-zero elements in $P_{\uparrow}^{\phi}$. Value iteration for BMDPs and BSMDPs is very similar to (13). Initialize $v^{(0)} \geq 0$ and $k = 0$. Then compute for each $i \in \mathcal{S}$

$$v^{(k+1)}(i) = \max_{a \in \mathcal{A}} \left(r_{\updownarrow}^{a}(i) + \gamma f_{\downarrow}\left(P_{\updownarrow}^{a}(i\bullet), v^{(k)}\right) \right) \qquad (21)$$

until the error bound of (13) is met. We can apply the same error bound as for MDPs because (21) like (13) can be shown to be a contraction mapping with factor γ; the proof can be found in [10]. Using this property, Banach's fixed point theorem can be applied to show that the iteration converges to a unique fixed point which corresponds to the value of an optimal policy.

For policy iteration let $\phi^{(1)} \in \Pi$ be some initial policy, $v^{(0)} = r^{\phi^{(1)}}$ and $k = 1$. Then

$$\text{Solve } r_{\downarrow}^{\phi^{(k)}} = \left(I - \gamma M_{\downarrow}\left(P_{\updownarrow}^{\phi^{(k)}}, v^{(k-1)}\right)\right) v^{(k)}$$
$$\phi^{(k+1)}(i) = \arg\max_{a \in \mathcal{A}} \left(r_{\updownarrow}^{a}(i) + \gamma f_{\downarrow}(P_{\updownarrow}^{a}(i\bullet), v^{(k)})\right) \text{ for all } i \in \mathcal{S} \qquad (22)$$
$$\text{choosing } \phi^{(k+1)}(i) = \phi^{(k)}(i) \text{ when possible.}$$

There are two variants of policy iteration in this case. In the first variant the solution (first line in (22)) and policy selection (second line in (22)) are performed successively until $\phi^{(k)} = \phi^{(k-1)}$. In the second variant the solution is iterated until $M_{\downarrow}\left(P_{\updownarrow}^{\phi^{(k)}}, v^{(k-1)}\right) = M_{\downarrow}\left(P_{\updownarrow}^{\phi^{(k-1)}}, v^{(k-2)}\right)$ and then a new policy is selected. Again the algorithm terminates when the policy remains the same. This variant of policy iteration is also proposed in [16].

The convergence of both variants of policy iteration follows, again, from the fixed point theorem, as the steps in (22) are all contraction mappings: the policy selection step is analogous to one step of value iteration, and the policy evaluation step can be written as an iterative application of $v_{i+1}^{(k+1)} = r_{\downarrow}^{\phi^{k}} + \gamma M_{\downarrow}\left(P_{\updownarrow}^{\phi^{(k)}} v^{(k)}\right) v_{i}^{(k+1)}$, which also is a contraction mapping with factor γ and has $v^{(k+1)}$ as its fixed point. This implies, again, the existence of a unique fixed point which corresponds to the value of the optimal policy.

Value and both variants of policy iteration can be applied for BMDPs and BSMDPs. For BSMDPs we observe that the main difference from BMDPs is the consideration of transition times, which affects the discount factor. Therefore, for BSMDPs, we have to substitute γ by $\boldsymbol{\Gamma}_\downarrow^\phi(i,i)$ and $\gamma_\uparrow^a(i)$ in (21) and (22). This, in turn, affects the contraction mapping statement and replaces the contraction factor with a term from $\gamma_\downarrow^a$; however, as it always is $\gamma_\downarrow^a(i) < 1$, since $\gamma_\downarrow^a(i)$ is the expected value of $\exp(-\beta X)$ with a non-negative random variable X, the contraction mapping property still holds.

In contrast to value and policy iteration for plain MDPs, the functions $f_\downarrow$ or $\boldsymbol{M}_\downarrow$ have to be evaluated in each iteration and it is unclear how this affects the performance of both algorithms and how much more effort has to be spent in comparison to the optimization of MDPs.

4.3   Average Rewards for Bounded Parameter (Semi-) Markov Decision Processes

As already mentioned, computation of the average reward has rarely been considered in the literature, and for BMDPs only the value iteration algorithm from [19] is available. We develop new algorithms based on relative value iteration and on policy iteration. It is important to note that our algorithms are designed for unichain models; multichain models are subject to future research. The first step is to transform the BSMDP into an, according to average reward, equivalent BMDP using uniformization as in (19).

Each BSMDP defines a set of SMDPs and each SMDP from the set can be transformed into an equivalent MDP. We begin with rewards and obtain for $r^a \in r_\uparrow^a$ and $y^a \in y_\downarrow^a$ the bounds $\bar{s}_\downarrow^a(i) = \frac{r_\downarrow^a(i)}{y_\downarrow^a(i)} \leq \frac{r^a(i)}{y^a(i)} \leq \frac{r_\uparrow^a(i)}{y_\downarrow^a(i)} = \bar{s}_\uparrow^a(i)$ for all $i \in \mathcal{S}$. The uniformization rate is then given by $\eta = \min_{i \in \mathcal{S}} \min_{a \in \mathcal{A}} y_\downarrow^a(i)/(1 - \boldsymbol{P}_\downarrow^a(i,i))$ and matrices $\bar{\boldsymbol{Q}}_\downarrow^a, \bar{\boldsymbol{Q}}_\uparrow^a$ are defined as

$$\bar{\boldsymbol{Q}}_\downarrow^a = \boldsymbol{I} + \eta \cdot \operatorname{diag}(y_\downarrow^a)^{-1}(\boldsymbol{P}_\downarrow^a - \boldsymbol{I}), \bar{\boldsymbol{Q}}_\uparrow^a = \boldsymbol{I} + \eta \cdot \operatorname{diag}(y_\uparrow^a)^{-1}(\boldsymbol{P}_\uparrow^a - \boldsymbol{I}) \qquad (23)$$

The transformation can also be applied for BMDPs. In this case $y_\downarrow^a = y_\uparrow^a = \boldsymbol{1}$ and the transformation has no effect if $\boldsymbol{P}^a(i,i) = 0$ for some $i \in \mathcal{S}$, $a \in \mathcal{A}$ exists. $\bar{s}_\updownarrow^a = (\bar{s}_\downarrow^a, \bar{s}_\uparrow^a)$ and $\bar{\boldsymbol{Q}}_\updownarrow^a = (\bar{\boldsymbol{Q}}_\downarrow^a, \bar{\boldsymbol{Q}}_\uparrow^a)$ define the reward and transition probability bounds of a BMDP. Optimization according to the average reward can be done with relative value or policy iteration.

The relative value iteration algorithm performs the following iteration step starting with some initial vector $\boldsymbol{v}^{(0)} \geq \boldsymbol{0}$ and $k = 0$ for some previously fixed constant i_0. Intuitively, i_0 is the state where the relative value iteration is applied and from which the changes made by the Bellman updates are propagated.

$$\begin{aligned} \boldsymbol{w}_\downarrow^{(k)} &= \boldsymbol{v}_\downarrow^{(k)} - \boldsymbol{e}_{i_0}^T \boldsymbol{v}_\downarrow^{(k)}(i_0) \\ \boldsymbol{v}_\downarrow^{(k+1)}(i) &= \max_{a \in \mathcal{A}} \left(r_\downarrow^a(i) + f_\downarrow \left(\bar{\boldsymbol{Q}}_\updownarrow^a, \boldsymbol{w}^{(k)} \right) \right) \text{ for all } i \in \mathcal{S}. \end{aligned} \qquad (24)$$

until the error bound given in (17) is met. We expect convergence from this algorithm as this is a straightforward application of the relative value iteration algorithm from [15] to BMDPs.

The policy iteration algorithm is derived from the algorithm of Hoffman and Karp [12] and has the same convergence properties. It starts with some initial policy $\bar{\phi}^{(1)}$, $k = 1$, $\bar{h}_\downarrow^{(0)} = r_\downarrow^{\bar{\phi}^{(1)}}$ and evaluates then

$$\text{Solve } \bar{s}^{\bar{\phi}^{(k)}} = \left(I - M_\downarrow \left(\bar{Q}_\uparrow^{\phi^{(k)}}, \bar{h}_\downarrow^{(k-1)}\right)\right) \bar{h}_\downarrow^{(k)} + \bar{H}_\downarrow^{(k)} \mathbb{1} \text{ with } \bar{h}_\downarrow^{(k)}(i_0) = 0$$

$$\bar{\phi}^{(k+1)}(i) = \arg\max_{a \in \mathcal{A}} \left(\bar{s}_\downarrow^a(i) + f_\downarrow \left(\bar{Q}_\uparrow^a(i\bullet), \bar{h}_\downarrow^{(k)}\right)\right) \text{ for all } i \in \mathcal{S}$$

choosing $\bar{\phi}^{(k+1)}(i) = \bar{\phi}^{(k)}(i)$ when possible.

$$(25)$$

Here, the scalar $\bar{H}_\downarrow^{(k)}$ corresponds to the average gain in step k. We also note that as in the discounted case, two variants of policy iteration can be defined.

5 Experimental Results

We use two series of experiments to evaluate the presented algorithms. First, randomly generated (B)(S)MDPs are analyzed and afterwards a BSMDP resulting from a maintenance model is analyzed. All computations were carried out on a PC with a 3.0 GHz 20-Core processor and 126 GB main memory running Debian Linux. All algorithms are implemented in *octave*. *Matlab* implementations of the algorithms are currently under way. We compare the different algorithms for MDPs and BMDPs. Computational times for SMDPs and BSMDPs are similar since they are transformed into MDPs and BMDPs. The transformation usually requires only a small amount of time.

Models with expected average criterion: In a first series of experiments we use randomly generated dense matrices and vectors to define (B)MDPs with a number of states varied from 100 to 1200 and a number of actions $|\mathcal{A}| \in \{5, 10, 15, 20\}$. We repeat each run 10 times with newly generated matrices and determine the mean and variance of the solution time. In all experiments, relative value iteration runs with precision parameter $\epsilon \in \{10^{-4}, 10^{-9}\}$. The left plot in Fig. 1 shows computation times for all tested algorithms for MDPs as a function of the number of states. As one can see, policy iteration outperforms value iteration with precision parameter $\epsilon = 10^{-9}$. For $\epsilon = 10^{-4}$ value iteration is faster than policy iteration. The computation time for the MDPs is small, even the largest models can be solved in less than 10 s. The plot, shows only the mean values because variances are so small that confidence intervals would become almost invisible. This indicates that at least for randomly generated dense matrices, computational times are mainly determined by the number of states and actions and not by the structure. This will be different for sparse matrices, where the structure and size of the non-zero elements determines the runtimes of value and policy iteration much more.

For BMDPs we compare both variants of policy iteration and relative value iteration. The experimental results for the long term average reward are shown in the right plot in Fig. 1. As one can see, the second variant of policy iteration yields the smallest solution time, whereas the first variant of policy iteration

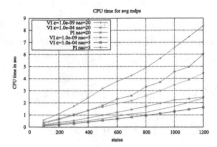
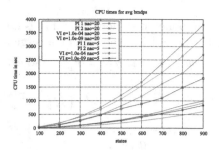

Fig. 1. Average CPU time of average reward criterion solution algorithms. The plotted results are obtained for models with 5 and 20 actions. *nac* is the number of actions, PI 1 and PI 2 denote two variants of policy iteration, and VI stands for value iteration.

requires the longest solution time for BMDPs. Again the variance is relatively small such that the mean value bears enough information. For instance, for $|\mathcal{S}| = 400, |\mathcal{A}| = 20$, value iteration requires 682 s on average with $\sigma = 5.4$, the first variant of policy iteration requires 755 s in average with $\sigma = 40.07$, and the second variant takes 486 s on average with $\sigma = 55.7$. For larger state spaces, e.g., $|\mathcal{S}| = 900$, value iteration requires 3327 s on average with relatively small $\sigma = 9.33$, the first variant of policy iteration has mean solution time of 3801 s with large $\sigma = 283.08$, and the second variant takes 2689 s with $\sigma = 371.2$. The effort to solve a BMDP compared to an MDP of the same size and with the same number of actions is between two and three orders of a magnitude larger. This is caused by the effort to evaluate the functions $M_{\downarrow}$ and $f_{\downarrow}^{a}$ and a larger number of iterations which is caused by the two levels, namely the change of the policy and the matrix belonging to a policy. The proposed algorithms for computing the long run average reward for BMDPs are original. We are aware of only one value iteration based algorithm presented in [19]. This algorithm is based on the value iteration for the discounted reward where the discount factor is slowly increased towards 1. The paper presenting the algorithm contains no numerical results. We tested the algorithm on the randomly generated test examples where the iteration is stopped if the estimated error is below 10^{-4} or 10.000 iterations have been performed. The value iteration variant from [19] is denoted as *CB*. Results are shown in Fig. 2. It can be seen that the runtimes of the algorithm from [19] are several orders of a magnitude longer than the runtime of our value iteration variant and that most times the algorithm stops due to the number of iterations resulting in an unsatisfactory error.

The expected discounted reward criterion: We now analyze the performance of the algorithms for the expected discounted reward. In this case we have another free parameter, namely the discount factor which is chosen from $\{0.5, 0.9\}$. Results are shown in Fig. 3. In principle results are similar to the results for the long run average reward. The difference between the solution time for an MDP and a BMDP is now less than two orders of magnitude which is smaller than in the average reward case. This is caused by the decreasing

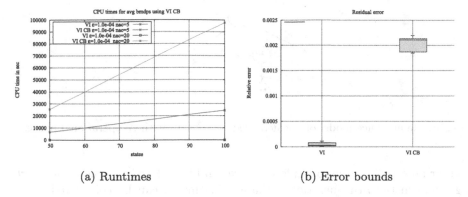

(a) Runtimes   (b) Error bounds

Fig. 2. Comparison between average reward criterion value iteration variants. *nac* is the number of actions, VI denotes value iteration and VI CB is the value iteration variant from [10].

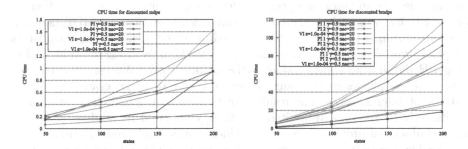

Fig. 3. CPU time of discounted reward criterion solution algorithms. The plotted results are obtained for models with 5 and 20 actions. *nac* is the number of actions, PI 1 and PI 2 denote two variants of policy iteration, and VI stands for value iteration.

influence of the future due to discounting which makes the evaluation of $M_\downarrow$ more stable.

Case study: Dependability models: To analyze algorithmic performance on more realistic processes, we consider a maintenance and repair (M&R) model. The model describes two components that may degrade over time and can be repaired or replaced; the states of an individual component are visualized in Fig. 4. Degrading is modeled as a stepwise process where a component can be in one of n *operational phases*, and the change from the kth to the $k-1$st operational phase occurs according to a PHD $(\pi_k^o, \boldsymbol{D}_{0,k}^o)$; analogously, repair and replacement are modeled as well by PHDs $(\pi_k^r, \boldsymbol{D}_{0,k}^r)$ for the repair process in the k-th operational phase and a PHD with representation $(\pi^f, \boldsymbol{D}_0^f)$ for the replacement process in the case of a failure. Two actions are possible, either letting the system run or performing maintenance on a component. Without maintenance, the component either eventually degrades to the first operational phase and fails

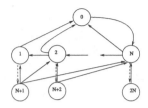

Fig. 4. Maintenance model of a generalized component with N operational phases

then or fails earlier (due to random failure) and is replaced by a new one, starting again in the nth operational phase. This model can be considered on an event-based level, by associating the states with operational phases and allowing actions when a component changes its operational phase. In our application, we consider two components with a shared repair worker to model scarcity of repair resources. Depending on which sojourn times are used in operational phases (we may consider mean rates or upper and lower exponential distribution bounds for the PHDs), we can derive, after uniformization is applied, an MDP (sojourn times are assumed to be exponentially distributed) or a BMDP (sojourn times are phase type distributed and lower/upper rate bounds are used). The underlying PH distributions are random PHDs of order two. We vary the number of operational phases in the range between 4 and 10 to generate MDPs and BMDPs of different sizes. The results can be seen in Fig. 5. Policy iteration is significantly better than value iteration here. Compared to random instances, one can observe a drop in performance which can be explained by a more complex problem structure. However, also for the example it becomes clear that the introduction of uncertainty (i.e., going from MDPs to BMDPs) has its price which is a significantly higher effort to compute optimal policies.

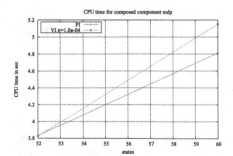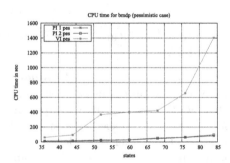

Fig. 5. CPU time in dependence of state space size for dependability models. PI 1, 2 pes denotes two variants of policy iteration for pessimistic case; VI denotes value iteration.

6   Conclusions

In this work we develop variants of value and policy iteration to compute an optimal policy and value vector according to the long run average and the discounted reward for BMDPs. Furthermore, the class of BMDPs is extended to BSMDPs. The algorithms are extended to this class of processes. The different algorithms are compared on a set of examples such that the price of uncertainty, namely the additional effort to analyze a BMDP compared to a MDP, becomes visible. Although the algorithms have been proposed here only for BMDPs, they can be used for other classes of MDPs with imprecise or uncertain parameters that appeared in the literature [9,16,20]. Furthermore, the algorithms can be adopted for multichain processes [15]. To allow efficient optimization of large processes it is necessary to implement variants of the algorithms that work on sparse data structures and possibly exploit the inherent parallelism of several operations. This will be a topic for future research.

References

1. Analysis of Markov decision processes under parameter uncertainty online companion. http://ls4-www.cs.tu-dortmund.de/cms/de/home/dohndorf/publications/
2. Bertsekas, D.P.: Dynamic Programming and Optimal Control, vol. 2, 3rd edn. Athena Scientific (2005, 2007)
3. Beutler, F.J., Ross, K.W.: Uniformization for Semi-Markov decision processes under stationary policies. J. Appl. Probab. **24**, 644–656 (1987)
4. Buchholz, P., Kriege, J., Felko, I.: Input Modeling with Phase-Type Distributions and Markov Models. SM. Springer, Cham (2014)
5. Chen, T., Hahn, E.M., Han, T., Kwiatkowska, M.Z., Qu, H., Zhang, L.: Model repair for Markov decision processes. In: TASE, pp. 85–92 (2013)
6. Cubuktepe, M., Jansen, N., Junges, S., Katoen, J., Papusha, I., Poonawala, H.A., Topcu, U.: Sequential convex programming for the efficient verification of parametric MDPs. CoRR, abs/1702.00063 (2017)
7. Delgado, K.V., de Barros, L.N., Cozman, F.G., Sanner, S.: Using mathematical programming to solve factored Markov decision processes with imprecise probabilities. Int. J. Approx. Reasoning **52**(7), 1000–1017 (2011)
8. Delgado, K.V., Sanner, S., de Barros, L.N.: Efficient solutions to factored MDPs with imprecise transition probabilities. Artif. Intell. **175**, 1498–1527 (2011)
9. Filho, R.S., Cozman, F.G., Trevizan, F.W., de Campos, C.P., de Barros, L.N.: Multilinear and integer programming for Markov decision processes with imprecise probabilities. In: 5th Int. Symposium on Imprecise Porbability: Theories and Applications, Prague, Czech Republic, pp. 395–404 (2007)
10. Givan, R., Leach, S.M., Dean, T.L.: Bounded-parameter Markov decision processes. Artif. Intell. **122**(1–2), 71–109 (2000)
11. Gross, D., Miller, D.: The randomization technique as a modeling tool and solution procedure for transient Markov processes. Oper. Res. **32**, 343–361 (1984)
12. Hoffman, A.J., Karp, R.M.: On nonterminating stochastic games. Manage. Sci. **12**(5), 359–370 (1966)
13. Kallenberg, L.: Markov decision processes. Lecture Notes, University Leiden (2011). https://www.math.leidenuniv.nl/~kallenberg/Lecture-notes-MDP.pdf

14. Müller, A., Stoyan, D.: Comparison Methods for Stochastic Models and Risks. Wiley, Chichester (2002)
15. Puterman, M.L.: Markov Decision Processes. Wiley, New York (2005)
16. Satia, J.K., Lave, R.E.: Markovian decision processes with uncertain transition probabilities. Oper. Res. **21**(3), 728–740 (1973)
17. Serfozo, R.F.: An equivalence between continuous and discrete time Markov decision processes. Oper. Res. **27**(3), 616–620 (1979)
18. Sigaud, O., Buffet, O. (eds.): Markov Decision Processes in Artificial Intelligence. Wiley-ISTE (2010)
19. Tewari, A., Bartlett, P.L.: Bounded parameter Markov decision processes with average reward criterion. In: Bshouty, N.H., Gentile, C. (eds.) COLT 2007. LNCS (LNAI), vol. 4539, pp. 263–277. Springer, Heidelberg (2007). doi:10.1007/978-3-540-72927-3_20
20. White, C.C., Eldeib, H.K.: Markov decision processes with imprecise transition probabilities. Oper. Res. **42**(4), 739–749 (1994)

Bounded Aggregation for Continuous Time Markov Decision Processes

Peter Buchholz, Iryna Dohndorf, Alexander Frank,
and Dimitri Scheftelowitsch[✉]

Department of Computer Science, TU Dortmund, Dortmund, Germany
{peter.buchholz,iryna.dohndorf,
alexander.frank,dimitri.scheftelowitsch}@cs.tu-dortmund.de

Abstract. Markov decision processes suffer from two problems, namely the so-called *state space explosion* which may lead to long computation times and the *memoryless property of states* which limits the modeling power with respect to real systems. In this paper we combine existing state aggregation and optimization methods for a new aggregation based optimization method. More specifically, we compute reward bounds on an aggregated model by exchanging state space size with uncertainty. We propose an approach for continuous time Markov decision models with discounted or average reward measures.

The approach starts with a portioned state space which consists of blocks that represent an abstract, high-level view on the state space. The sojourn time in each block can then be represented by a phase-type distribution (PHD). Using known properties of PHDs, we can then bound sojourn times in the blocks and also the accumulated reward in each sojourn by constraining the set of possible initial vectors in order to derive tighter bounds for the sojourn times, and, ultimatively, for the average or discounted reward measures. Furthermore, given a fixed policy for the CTMDP, we can then further constrain the initial vector which improves reward bounds. The aggregation approach is illustrated on randomly generated models.

Keywords: Markov Decision Process · Aggregation · Discounted reward · Average reward · Bounds

1 Introduction

Continuous time Markov decision processes (CTMDPs) are a well-established class of stochastic processes which are widely applied in performance and dependability analysis. A significant problem for Markov models are uncomfortably high-dimensional state spaces which lead to high runtime complexities. One way to handle this issue is bounded state aggregation, where each state results from aggregation of several detailed states, as discussed in [3,7]. Another problem when decisions are added to Markov models are the inherent memoryless property of states. In Markov models this problem is alleviated by using phase type

© Springer International Publishing AG 2017
P. Reinecke and A. Di Marco (Eds.): EPEW 2017, LNCS 10497, pp. 19–32, 2017.
DOI: 10.1007/978-3-319-66583-2_2

distributions (PHDs) [6]. However, this step can not be used in Markov Decision Processes (MDPs) because decisions are made in states and if PHDs are used for generally distributed sojourn times in states, each state of the PHD becomes a decision state which does not correspond to the state of the system where decisions can be made. By bounded aggregation also this problem can be handled.

The processes resulting from bounded aggregation is a so called bounded parameter MDP (BMDP), where for some parameters only intervals and not exact values are known. When lower and upper bounds are known, a set of CTMDPs, rather than a single CTMDP, is described. The goal of an optimization is then the minimization or maximization of the worst result value. The available papers on bounded aggregation in Markov processes or Markov decision processes consider only discrete time models. In this paper, the approach is extended to continuous time models coincidently computing improved bounding parameters based on recent work and available results for PHDs.

An extensive discussion of MDP theory and its applications is given in [13], where different optimality criteria for the unaltered MDP formalism are discussed. Given some uncertainty in transitions and rewards, a more general Markov model is necessary, like BMDPs described in [11]. In [4] some variants of policy and value iteration approaches to compute an optimal policy and value vector for the case of average and discounted reward optimality measures are evaluated with respect to their runtime.

There exist several papers concerned with aggregation of MDPs. In [10], a factored MDP is reduced to a MDP with an exponentially smaller state space MDP by stochastic bisimulation, such that an optimal policy for the reduced one is also optimal for the original MDP. The authors in [12] introduce different abstraction schemes for the states of a MDP. Some other techniques for state aggregation are given by $(\varepsilon_p, \varepsilon_f)$-lumpable partitions [14] and ε-homogeneous partitions [8]. In [15] numerical methods for bounding the stationary distribution for large state spaces are given which can be extended to obtain better bounds for BMDP models.

For the discrete time Markov models with state aggregation some approximations are studied: [1] treats approximate policy iteration for the described problem with discounted rewards and [18] attends to a value iteration algorithm.

In [8], it is shown how the reduced MDP with states corresponding to blocks of a partition of the state space can be generated. Furthermore, upper and lower bounds on the transition probabilities and rewards in the resulting BMDP model correspond to bounds on the transition probabilities for states that are grouped in the same partition. However, the mentioned approach operates only on discrete time models, and it computes simple bounds using minimal and maximal exit probabilities out of aggregated blocks.

In this paper, the approach of [8] is extended. Bounded aggregation for continuous time Markov decision models and the differences compared to available methods for discrete time Markov formalisms are considered. For discounted CTMDPs, continuous time introduces a different bounded aggregation method. Another goal is to improve upper and lower bounds for continuous time bounded (semi) Markov decision problems and compare the results to the previous work [8].

The paper is organized as follows. In the following section, we give an overview of the mathematical foundations for (semi) Markov decision processes and their extensions to uncertain transition probabilities as well as uniformization and optimization techniques for computation of optimal value vectors for Markov processes. Then, in Sect. 3, we briefly summarize known aggregation results for MDPS and develop an extended bounded aggregation approach for CTMDPs to derive a reduced state space model and make computing improved bounds and optimal policies tractable even for large state spaces. Finally, we continue with some examples and discuss the results in Sect. 5.

2   Background and Definitions

Here, we introduce basic definitions and notation. Vector and matrix identifiers are written in bold script, and individual elements of a vector v or a matrix M are accessed by $v(i)$ and $M(i,j)$. A column vector of ones is designated by $\mathbb{1}$.

2.1   Markov Decision Processes

A continuous-time Markov decision process (CTMDP) is defined as a tuple $\left(\mathcal{S}, \mathcal{A}, (Q^a)_{a \in \mathcal{A}}, (r^a)_{a \in \mathcal{A}}, p\right)$ where $\mathcal{S}$ is a finite set of *states* of a given order n, $\mathcal{A}$ is a finite set of *actions* of order m, $Q^a \in \mathbb{R}^{n \times n}$ is a transition rate matrix with $Q^a(i,j)$ giving the transition rate of moving from the state i to some state j when action a has been chosen. For the transition rate matrix Q^a it has to hold that $Q^a \mathbb{1} = 0$ and $Q^a(i,j) \geq 0$ if $i \neq j$ for all actions $a \in \mathcal{A}$. Furthermore, the initial probability distribution vector $p \in \mathbb{R}^{1 \times n}$ and the reward rate vector $r^a \in \mathbb{R}^{n \times 1}$ define a MDP. In the following, the states are numbered as $\mathcal{S} = \{1, \ldots, n\}$, and the actions are numbered as $\mathcal{A} = \{1, \ldots, m\}$.

To optimize some performance criteria of a CTMDP decision rules and policies are specified. A decision rule is a mapping $u_t : \mathcal{S} \to \mathcal{A}$ which is an assignment of actions to states at some point in time t. A policy can then be defined as a sequence of decision rules $\pi = (u_0, u_1, \ldots, u_T)$ for some $T \leq \infty$. A deterministic policy which is independent of time t is called *pure*. We consider here only *pure* policies and denote them for simplicity as *policies*. A pure policy can be described by a vector $\pi \in \mathcal{A}^{\mathcal{S}}$. We use the notation π to denote the policy whereas the vector notation π is applied if specific elements of the policy are accessed, i.e. $\pi(s)$ is the action chosen in state s under policy π. We designate by $Q^{\pi(t)}$ and $r^{\pi(t)}$ the matrices and vectors that are constructed from $Q^{\pi(t,s)}$ and $r^{\pi(t,s)}$ in row s. If the MDP is in state s and action a is selected, then it accumulates reward with rate $r^{\pi(t,s)}(s)$, its sojourn time in this state is exponentially distributed with rate $-Q^{\pi(t,s)}(s,s)$, and the transition probability to a different state s' is $\frac{Q^{\pi(t,s)}(s,s')}{-Q^{\pi(t,s)}(s,s)}$. For the definition of optimal policies and their values in CTMDPs as the methods for computing them we refer to the literature [5,13,17].

Uniformization: For long-term, "stationary" reward measures such as expected average reward and expected discounted reward, it is possible to transform continuous-time MDPs into discrete-time models with the same optimality behavior, which is sufficient for finding optimal policies [2,13,16]. Intuitively, a CTMDP is transformed into a discrete-time MDP in two steps: First, the sojourn time distributions are made identical for all states by introducing virtual self-transitions. Second, with uniform sojourn time distribution, the CTMDP is transformed to an equivalent discrete-time MDP.

Given a CTMDP $\left(\mathcal{S}, \mathcal{A}, (\boldsymbol{Q}^a)_{a \in \mathcal{A}}, (\boldsymbol{r}^a)_{a \in A}, \boldsymbol{p}\right)$, we can transform it into a discrete-time MDP with the following operations. First, we choose $\lambda \geq -\boldsymbol{Q}^a(s, s)$ for all $s \in \mathcal{S}, a \in \mathcal{A}$; λ is the so-called *uniformization rate*. Define matrices $(\boldsymbol{P}^a)_{a \in \mathcal{A}}$ and vectors $(\boldsymbol{z}^a)_{a \in \mathcal{A}}$ with $\boldsymbol{P}^a = \boldsymbol{I} + \frac{1}{\lambda}\boldsymbol{Q}^a$. The reward vectors are modified depending on the reward measure. For expected discounted rewards with discount rate β, we define reward vectors $\boldsymbol{z}_\beta^a$ with $\boldsymbol{z}_\beta^a(s) = \frac{r^a(s)}{\lambda + \beta}$ and assume a discount factor $\gamma = \frac{\lambda}{\lambda + \beta}$. For the expected average reward measure, the reward vectors are $\bar{\boldsymbol{z}}^a$ with $\bar{\boldsymbol{z}}^a(s) = \frac{r^a(s)}{\lambda}$.

This construction yields a discrete-time MDP $\left(\mathcal{S}, \mathcal{A}, (\boldsymbol{P}^a)_{a \in \mathcal{A}}, (\boldsymbol{z}^a)_{a \in \mathcal{A}}\right)$, where, depending on the reward measure selected, the reward vectors $\boldsymbol{z}^a$ are either $\bar{\boldsymbol{z}}^a$ or $\boldsymbol{z}_\beta^a$. The uniformization method is summarized in Algorithm 1.

Algorithm 1. Uniformization method for CTMDPs

Require: CTMDP $\left(\mathcal{S}, \mathcal{A}, (\boldsymbol{Q}^a)_{a \in \mathcal{A}}, (\boldsymbol{r}^a)_{a \in A}\right)$, discount rate β, *discounted* is true for
    the discounted reward measure and false else.
1:   $\lambda = \max_{\forall i \in \mathcal{S}, \forall a \in \mathcal{A}} |\boldsymbol{Q}^a(i, i)|$;
2: **for** $a \in \mathcal{A}$ **do**
3:      $\boldsymbol{P}^a = \boldsymbol{I} + \frac{1}{\lambda}\boldsymbol{Q}^a$;
4:      **if** *discounted* **then**
5:          $\boldsymbol{z}^a(i) = \frac{r^a(i)}{\lambda + \beta}, \quad \forall i \in \mathcal{S}$;
6:      **else**
7:          $\boldsymbol{z}^a(i) = \frac{r^a(i)}{\lambda}, \quad \forall i \in \mathcal{S}$;
8: **if** *discounted* **then**
9:      $\gamma = \frac{\lambda}{\lambda + \beta}$;
10: **return** Discrete-time MDP $\left(\mathcal{S}, \mathcal{A}, (\boldsymbol{P}^a)_{a \in \mathcal{A}}, (\boldsymbol{z}^a)_{a \in \mathcal{A}}\right)$, discount factor γ if *discounted* is true;

2.2 Bounded-Parameter Markov Decision Processes

In most cases, the parameters of a stochastic model are not known exactly. Consequently, they can be given by intervals rather than point estimates. The formalism of *bounded-parameter MDPs* [8,11] captures this concept. Bounded-parameter MDPs have been often defined in the literature. We review their definition and some optimality results here.

A bounded-parameter MDP is a tuple $\{\mathcal{S}, \mathcal{A}, \left(\boldsymbol{P}_{\updownarrow}^a\right)_{a \in \mathcal{A}}, \left(\boldsymbol{r}_{\updownarrow}^a\right)_{a \in \mathcal{A}}\}$ containing a set of discrete-time MDPs defined by a state and action space $\mathcal{S}, \mathcal{A}$. The discounting is performed with a discount factor $\gamma \in [0, 1)$. For each action $a \in \mathcal{A}$ lower and upper bounds on the transition probability parameters $\boldsymbol{P}_{\updownarrow}^a = (\boldsymbol{P}_{\downarrow}^a, \boldsymbol{P}_{\uparrow}^a)$ are defined with matrices $\boldsymbol{P}_{\downarrow}^a, \boldsymbol{P}_{\uparrow}^a$ satisfying the conditions

$$\boldsymbol{P}_{\downarrow}^a \leq \boldsymbol{P}_{\uparrow}^a,$$

where $\boldsymbol{P}_{\downarrow}^a, \boldsymbol{P}_{\uparrow}^a \in \mathbb{R}_{\geq 0}^{n \times n}$ and $\boldsymbol{P}_{\downarrow}^a \mathbf{1} \leq \mathbf{1} \leq \boldsymbol{P}_{\uparrow}^a \mathbf{1}$. Similarly, lower and upper bounds for the rewards are defined as $\boldsymbol{r}_{\updownarrow}^a = \left(\boldsymbol{r}_{\downarrow}^a, \boldsymbol{r}_{\uparrow}^a\right)$ with $\boldsymbol{r}_{\downarrow}^a, \boldsymbol{r}_{\uparrow}^a \in \mathbb{R}_{\geq 0}^{n \times 1}$ where the condition

$$\boldsymbol{r}_{\downarrow}^a \leq \boldsymbol{r}_{\uparrow}^a$$

is satisfied for all actions $a \in \mathcal{A}$.

The BMDP model defines a set of discrete-time MDPs with parameters varying according to this bounds. One is often interested in the set of policies that optimize the lower and upper bounds for the reward measures from the set $\boldsymbol{r}_{\updownarrow}^a$. These policies are permissible for the whole set of MDPs contained in $\{\mathcal{S}, \mathcal{A}, \left(\boldsymbol{P}_{\updownarrow}^a\right)_{a \in \mathcal{A}}, \left(\boldsymbol{r}_{\updownarrow}^a\right)_{a \in \mathcal{A}}\}$, the optimistic policy optimizing the upper bound of $\boldsymbol{r}_{\uparrow}^a$, and the pessimistic policy optimizing the lower bound for the rewards. In the following we consider only the lower bound computation, since the upper bound case is analogous. As in the area of robust optimization, the objective is to obtain the optimal solution for the whole uncertainty set. For BMDPs with discounted reward criterion one is interested in policy that maximizes the lower bound. In the pessimistic case the policy should fulfill

$$\pi_{\downarrow} = \arg\max_{\pi \in \Pi} \min_{\boldsymbol{P}^\pi \in \boldsymbol{P}_{\updownarrow}^\pi} g_{\gamma \downarrow}^\pi \tag{1}$$

for a value function $g_{\gamma \downarrow}^\pi$ which maps the policy π and the γ-discounted MDP $\boldsymbol{P}^\pi$ to the value of π in the MDP $\boldsymbol{P}^\pi$. To obtain the optimal pessimistic gain vector the Bellman-like equation has to be solved for each state $i \in \mathcal{S}$

$$g_{\gamma \downarrow}(i) = \max_{\pi \in \Pi} \min_{\boldsymbol{P}^\pi \in \boldsymbol{P}_{\updownarrow}^\pi} \left(\boldsymbol{r}_{\downarrow}^\pi(i) + \gamma \sum_{j \in \mathcal{S}} \boldsymbol{P}^\pi(i, j) g_{\gamma \downarrow}(j)\right). \tag{2}$$

The analysis of a BMDP can be performed efficiently regarding the nested $\max\min$ operator [4]. For the BMDPs with average reward criterion, define the expected reward in the k-th step in the future $R(\boldsymbol{r}, \boldsymbol{P}, k) = \boldsymbol{P}^k \boldsymbol{r}$. Then an optimal policy and the associated gain vector is the solution of the following equations.

$$\bar{g}_{\downarrow} = \max_{\pi \in \Pi} \min_{\boldsymbol{P} \in \boldsymbol{P}_{\updownarrow}^\pi} \lim_{K \to \infty} \left(\frac{1}{K} \sum_{k=1}^{K} R(\boldsymbol{r}_{\downarrow}^\pi, \boldsymbol{P}, k)\right),$$

$$\pi_{\downarrow} = \arg\max_{\pi \in \Pi} \min_{\boldsymbol{P} \in \boldsymbol{P}_{\updownarrow}^\pi} \lim_{K \to \infty} \left(\frac{1}{K} \sum_{k=1}^{K} R(\boldsymbol{r}_{\downarrow}^\pi, \boldsymbol{P}, k)\right). \tag{3}$$

For further reference on analysis algorithms for BMDPs, we refer to [4,11].

3   Bounded Aggregation Approach

Now that the backgrounds and syntax are clear, we are able to deal with aggregation. In the first part of this section we describe a common method for state space aggregation. Then in the second part we present our main contribution by introducing a new aggregation method for CTMDPs with specific constraints. Our refinement is in the calculation of the rates λ_i^{a-} and λ_i^{a+}, which are the bounds for the diagonal elements of the aggregated transition rate matrices. The last part consists of an application example.

3.1   Aggregation of MDPs

In this subsection we describe existing concepts of state space aggregation and of bounds and apply known results to continuous Markov processes. The state space S can be clustered into blocks with states from the set S_{ij} which exhibit nearly the same stochastic behavior with respect to other blocks [7–9]. There are different motivations and approaches to define or compute the state space partition. In general, the computation of an *optimal* partition with respect to minimal rates between blocks is NP-hard [8]. This implies that only heuristic approaches for computing a partition are useful. In our setting an additional motivation exists, namely the combination of states where decisions should be identical (e.g. due to physical restrictions like the unobservability of the detailed state). In the aggregated process all states in a block are represented by a single state such that a single decision is naturally chosen in this state.

Typically, parameter bounds for the aggregated process are obtained due to the bounds on the transition probabilities of separated blocks. Consider a continuous time Markov decision model. Assume that the generator matrix Q^a can be structured into k submatrices Q_{ij}^a of dimension $n_i \times n_j$ belonging to some block of states B_{ij}.

$$Q^a = \begin{bmatrix} Q_{11}^a & \cdots & Q_{1k}^a \\ \vdots & \ddots & \vdots \\ Q_{k1}^a & \cdots & Q_{kk}^a \end{bmatrix}$$

Then the aggregated Markov process can be generated by substituting each block B_{ij} by a single macro state $s \in \{1, \ldots, k\}$ thus shrinking the initial state space to overall k aggregates. Let $\tilde{S}$ denote the state space structured into macro states. Let now $0 \leq q_{ij}^{a-} \leq q_{ij}^{a+} < \infty$ be the upper and lower bounds for transition rates between two macro states $i, j \in \tilde{S}$ as described in [3]. The bounds can then be computed with

$$\begin{aligned} q_{ij}^{a-} &= \min_{m=1,\ldots,n_i} \left(\sum_{l=1}^{n_j} Q_{ij}^a(m,l) \right) \\ q_{ij}^{a+} &= \max_{m=1,\ldots,n_i} \left(\sum_{l=1}^{n_j} Q_{ij}^a(m,l) \right) \end{aligned} \tag{4}$$

such that intervals bounding the uncertain transition rates between two macro states i and j can be easily obtained as $q_{ij\updownarrow}^a = \{q^a \mid q_{ij}^{a-} \leq q^a \leq q_{ij}^{a+}\}$.

3.2   New Aggregation Method for CTMDPs

If we assume that the system starts in some macro state $i \in \tilde{S}$, then the expected sojourn time in i under decision a can be obtained using phase-type distribution (PHD) with subgenerator matrix $\boldsymbol{D}_i^a = \boldsymbol{Q}_{ii}^a$. Then, the process stays in macro state i a time which is distributed according to the PHD with parameters $(\boldsymbol{\phi}, \boldsymbol{D}_i^a)$ and afterwards moves to the next macro state j with rate $q_{ij}^a \in q_{ij\uparrow}^a$.

Let us consider some macro state $i \in \tilde{S}$. Rewriting the generator matrix $\boldsymbol{Q}^a$ for some action $a \in \mathcal{A}$ as

$$
\boldsymbol{Q}^a = \begin{bmatrix} \boldsymbol{Q}_{ii}^a & \boldsymbol{E}_{i\rightarrow}^a \\ \boldsymbol{F}_{i\leftarrow}^a & \begin{pmatrix} \boldsymbol{Q}_{jj}^a \cdots \\ \vdots \ddots \vdots \\ \cdots \boldsymbol{Q}_{kk}^a \end{pmatrix} \end{bmatrix}, \tag{5}
$$

where the transition rate matrix $\boldsymbol{E}_{i\rightarrow}^a$ is of dimension $n_i \times \sum_{l \in \tilde{S} \setminus \{i\}} n_l$ and the matrix $\boldsymbol{F}_{i\leftarrow}^a$ is of dimension $\sum_{l \in \tilde{S} \setminus \{i\}} n_l \times n_i$, the initial vector of the PHD with subgenerator $\boldsymbol{D}_i^a$ can be approximated using rows of the matrix $\boldsymbol{F}_{i\leftarrow}^a$ as follows.

Let $q = \sum_{l=1, l \neq i}^{k} n_l$ be the number of rows of the matrix $\boldsymbol{F}_{i\leftarrow}^a$. Note that PHD with subgenerator $\boldsymbol{D}_i^a$ describes the sojourn time distribution of a macro state i corresponding to the submatrix $\boldsymbol{Q}_{ii}^a$. The initial vector $\boldsymbol{\phi}$ of the PHD with subgenerator $\boldsymbol{D}_i^a$ can be guessed using rows of the matrix $\boldsymbol{F}_{i\leftarrow}^a$ in order to bound the sojourn time as follows. We obtain an initial vector $\boldsymbol{\phi}_l^a$, for each row $l \in \{1, \ldots, q\}$ of $\boldsymbol{F}_{i\leftarrow}^a$, $\forall a \in \mathcal{A}$ where $\boldsymbol{F}_{i\leftarrow}^a(l\bullet)$ is the lth row of the matrix $\boldsymbol{F}_{i\leftarrow}^a$ and $\boldsymbol{F}_{i\leftarrow}^a(l\bullet) \neq \boldsymbol{0}$ as

$$
\boldsymbol{\phi}_l^a = \boldsymbol{F}_{i\leftarrow}^a(l\bullet) / \|\boldsymbol{F}_{i\leftarrow}^a(l\bullet)\|_1. \tag{6}
$$

Note that Eq. 6 is in fact a normalization of the vector $\boldsymbol{F}_{i\leftarrow}^a(l\bullet)$. Evaluating the Eq. 6 for all non-zero rows q of $\boldsymbol{F}_{i\leftarrow}^a$ and for all $a \in \mathcal{A}$, the initial vectors resulting in minimal and maximal expected sojourn times of the PHD with subgenerator $\boldsymbol{D}_i^a$ can be computed. Then, the sojourn time bounds can be obtained as

$$
\begin{aligned}
\nu_i^{a-} &= \min_{\forall \boldsymbol{\phi}_j \in \Phi} \left(\boldsymbol{\phi}_j (-\boldsymbol{D}_i^a)^{-1} \boldsymbol{1} \right), \forall a \in \mathcal{A} \\
\nu_i^{a+} &= \max_{\forall \boldsymbol{\phi}_j \in \Phi} \left(\boldsymbol{\phi}_j (-\boldsymbol{D}_i^a)^{-1} \boldsymbol{1} \right), \forall a \in \mathcal{A}
\end{aligned} \tag{7}
$$

where $\boldsymbol{1}$ is a vector of dimension $n_i \times 1$. In Eq. 7, Φ is the set containing probability distribution vectors computed using (6) for all non-zero rows $l \in \{1, \ldots, q\}$ of $\boldsymbol{F}_{i\leftarrow}^a$, and all $a \in \mathcal{A}$. The rate bounds for the sojourn time distributions can then be estimated by $\lambda_i^{a-} = \frac{1}{\nu_i^{a-}}$ and $\lambda_i^{a+} = \frac{1}{\nu_i^{a+}}$ for all macro states $i \in \tilde{S}$.

Now we turn our attention to the non-diagonal elements $q_{ij}^{a\pm}$. As the value of $q_{ij}^{a\pm}$ is a bound on the exit rate from macro state i to macro state j, we bound it by computing the probability to enter macro state j from macro state i. This probability is $\boldsymbol{\phi}_k (-\boldsymbol{D}_i^a)^{-1} \boldsymbol{Q}_{ij}^a \boldsymbol{1}$. By multiplying it with the rate bounds we get the bounds

$$q_{ij}^{a-} = \lambda_i^{a+} \min_{\forall \phi_k \in \Phi} \left(\phi_k (-D_i^a)^{-1} Q_{ij}^a \mathbb{1} \right), \forall a \in \mathcal{A}$$

$$q_{ij}^{a+} = \lambda_i^{a-} \max_{\forall \phi_k \in \Phi} \left(\phi_k (-D_i^a)^{-1} Q_{ij}^a \mathbb{1} \right), \forall a \in \mathcal{A}. \tag{8}$$

as $\nu_i^{a+} \geq \nu_i^{a-} \Leftrightarrow \lambda_i^{a+} \leq \lambda_i^{a-}$.

Together, we obtain a continuous time BMDP model where λ_i^{a-} and λ_i^{a+} from (7) specify bounds of an exponential distribution for each macro state and (8) specifies bounds for transition rates between macro states. The resulting aggregated process is shown below.

$$\boldsymbol{Q}^{a-} = \begin{bmatrix} -\lambda_1^{a-} & q_{12}^{a-} & \cdots & q_{1k}^{a-} \\ q_{21}^{a-} & -\lambda_2^{a-} & \cdots & q_{2k}^{a-} \\ \vdots & \ddots & \ddots & \vdots \\ q_{k1}^{a-} & q_{k2}^{a-} & \cdots & -\lambda_k^{a-} \end{bmatrix}, \boldsymbol{Q}^{a+} = \begin{bmatrix} -\lambda_1^{a+} & q_{12}^{a+} & \cdots & q_{1k}^{a+} \\ q_{21}^{a+} & -\lambda_2^{a+} & \cdots & q_{2k}^{a+} \\ \vdots & \ddots & \ddots & \vdots \\ q_{k1}^{a+} & q_{k2}^{a+} & \cdots & -\lambda_k^{a+} \end{bmatrix}. \tag{9}$$

As for bounding matrices $\boldsymbol{Q}^{a-}(i,j) \leq \boldsymbol{Q}^{a+}(i,j)$ has to hold, the diagonal elements are $-\lambda_i^{a-}$ resp. $-\lambda_i^{a+}$ since $-\lambda_i^{a-} \leq -\lambda_i^{a+}$. Afterwards, the obtained bounds can be further improved as follows. First, we apply the uniformization technique described in Sect. 2.1 and solve Eq. 3 for the aggregated discrete-time BMDP model resulting from the uniformization. Then we use the pessimistic optimal policy $\boldsymbol{\pi}_\downarrow$ to update the bounds, but, in principle, also the optimistic optimal policy can be used to obtain tighter bounds.

Assume that the optimal policy $\boldsymbol{\pi}_\downarrow$ obtained for an aggregated process holds for all states partitioned in a block corresponding to the macro state for which the optimal action has been determined. We compute Eq. 7 where possible initial vectors in the set Φ are obtained using optimal policy $\boldsymbol{\pi}_\downarrow$ as follows

$$\phi_l^{\pi_\downarrow(l)} = F_{i\leftarrow}^{\pi_\downarrow(l)}(l\bullet)/\|F_{i\leftarrow}^{\pi_\downarrow(l)}(l\bullet)\|_1, \tag{10}$$

for all non-zero rows l of the policy matrix $\boldsymbol{Q}^{\pi_\downarrow}$ which is assembled by picking $\boldsymbol{Q}^{\pi_\downarrow}(l\bullet) = \boldsymbol{Q}^{\pi_\downarrow(l)}(l\bullet)$. At first we optimize the sojourn time bounds ν_i^{a-} and ν_i^{a+} over the whole set of initial vectors Φ. The update step supplies us a reduced subset of Φ that leads to an improved optimization. In general we get tighter bounds for the sojourn times by recalculating. We can now summarize our approach in Algorithm 2.

---

function COMPUTE_INITIAL_VECTORS(Set $\mathcal{F}$ containing matrices $\boldsymbol{F}$)
    $\Phi = \emptyset$;
    for $\boldsymbol{F} \in \tilde{\mathcal{F}}$ **do**
        q = rows($\boldsymbol{F}$);
        for $i = 1 \rightarrow q$ **do**
            Compute ϕ_i as given in Eq. 6;
            $\Phi = \Phi \cup \phi_i$;
    return Set Φ containing guessed initial vectors;

---

function COMPUTE_SOJOURN_TIME_BOUNDS(Φ, Set $\mathcal{D}_i$ containing matrices $\boldsymbol{D}_i^a$, $a \in \{1, \ldots, m\}$)
  $\Lambda = \emptyset$;
  Evaluate Eq. 7 using sets Φ and $\mathcal{D}_i$; Save results in the set Λ;
  return Set Λ containing λ_i^{a-}, λ_i^{a+} for all $a \in \mathcal{A}$;

Algorithm 2. Aggregation algorithm for CTMDPs

Require: Block structured CTMDP process with decision matrices $\boldsymbol{Q}^a$, $\forall a \in \mathcal{A}$. For each generator $\boldsymbol{Q}^a$, b blocks of dimension $n_i \times n_j$ and corresponding submatrices $\boldsymbol{Q}_{ij}^a$. Independend of a we define k as the number of blocks in a row of $\boldsymbol{Q}^1$.

1: $\mathcal{F} = \emptyset$; $\mathcal{D}_1 = \emptyset, \ldots, \mathcal{D}_k = \emptyset$; $\Lambda_i = \emptyset, \ldots, \Lambda_k = \emptyset$;
2: **for** $i = 1 \to k$ **do**
3:     **for** $\forall a \in \mathcal{A}$ **do**
4:         Compute a PHD with subgenerator $\boldsymbol{D}_i^a = \boldsymbol{Q}_{ii}^a$; $\mathcal{D}_i = \mathcal{D}_i \cup \boldsymbol{D}_i^a$;
5:         Compute matrix $\boldsymbol{F}_{i\leftarrow}^a$ as given in Eq. 5; $\mathcal{F} = \mathcal{F} \cup \boldsymbol{F}_{i\leftarrow}^a$;
6:         $\Phi = \text{compute\_initial\_vectors}(\mathcal{F})$;
7:         $\Lambda_i = \text{compute\_sojourn\_time\_bounds}(\Phi, \mathcal{D}_i)$;
8:         $\tilde{r}_i^{a+}, \tilde{r}_i^{a-} = $ maximum/minimum of all rewards in this block and action;
9:         **for** $j = 1 \to k$ **do**
10:            **if** $i \neq j$ **then**
11:                Compute transition rate bounds q_{ij}^{a-} and q_{ij}^{a+} by the given set Φ using Eq. 8 $\forall a \in \mathcal{A}$;
12:     $\mathcal{F} = \emptyset$;
13: Compute bounded discrete-time MDP model $\left(\tilde{\mathcal{S}}, \mathcal{A}, \left(\boldsymbol{P}_{\updownarrow}^a\right)_{a\in\mathcal{A}}, \left(r_{\updownarrow}^a\right)_{a\in\mathcal{A}}, p_{\updownarrow}\right)$ using Algorithm 1;
14: Compute optimal policy $\boldsymbol{\pi}_\downarrow$ and gain vector $\boldsymbol{g}_\downarrow$ using Eq. 1 and Eq. 2 or Eq. 3 ;
15: Determine policy matrix $\boldsymbol{Q}^{\boldsymbol{\pi}_\downarrow}$ with $\boldsymbol{Q}^{\boldsymbol{\pi}_\downarrow}(l\bullet) = \boldsymbol{Q}^{\boldsymbol{\pi}_\downarrow(l)}(l\bullet)$ for each row l of $\boldsymbol{Q}^{\boldsymbol{\pi}_\downarrow}$;
16: **for** $i = 1 \to k$ **do**   ▷ Update bounds according to the optimal policy
17:     Compute matrix $\boldsymbol{F}_{i\leftarrow}^{\boldsymbol{\pi}_\downarrow}$ as given in Eq. 5; $\mathcal{F} = \mathcal{F} \cup \boldsymbol{F}_{i\leftarrow}^{\boldsymbol{\pi}_\downarrow}$;
18:     $\Phi = \text{compute\_initial\_vectors}(\mathcal{F})$;
19:     $\Lambda_i = \text{compute\_sojourn\_time\_bounds}(\Phi, \mathcal{D}_i)$;
20:     **for** $j = 1 \to k$ **do**
21:         **if** $i \neq j$ **then**
22:             Compute transition rate bounds q_{ij}^{a-} and q_{ij}^{a+} by the given set Φ using Eq. 8 $\forall a \in \mathcal{A}$;
23: $\mathcal{F} = \emptyset$;
24: Compute set of aggregated transition rate matrices $\left(\tilde{\boldsymbol{Q}}_{\updownarrow}^a\right)_{a\in\mathcal{A}}$ like in 9;
25: **return** $\left(\tilde{\boldsymbol{Q}}_{\updownarrow}^a\right)_{a\in\mathcal{A}}, \left(\tilde{r}_{\updownarrow}^a\right)_{a\in\mathcal{A}}$

---

4 Experiments

We perform different experiments with randomly generated CTMDP instances with state space sizes ranging from 100 to 500 to compare the different aggregation approaches. All computations were performed on a machine with a 3.0 GHz

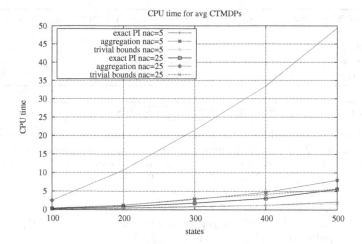

Fig. 1. CPU times for average reward criterion aggregation algorithms. The plotted results are obtained for random CTMDP models with 5 and 25 actions, an average reward of 5, an average sojourn time of $1/5 \cdot |S|$ and a block size of mean $10/3$.

20-Core processor and 126 GB main memory running Debian Linux. We used *Matlab* implementation of our algorithms.

First, we analyzed randomly generated CTMDP models with dense matrices and reward vectors with a number of states varied from 100 to 500 and a number of actions $\mathcal{A} = \{5, 25\}$. The average sojourn times (the entrees on the diagonal of a transition rate matrix) depends on the size of states and is $1/5 \cdot |S|$ and all nondiagonal elements are randomly and normalized to the sojourn time. Also the number of blocks and their size depend on $|S|$, cause there are $3/10 \cdot |S|$ blocks given with a mean size of $10/3$. In both cases, the discounted and the average, the rewards are expected 5. For the discounted problem we test dieffernt values for β, but the results are similar enough to show you only one case for $\beta = 2$. For every combination of state space and actions we build ten examples, compare the seperate results of the aggregation methods, and then we compute the mean of them. To obtain the average or discounted reward value and optimal strategy policy iteration method has been applied.

The plots in Fig. 1 show computation times for exact and aggregation algorithms for CTMDPs as a function of the number of states. We compared results obtained using the exact solution method, trivial aggregation and the improved aggregation methods. In the trivial aggregation method bounds for block sojourn times are obtained using minimal and maximal exit rates out of block. As one can see, the improved aggregation algorithm requires much more computation time. The reason is the computational effort required to compute Eq. 7 in order to derive tighter bounds.

In Tables 1 and 2 we list the compared values computed by an intuitive aggregation algorithm and by our new aggregation method. The exact solution is only given to show you the quality of our results. The main consequence is the relative

Table 1. Results for average CTMDPs with 25 actions

Number of states (States)	Trivial lower bounds (TrivL)	Improved lower bounds (ImprL)	Exact solution (Exact)	Improved upper bounds (ImprL)	Trivial upper bounds (TrivU)	Relative ratio (Ratio)
100	0.14378	0.18362	0.24946	0.24040	0.24954	1.71247
200	0.06657	0.08501	0.12466	0.11924	0.12469	1.50527
300	0.04580	0.05705	0.08313	0.07958	0.08314	1.46999
400	0.03455	0.04293	0.06118	0.05954	0.06237	1.47638
500	0.02868	0.03531	0.04904	0.04768	0.04991	1.52802

Table 2. Results for discounted CTMDPs with 25 actions and discount factor 2

(States)	(TrivL)	(ImprL)	(Exact)	(ImprU)	(TrivU)	(Ratio)
100	2.87161	3.49862	5.08380	5.19395	5.22686	1.38927
200	2.66808	3.34325	4.97368	5.07058	5.10530	1.41288
300	2.73046	3.37402	4.94414	5.03400	5.06653	1.40827
400	2.69198	3.32580	4.92630	5.01026	5.04692	1.39925
500	2.64740	3.31049	4.91676	5.00030	5.03700	1.41653

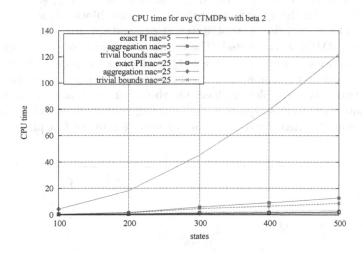

Fig. 2. CPU times for average reward criterion aggregation algorithms. The plotted results are obtained for random CTMDP models with 5 and 25 actions, an average reward of 5, an average sojourn time of $1/5 \cdot |S|$, a block size of mean $10/3$ and $\beta = 2$.

ratio, which is calculated as the quotient of the difference between the trivial bounds and improved bounds. With our aggregation method we gain a relativ improvement of round about 50% in comparison to the intuitive algorithm. A relativ ratio of 2 would mean, that the span of upper and lower bounds is halved.

We discuss the applicability of the aggregation approach using a small closed central server queueing network where jobs can be alternatively routed to one of two peripheral servers. The queueing network is illustrated in Fig. 3. The central queue is a FCFS station with exponentially distributed service times with rate $\mu = 2$. After leaving the central station, a job enters one of the two peripheral stations. The choice of the station is a decision which can be made upon leaving the central station. Service times at the peripheral stations are distributed according to an Erlang 2 distribution with mean 1 at station Q_2 and according to a hyperexponential distribution with parameters $\mu_{31} = 4$, $\mu_{32} = 1/4$ and $p_{21} = 0.2$ for queue Q_3. Thus, both peripheral queues have the same mean service time.

The decision to choose Q_2 or Q_3 can be made according to the current population at the queues but cannot be based on the state of the service time distribution which is introduced in the Markov model to describe non-exponential times. If the whole model is interpreted as a CTMDP, then the optimal policy will consider the internal state of the service time and it might be better to choose a longer queue. E.g., if in Q_3 a single customer is in phase 2, then the mean service time of this customer is 4, whereas the mean service time of a customer in phase 2 of Q_2 is 0.5. Thus, in this situation Q_2 is the better choice as long as it contains at most 3 customers more than Q_3.

If decision have to be made based on the population only, states with the same population in the queues are collected in one block. This implies that in our case blocks contain up to 4 states, if Q_2 and Q_3 are non-empty. Using aggregation a BMDP is computed. The robust and therefore pessimistic policy for this BMDP avoids routing into queue Q_2 as long as the population difference between Q_3 does not become too large because in the worst case, the service time distribution is in the slower phase. On the other hand, an optimistic policy tries to route customers to queue Q_3 because the service time might be much smaller than in Q_2 whenever the customer in service is in the fast phase (Fig. 3).

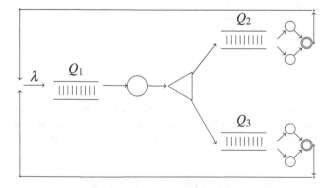

Fig. 3. Queueing network example.

Table 3. Results for the described example in the average and discounted case

(States)	(TrivL)	(ImprL)	(Exact)	(ImprU)	(TrivU)	(Ratio)
25 avg	0.6290	0.6743	1.0053	1.1535	1.1866	1.1636
61 avg	0.5756	0.6562	0.9912	1.1047	1.1779	1.3429
113 avg	0.6115	0.6707	0.8467	1.1384	1.2247	1.3111
25 disc	2.6608	2.9835	4.2256	5.4417	5.6080	1.1989
61 disc	2.5388	2.8003	4.3079	5.2663	5.4300	1.1724
113 disc	1.9375	2.2571	4.1899	5.4533	5.6438	1.1596

We analyze the model for the populations 3, 5 and 7 resulting in CTMDPs with 25, 61 and 113 states. The state spaces of the corresponding BMDPs contain 10, 21 and 36 states. For the results look at Table 3.

5   Conclusions

In this paper, we propose a state aggregation method for CTMDPs. In general, state aggregation enables one to reduce the number of states in a given CTMDP by deriving a bounded-parameter Markov model. The paper presents an improved aggregation approach to compute upper and lower reward bounds for CTMDPs for groups of similar states which are treated equally by a decision maker. It is shown that continuous time models can be efficiently aggregated when sojourn times in blocks are approximated using phase-type distributions. In the proposed method, one can refine the obtained bounds after an optimal policy has been computed. Comparing our results to established aggregation methods, we show that the proposed algorithm computes better bounds by an acceptable extra computational effort. Though the required CPU time is increased by a factor close to two, the difference between upper and lower bounds is reduced by nearly one half compared to a trivial aggregation algorithm.

The approach can be extended to refine the reward aggregation on the basis of stationary quantities in CTMDPs as presented in [3]. We have evaluated our aggregation method on randomly generated CTMDPs and a small queueing model. A special case that is in our opinion most interesting occurs when the states that are lumped into one are similar in behavior (with respect to transition and reward rates); we conjecture that in this case, our approach would show an even further improvement over the standard bounded-parameter aggregation approach. Furthermore, it is possible to improve the bounds for a fixed policy further by using the iterative bounding approach form [15].

References

1. Abate, A., Česka, M., Kwiatkowska, M.: Approximate policy iteration for Markov decision processes via quantitative adaptive aggregations. In: Artho, C., Legay, A., Peled, D. (eds.) ATVA 2016. LNCS, vol. 9938, pp. 13–31. Springer, Cham (2016). doi:10.1007/978-3-319-46520-3_2
2. Beutler, F.J., Ross, K.W.: Uniformization for semi-Markov decision processes under stationary policies. J. Appl. Probability **24**, 644–656 (1987)
3. Buchholz, P.: Bounding reward measures of Markov models using the Markov decision processes. Numerical Lin. Alg. with Applic. **18**(6), 919–930 (2011)
4. Buchholz, P., Dohndorf, I., Scheftelowitsch, D.: Analysis of Markov decision processes under parameter uncertainty. In: Reinecke, P., Di Marco, A. (eds.) EPEW 2017. LNCS, vol. 10497, pp. 3–18. Springer, Cham (2017). doi:10.1007/978-3-319-66583-2_1
5. Buchholz, P., Hahn, E.M., Hermanns, H., Zhang, L.: Model checking algorithms for CTMDPs. In: Computer Aided Verification - 23rd International Conference, CAV 2011, Snowbird, UT, USA, 14–20 July 2011, Proceedings, pp. 225–242 (2011)
6. Buchholz, P., Kriege, J., Felko, I.: Input Modeling with Phase-Type Distributions and Markov Models. SM. Springer, Cham (2014)
7. Courtois, P., Semal, P.: Bounds for the positive eigenvectors of nonnegative matrices and for their approximations by decomposition. J. ACM **31**(4), 804–825 (1984)
8. Dean, T.L., Givan, R., Leach, S.M.: Model reduction techniques for computing approximately optimal solutions for Markov decision processes. In: Geiger, D., Shenoy, P.P. (eds.) UAI 1997: Proceedings of the Thirteenth Conference on Uncertainty in Artificial Intelligence, Brown University, Providence, Rhode Island, USA, 1–3 August 1997, pp. 124–131. Morgan Kaufmann (1997)
9. Franceschinis, G., Muntz, R.R.: Bounds for quasi-lumpable Markov chains. Perform. Eval. **20**(1–3), 223–243 (1994)
10. Givan, R., Dean, T.L., Greig, M.: Equivalence notions and model minimization in Markov decision processes. Artif. Intell. **147**(1–2), 163–223 (2003)
11. Givan, R., Leach, S.M., Dean, T.L.: Bounded-parameter Markov decision processes. Artif. Intell. **122**(1–2), 71–109 (2000)
12. Li, L., Walsh, T.J., Littman, M.L.: Towards a unified theory of state abstraction for MDPs. In: International Symposium on Artificial Intelligence and Mathematics, ISAIM 2006, Fort Lauderdale, Florida, USA, 4–6 January 2006 (2006)
13. Puterman, M.L.: Markov Decision Processes. Wiley, New York (2005)
14. Ren, Z., Krogh, B.: State aggregation in Markov decision processes. In: Proceedings of the 41st IEEE Conference on Decision and Control, vol. 4, pp. 3819–3824. IEEE (2002)
15. Semal, P.: Refinable bounds for large Markov chains. IEEE Trans. Computers **44**(10), 1216–1222 (1995)
16. Serfozo, R.F.: An equivalence between continuous and discrete time Markov decision processes. Oper. Res. **27**(3), 616–620 (1979)
17. Tewari, A., Bartlett, P.L.: Bounded parameter Markov decision processes with average reward criterion. In: Bshouty, N.H., Gentile, C. (eds.) COLT 2007. LNCS, vol. 4539, pp. 263–277. Springer, Heidelberg (2007). doi:10.1007/978-3-540-72927-3_20
18. Van Roy, B.: Performance loss bounds for approximate value iteration with state aggregation. Math. Oper. Res. **31**(2), 234–244 (2006)

Interactive Markovian Equivalence

Arpit Sharma[✉]

Department of Electrical Engineering and Computer Science,
Indian Institute of Science Education and Research Bhopal, Bhopal, India
arpit@iiserb.ac.in

Abstract. Behavioral equivalences relate states which are indistinguishable for an external observer of the system. This paper defines two equivalence relations, interactive Markovian equivalence (IME) and weak interactive Markovian equivalence (WIME) for closed IMCs. We define the quotient system under these relations and investigate their relationship with strong bisimulation and weak bisimulation, respectively. Next, we show that both IME and WIME can be used for repeated minimization of closed IMCs. Finally we prove that time-bounded reachability properties are preserved under IME and WIME quotienting.

Keywords: Markov chains · Scheduler · Equivalence · Bisimulation · Reachability

1 Introduction

Interactive Markov chains (IMCs) [15,16] extend labeled transition systems (LTS) with stochastic aspects. IMCs thus support both reasoning about non-deterministic behaviors as in LTSs and stochastic phenomena as in continuous-time Markov chains (CTMCs). IMCs are compositional, i.e., a parallel composition operator allows one to construct a complex IMC from several component IMCs running in parallel.

IMCs are widely used for performance and dependability analysis of complex distributed systems, e.g., shared memory mutual exclusion protocols [20]. They have been used as semantic model for amongst others dynamic fault trees [9,10], architectural description languages such as AADL [8,11], generalized stochastic Petri nets [17], and Statemate [7]. They are also used for modeling and analysis of GALS (Globally Asynchronous Locally Synchronous) hardware design [12]. For analysis, model checking algorithms [14,21] are applied on *closed*[1] IMC models to compute the probability of linear or branching real-time objectives, e.g., extremal time-bounded reachability probabilities [16,21] and expected time [14].

Equivalence relations are used to compare the behavior of IMC models [15]. Abstraction techniques based on equivalence relations reduce the state space of IMCs, by aggregating equivalent states into a single state. The reduced state space obtained under an equivalence relation, called a quotient, can then be used

---

[1] An IMC is said to be closed if it is not subject to any further synchronization.

© Springer International Publishing AG 2017
P. Reinecke and A. Di Marco (Eds.): EPEW 2017, LNCS 10497, pp. 33–49, 2017.
DOI: 10.1007/978-3-319-66583-2_3

for analysis provided it preserves a rich class of properties of interest. Strong and weak bisimulation [15,16] are two well known equivalence relations for IMCs. Both these equivalences preserve time-bounded reachability probabilities [16].

This paper proposes a novel theoretical framework for the state space reduction of closed IMCs. We define interactive Markovian equivalence and weak interactive Markovian equivalence for closed IMCs. Unlike bisimulation which compares states on the basis of their direct successors, IME considers a *two-step* perspective. Before explaining the idea of IME, let us recall that every state of a closed IMC can either have Markovian transitions or τ-labeled interactive transitions. Every Markovian transition is labeled with a positive real number λ. This parameter indicates the rate of the exponential distribution, i.e., the probability of a λ-labeled transition to be enabled within t time units equals $1 - e^{-\lambda \cdot t}$. Two Markovian states s and s' are IME equivalent if for each pair of their direct predecessors *weighted rate* to directly move to any equivalence class via the equivalence class $[s] = [s']$ coincides. Similarly, two interactive states are IME equivalent if for each pair of their direct predecessors it is possible to reach the same set of equivalence classes in two steps. For WIME, we abstract from stutter steps and thus each predecessor of equivalence class C should reach the same set of equivalence classes in two or more steps such that all the extra steps are taken within C.

Contributions. The main contributions of this paper are as follows:

- We provide a structural definition of IME on closed IMCs, define quotient under IME and investigate its relationship with strong bisimulation.
- We provide a structural definition of weak IME (WIME) on closed IMCs, define quotient under WIME and investigate its relationship with weak bisimulation.
- Finally, we prove that time-bounded reachability probabilities are preserved under IME and WIME quotienting.

1.1 Related Work

For continuous-time Markov chains (CTMCs), several variants of weak and strong bisimulation equivalence and simulation pre-orders have been defined in [4]. Their compatibility to (fragments of) stochastic variants of computation tree logic (CTL) has been thoroughly investigated, cf. [4]. In [5], Bernardo considered Markovian testing equivalence over sequential Markovian process calculus (SMPC), and coined the term T-lumpability [6] for the induced state-level aggregation where T stands for testing. His testing equivalence is a congruence w.r.t. parallel composition, and preserves transient as well as steady-state probabilities. Bernardo's T-lumpability has been reconsidered in [26] where weighted lumpability (WL) is defined as a structural notion on CTMCs. Note that DTA and MTL specifications are preserved under WL [26]. In [27], several linear-time equivalences (Markovian trace equivalence, failure and ready trace equivalence) for CTMCs have been investigated. Testing scenarios based on push-button experiments have been used for defining these equivalences.

In [22], authors have defined strong bisimulation relation for CTMDPs. This paper also proves that CSL properties are preserved under bisimulation for CTMDPs. Trace semantics for interactive Markov chains (IMCs) have been defined in [28]. In this paper testing scenarios using button pushing experiments have been used to define several variants of trace equivalences that arise by varying the type of schedulers. Note that the relationship of IME and trace semantics for IMCs is not clear. In the branching-time setting, strong and weak bisimulation relations for IMCs have been defined in [15,16]. IME and strong bisimulation are incomparable. Similarly, in the weak setting, WIME and weak bisimulation are incomparable. Our definition of IME here builds on that investigated in [26] for CTMCs.

Organisation of the paper. Section 2 briefly recalls the main concepts of IMCs. Section 3 defines interactive Markovian equivalence (IME) and investigates its relationship with strong bisimulation. Section 4 defines the weaker variant of IME (WIME) and investigates its relationship with weak bisimulation. Section 5 proves the preservation of time-bounded reachability properties. Finally, Sect. 6 concludes the paper and discusses directions for future research.

2 Preliminaries

This section presents the necessary definitions and basic concepts related to interactive Markov chains (IMCs) that are needed for the understanding of the rest of this paper.

Definition 1 (IMC). *An* interactive Markov chain *(IMC) is a tuple* $\mathcal{I} = (S, s_0, Act, AP, \rightarrow, \Rightarrow, L)$ *where:*

- *S is a finite set of states,*
- *s_0 is the initial state,*
- *Act is a finite set of actions,*
- *AP is a finite set of atomic propositions,*
- *$\rightarrow \subseteq S \times Act \times S$ is a set of interactive transitions,*
- *$\Rightarrow \subseteq S \times \mathbb{R}_{>0} \times S$ is a set of Markovian transitions, and*
- *$L : S \rightarrow 2^{AP}$ is a labeling function.*

We abbreviate $(s, a, s') \in \rightarrow$ as $s \xrightarrow{a} s'$ and similarly, $(s, \lambda, s') \in \Rightarrow$ by $s \xRightarrow{\lambda} s'$. Let $IT(s)$ and $MT(s)$ denote the set of interactive and Markovian transitions that leave state s. A state s is *Markovian* iff $MT(s) \neq \varnothing$ and $IT(s) = \varnothing$; it is *interactive* iff $MT(s) = \varnothing$ and $IT(s) \neq \varnothing$. Further, s is a *hybrid* state iff $MT(s) \neq \varnothing$ and $IT(s) \neq \varnothing$; finally s is a *deadlock* state iff $MT(s) = \varnothing$ and $IT(s) = \varnothing$. In this paper we only consider those IMCs that do not have any deadlock states. Let $MS \subseteq S$ and $IS \subseteq S$ denote the set of Markovian and interactive states in IMC $\mathcal{I}$. For any Markovian state $s \in MS$ let $R(s, s') = \sum \{\lambda | s \xRightarrow{\lambda} s'\}$ be the rate to move from state s to state s'. The exit rate for state s is defined by: $E(s) = \sum_{s' \in S} R(s, s')$.

It is easy to see that an IMC where $MT(s) = \varnothing$ for any state s is an LTS. An IMC where $IT(s) = \varnothing$ for any state s is a CTMC. The semantics of IMCs can thus be given in terms of the semantics of CTMCs (for Markovian transitions) and LTSs (for interactive transitions).

The meaning of a Markovian transition $s \overset{\lambda}{\Rightarrow} s'$ is that the IMC moves from state s to s' within t time units with probability $1 - e^{\lambda \cdot t}$. If s has multiple outgoing Markovian transitions to different successors, then we speak of a race between these transitions, known as the *race condition*. In this case, the probability to move from s to s' within t time units is $\frac{R(s,s')}{E(s)} \cdot (1 - e^{E(s) \cdot t})$.

Example 1. Consider the IMC $\mathcal{I}$ shown in Fig. 1, where $S = \{s_0, s_1, s_2, s_3, s_4, s_5, s_6, s_7, s_8, s_9\}$, $AP = \{a, b, c\}$, $Act = \{\alpha, \beta, \gamma\}$ and s_0 is the initial state. The set of interactive states is $IS = \{s_0, s_1, s_2\}$; MS contains all the other states. Note that there is no hybrid state in IMC $\mathcal{I}$. Non-determinism between action transitions appears in state s_0. Similarly, race condition due to multiple Markovian transitions appears in s_3 and s_4.

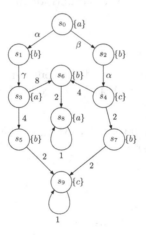

Fig. 1. An example IMC $\mathcal{I}$

We assume that in closed IMCs all outgoing interactive transitions from every state $s \in S$ are labeled with $\tau \in Act$ (internal action).

Definition 2 *(Maximal progress [16]).* In any closed IMC, interactive transitions take precedence over Markovian transitions.

Intuitively, the maximal progress assumption states that in closed IMCs, τ labeled transitions are not subject to interaction and thus can happen immediately[2], whereas the probability of a Markovian transition to happen immediately is zero. Accordingly, we assume that each state s has either only outgoing

---

[2] We restrict to models without zenoness.

τ transitions or outgoing Markovian transitions. In other words, a closed IMC only has interactive and Markovian states.

Definition 3 (IMC timed paths). *Let* $\mathcal{I} = (S, s_0, Act, AP, \rightarrow, \Rightarrow, L)$ *be an IMC. An infinite path* π *in* $\mathcal{I}$ *is a sequence* $s_0 \xrightarrow{\sigma_0, t_0} s_1 \xrightarrow{\sigma_1, t_1} s_2 \ldots s_{n-1} \xrightarrow{\sigma_{n-1}, t_{n-1}} s_n \ldots$ *where* $s_i \in S$, $\sigma_i \in Act$ *or* $\sigma_i = \bot$, *and* $t_i \in \mathbb{R}_{\geq 0}$ *is the sojourn time in state* s_i. *A finite path* π *is a finite prefix of an infinite path. The length of an infinite path* π, *denoted* $|\pi|$ *is* ∞; *the length of a finite path* π *with* $n + 1$ *states is* n.

We use a distinguished action $\bot \notin Act$ to indicate Markovian transitions and extend the set of actions to $Act_\bot = Act \cup \{\bot\}$. Let $Paths^{\mathcal{I}} = Paths^{\mathcal{I}}_{fin} \cup Paths^{\mathcal{I}}_\omega$ denote the set of all paths in $\mathcal{I}$ that start in s_0, where $Paths^{\mathcal{I}}_{fin} = \bigcup_{n \in \mathbb{N}} Paths^{\mathcal{I}}_n$ is the set of all finite paths in $\mathcal{I}$ and $Paths^{\mathcal{I}}_n$ denote the set of all finite paths of length n that start in s_0. Let $Paths^{\mathcal{I}}_\omega$ is the set of all infinite paths in $\mathcal{I}$ that start in s_0. For infinite path $\pi = s_0 \xrightarrow{\sigma_0, t_0} s_1 \xrightarrow{\sigma_1, t_1} s_2 \ldots$ and any $i \in \mathbb{N}$, let $\pi[i] = s_i$, the $(i + 1)$th state of π. For any $t \in \mathbb{R}_{\geq 0}$, let $\pi@t$ denote the sequence of states that π occupies at time t. Note that $\pi@t$ is in general not a single state, but rather a sequence of several states, as an IMC may exhibit immediate transitions and thus may occupy various states at the same time instant. Let $Act(s)$ denote the set of enabled actions from state s. Note that in case s is a Markovian state then $Act(s) = \{\bot\}$.

Example 2. Consider an example timed path $\pi = s_0 \xrightarrow{\alpha, 0} s_1 \xrightarrow{\gamma, 0} s_3 \xrightarrow{\bot, 1.5} s_2 \xrightarrow{\gamma, 0} s_5$. Here we have $\pi[2] = s_3$ and $\pi@(1.5 - \epsilon) = \langle s_3 \rangle$, where $0 < \epsilon < 1.5$. Similarly, $\pi@1.5 = \langle s_2 s_5 \rangle$.

σ-*algebra.* In order to construct a measurable space over $Paths^{\mathcal{I}}_\omega$, we define the following sets: $\Omega = Act_\bot \times \mathbb{R}_{\geq 0} \times S$ and the σ-field $\mathcal{J} = (2^{Act_\bot} \times \mathcal{J}_R \times 2^S)$, where $\mathcal{J}_R$ is the Borel σ-field over $\mathbb{R}_{\geq 0}$ [2,3]. The σ-field over $Paths^{\mathcal{I}}_n$ is defined as $\mathcal{J}_{Paths^{\mathcal{I}}_n} = \sigma(\{S_0 \times M_0 \times \ldots \times M_{n-1} | S_0 \in 2^S, M_i \in \mathcal{J}, 0 \leq i \leq n-1\})$. A set $B \in \mathcal{J}_{Paths^{\mathcal{I}}_n}$ is a base of a cylinder set C if $C = Cyl(B) = \{\pi \in Paths^{\mathcal{I}}_\omega | \pi[0 \ldots n] \in B\}$, where $\pi[0 \ldots n]$ is the prefix of length n of the path π. The σ-field $\mathcal{J}_{Paths^{\mathcal{I}}_\omega}$ of measurable subsets of $Paths^{\mathcal{I}}_\omega$ is defined as $\mathcal{J}_{Paths^{\mathcal{I}}_\omega} = \sigma(\cup_{n=0}^{\infty}\{Cyl(B) | B \in \mathcal{J}_{Paths^{\mathcal{I}}_n}\})$.

2.1 Schedulers

Non-determinism in an IMC is resolved by a scheduler [16,29]. Schedulers are also known as adversaries or policies. More formally, schedulers[3] are defined as follows:

---

[3] We only consider total-time deterministic positional (TTDP) schedulers as they are sufficient for computing the maximum (resp. minimum) probability of time-bounded reachability properties [16].

Definition 4 (Scheduler). *A scheduler for IMC $\mathcal{I} = (S, s_0, Act, AP, \rightarrow, \Rightarrow, L)$ is a measurable function $\mathcal{D} : S \times \mathbb{R}_{\geq 0} \rightarrow Act_{\perp}$, such that for $n \in \mathbb{N}$ and $t \in \mathbb{R}_{\geq 0}$,*

$$\mathcal{D}(s_n, t) \in Act(s_n)$$

Intuitively, next action to be executed depends on the current state and total-time that passed up to the current state. This scheduler is deterministic as it always selects the next action with probability 1. Note that for Markovian states scheduler always selects $\perp$ with probability 1. Let $Adv_{TTDP}(\mathcal{I})$ denote the set of all total-time deterministic positional schedulers of $\mathcal{I}$.

Once the non-deterministic choices of an IMC $\mathcal{I}$ have been resolved by a scheduler, say $\mathcal{D}$, the induced model obtained is purely stochastic. To that end the unique probability measure for probability space $(Paths_\omega^{\mathcal{I}}, \mathcal{J}_{Paths_\omega^{\mathcal{I}}})$ can be defined [18,21]. Given a scheduler $\mathcal{D}$ and a set Π of infinite paths, then $Pr_{\mathcal{D}}(\Pi)$ denotes the probability of visiting all paths in Π under scheduler $\mathcal{D}$ starting from initial state s_0. The probability of the set of paths of length $(n+1)$ is defined as a product between the probability of the set of paths of length n and the one-step transition probability to go from $(n+1)$-th state to $(n+2)$-th state by executing action, say α, selected by the scheduler $\mathcal{D}$ [21].

Assumptions. Throughout this paper we make the following assumptions:

1. Every state of IMC $\mathcal{I}$ has at least one predecessor. This is not a restriction, as any IMC $\mathcal{I} = (S, s_0, Act, AP, \rightarrow, \Rightarrow, L)$ can be transformed into an equivalent IMC $(S', s_0', Act, AP', \rightarrow', \Rightarrow, L')$ which fulfills this condition. This is done by adding a new state $\hat{s}$ to S equipped with a self-loop and which has a transition to each state in S without predecessors. Let all the outgoing transitions from $\hat{s}$ be labeled with τ. To distinguish this state from the others we set $L'(\hat{s}) = \#$ with $\# \notin AP$ (All other labels, states and transitions remain unaffected). Let $s_0' = s_0$. It follows that all states in $S' = S \cup \{\hat{s}\}$ have at least one predecessor. Moreover, the reachable state space of both IMCs coincides.
2. We also assume that the initial state s_0 of an IMC is distinguished from all other states by a unique label, say \$. This assumption implies that for any equivalence that groups equally labeled states, $\{s_0\}$ constitutes a separate equivalence class.

Remark 1. Both assumptions do not affect the basic properties of an IMC such as linear or branching real-time properties. For convenience, we neither show the state $\hat{s}$ nor the label \$ in figures. These assumptions are required as Sect. 3 proposes an equivalence relation for closed IMCs that checks reachability from predecessors of every equivalence class to its successor equivalence classes.

3 Interactive Markovian Equivalence

Before defining interactive Markovian equivalence, we first define some auxiliary concepts. All the definitions presented in this section are relative to a closed IMC

$\mathcal{I} = (S, s_0, Act, AP, \rightarrow, \Rightarrow, L)$, where $Act = \{\tau\}$. For any state $s \in S$ and $Act = \{\tau\}$, the set of τ-predecessors of s is defined by: $Pred(s, \tau) = \{s' \in S | s' \xrightarrow{\tau} s\}$ and $Pred(s) = \{s' \in S | R(s', s) > 0\} \cup Pred(s, \tau)$. Let for $C \subseteq S$, $Pred(C) = \bigcup_{s \in C} Pred(s)$. Similarly, the set of τ-successors of any state s is defined by: $Post(s, \tau) = \{s' \in S | s \xrightarrow{\tau} s'\}$ and $Post(s) = \{s' \in S | R(s, s') > 0\} \cup Post(s, \tau)$. Let $Post(C) = \bigcup_{s \in C} Post(s)$ and $Post(s, \tau, C) = \{s' \in C | s \xrightarrow{\tau} s'\}$.

Definition 5. *Let $C \subseteq S$, then C is said to be* interactive *closed iff $C \subseteq IS \land Pred(C) \subseteq IS$.*

Definition 6. *Let $C \subseteq S$, then C is said to be* Markovian *closed iff $C \subseteq MS \land Pred(C) \subseteq MS$.*

Let $I(S)$ denote the set of all possible subsets of S that are interactive closed. Let $M(S)$ denote the set of all possible subsets of S that are Markovian closed.

Example 3. Consider the IMC shown in Fig. 2 (left). Let $C = \{s_1, s_2\}$ and $D = \{s_5, s_6, s_7\}$. Here C is interactive closed since $C \subseteq IS$ and $Pred(C) = \{s_0\} \subseteq IS$. Similarly, D is Markovian closed.

Definition 7 (Predecessor based reachability[4]). *For $s \in S$ and $C, D \subseteq S$, the function $Pbr : S \times 2^S \times 2^S \rightarrow \{0, 1\}$ is defined as:*

$$Pbr(s, C, D) = \begin{cases} 1 & \text{if } \exists s' \in Post(s, \tau, C) \text{ s.t. } Post(s', \tau, D) \neq \varnothing \\ 0 & \text{otherwise.} \end{cases}$$

Definition 8 (Weighted probability). *For $s, s' \in S$ and $C \subseteq S$, the function $P : S \times S \times 2^S \rightarrow \mathbb{R}_{\geq 0}$ is defined by:*

$$P(s, s', C) = \begin{cases} \frac{P(s,s')}{P(s,C)} & \text{if } s' \in C \text{ and } P(s, C) > 0 \\ 0 & \text{otherwise.} \end{cases}$$

where $P(s, s') = \frac{R(s,s')}{E(s)}$ and $P(s, C) = \sum_{s' \in C} P(s, s')$.

Intuitively, $P(s, s', C)$ is the probability to move from state s to s' under the condition that s moves to some state in C.

Example 4. Consider the IMC shown in Fig. 2 (left). Let $C = \{s_5, s_6, s_7\}$. Then $P(s_3, s_5, C) = \frac{1}{3}$, $P(s_3, s_6, C) = \frac{2}{3}$, $P(s_4, s_6, C) = \frac{2}{3}$ and $P(s_4, s_7, C) = \frac{1}{3}$.

Definition 9 (Weighted rate). *For $s \in S$, and $C, D \subseteq S$, the function $wr : S \times 2^S \times 2^S \rightarrow \mathbb{R}_{\geq 0}$ is defined by:*

$$wr(s, C, D) = \sum_{s' \in C} P(s, s', C) \cdot R(s', D)$$

where $R(s', D) = \sum_{s'' \in D} R(s', s'')$.

---

[4] Note that Pbr is not really a probability, it is a Boolean indicator of whether certain states are reached in a certain way or not.

Intuitively, $wr(s, C, D)$ is the (weighted) rate to move from s to some states in D in two steps via states of C.

Example 5. Consider the example in Fig. 2 (left). Let $C = \{s_5, s_6, s_7\}$ and $D = \{s_8\}$. Then $wr(s_3, C, D) = P(s_3, s_5, C) \cdot R(s_5, D) + P(s_3, s_6, C) \cdot R(s_6, D) = \frac{1}{3} \cdot 0 + \frac{2}{3} \cdot 2 = \frac{4}{3}$. Similarly, for $D = \{s_9\}$, $wr(s_3, C, D) = P(s_3, s_5, C) \cdot R(s_5, D) + P(s_3, s_6, C) \cdot R(s_6, D) = \frac{1}{3} \cdot 2 + \frac{2}{3} \cdot 0 = \frac{2}{3}$.

Definition 10 (IME). *Equivalence $\mathcal{R}$ on S is an interactive Markovian equivalence (IME) if we have:*

1. $\forall (s_1, s_2) \in \mathcal{R}$ *it holds:* $L(s_1) = L(s_2)$ *and* $E(s_1) = E(s_2)$,
2. $\forall C \in S/_{\mathcal{R}}$ *s.t.* $C \in I(S)$, $\forall D \in S/_{\mathcal{R}}$ *and* $\forall s', s'' \in Pred(C)$ *it holds:* $Pbr(s', C, D) = Pbr(s'', C, D)$,
3. $\forall C \in S/_{\mathcal{R}}$ *s.t.* $C \in M(S)$, $\forall D \in S/_{\mathcal{R}}$ *and* $\forall s', s'' \in Pred(C)$ *it holds:* $wr(s', C, D) = wr(s'', C, D)$,
4. $\forall C \in S/_{\mathcal{R}}$ *s.t.* $C \notin I(S) \wedge C \notin M(S)$, *we have* $|C| = 1$.

States s_1, s_2 are IM related, denoted by $s_1 \equiv s_2$, if $(s_1, s_2) \in \mathcal{R}$ for some IME $\mathcal{R}$.

The first condition asserts that s_1 and s_2 are equally labeled and have identical exit rates. The second condition asserts that for any interactive closed equivalence class C, the predecessor based reachability of going from any two predecessors of C to D via any state in C must be equal. Similarly, the third condition requires that for any Markovian closed equivalence class C, the weighted rate of going from any two predecessors of C to D via any state in C must be equal. Fourth condition says that if C is neither Markovian closed nor interactive closed then the number of states in set C is 1.

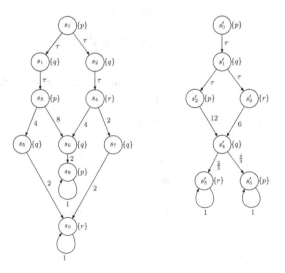

Fig. 2. An IMC $\mathcal{I}$ (left) and its quotient under an IME $\mathcal{R}$ (right)

Example 6. For the closed IMC in Fig. 2 (left), the equivalence relation induced by the partitioning $\{\{s_0\}, \{s_1, s_2\}, \{s_3\}, \{s_4\}, \{s_5, s_6, s_7\}, \{s_8\}, \{s_9\}\}$ is an IME relation.

Remark 2. It is easy to check that for any closed IMC where $MS = \varnothing$, the definition of IME coincides with that of Kripke minimization equivalence (KME) [25]. Similarly, for any closed IMC where $IS = \varnothing$, the definition of IME coincides with that of weighted lumpability (WL) [24,26].

3.1 Quotient IMC

Definition 11. *For an IME relation $\mathcal{R}$ on $\mathcal{I}$, the quotient IMC $\mathcal{I}/_\mathcal{R}$ is defined by $\mathcal{I}/_\mathcal{R} = (S/_\mathcal{R}, s'_0, Act, AP, \to', \Rightarrow', L')$ where:*

- *$S/_\mathcal{R}$ is the set of all equivalence classes under $\mathcal{R}$,*
- *$s'_0 = C$ where $s_0 \in C = [s_0]_\mathcal{R}$,*
- *$\to' \subseteq S/_\mathcal{R} \times Act \times S/_\mathcal{R}$ is defined as follows:*

$$\frac{C \in I(S) \wedge Pbr(s', C, D)=1,\ s' \in Pred(C)}{C \xrightarrow{\tau} D} \quad and \quad \frac{C \notin I(S) \wedge \exists s \in C, s' \in D : s \xrightarrow{\tau} s'}{C \xrightarrow{\tau} D},$$

- *$\Rightarrow' \subseteq S/_\mathcal{R} \times \mathbb{R}_{\geq 0} \times S/_\mathcal{R}$ is defined as follows:*

$$\frac{C \in M(S) \wedge \lambda = wr(s', C, D),\ s' \in Pred(C)}{C \xrightarrow{\lambda} D} \quad and \quad \frac{C \notin M(S) \wedge \lambda = R(s, D),\ s \in C}{C \xrightarrow{\lambda} D},$$

- *$L'(C) = L(s)$, where $s \in C$.*

Example 7. The quotient IMC for the Fig. 2 (left) under the IME relation with partition $\{\{s_0\}, \{s_1, s_2\}, \{s_3\}, \{s_4\}, \{s_5, s_6, s_7\}, \{s_8\}, \{s_9\}\}$ is shown in Fig. 2 (right).

Next, we show that any closed IMC $\mathcal{I}$ and its quotient under IME relation are $\equiv$-related.

Definition 12. *Any IMC $\mathcal{I}$ and its quotient $\mathcal{I}/_\mathcal{R}$ under IME $\mathcal{R}$ are $\equiv$-related, denoted by $\mathcal{I} \equiv \mathcal{I}/_\mathcal{R}$, if and only if there exists an IME relation $\mathcal{R}^*$ defined on the disjoint union of state space $S \uplus S/_\mathcal{R}$ such that*

$$\forall C \in S/_\mathcal{R}, \forall s \in C \implies (s, C) \in \mathcal{R}^*.$$

Theorem 1. *Let $\mathcal{I}$ be a closed IMC and $\mathcal{R}$ be an IME on $\mathcal{I}$. Then $\mathcal{I} \equiv \mathcal{I}/_\mathcal{R}$.*

Proposition 1. *Union of IMEs is not necessarily an IME.*

In simple words, it is possible that $\mathcal{R}_1, \mathcal{R}_2$ are two IMEs on S s.t. $\mathcal{R}_1 \cup \mathcal{R}_2$ is not an IME. Intuitively, it means that the original closed IMC $\mathcal{I}$ can be reduced in different ways.

3.2   Repeated Minimization

Next, we show that IME can be used for repeated minimization of a closed IMC. Intuitively, this means that if a quotient system $\mathcal{I}'$ has been obtained from a closed IMC $\mathcal{I}$ under IME $\mathcal{R}$, then it might still be possible to further reduce $\mathcal{I}'$ to $\mathcal{I}''$ under some IME $\mathcal{R}'$.

Example 8. Consider the example in Fig. 3 (left). IMC in Fig. 3 (middle) is the quotient for the IME induced by the partition $\{\{s_0\}, \{s_1, s_2\}, \{s_3, s_4\}, \{s_5\}, \{s_6\}, \{s_7\}, \{s_8\}\}$. IMC in Fig. 3 (right) is the quotient of the IMC Fig. 3 (middle) for the IME induced by the partition $\{\{s_0'\}, \{s_1'\}, \{s_2', s_3'\}, \{s_4'\}, \{s_5'\}, \{s_6'\}\}$. It is easy to check that s_3, s_4, s_5 in the original system cannot be merged in one shot, since s_1 can reach states labeled with atomic propositions a and b in two steps via s_3 and s_4 respectively, but s_2 cannot reach these states. This is no longer a problem once s_1 and s_2 are merged as shown in Fig. 3 (middle) as s_2', s_3' now have a single predecessor, i.e., s_1'.

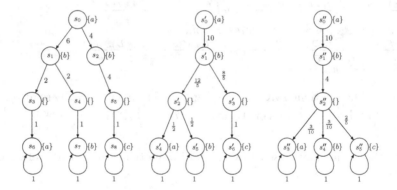

Fig. 3. Repeated minimization

Next, we investigate the relationship between IME and strong bisimulation for IMCs.

3.3   IME vs. Bisimulation

Definition 13 *(Strong bisimulation [15,16]). Let $\mathcal{I} = (S, s_0, Act, AP, \rightarrow, \Rightarrow, L)$ be a closed IMC. An equivalence relation $\mathcal{R} \subseteq S \times S$ is a strong bisimulation on $\mathcal{I}$ if for any $(s_1, s_2) \in \mathcal{R}$ and equivalence class $C \in S/_{\mathcal{R}}$ the following holds:*

- $L(s_1) = L(s_2)$,
- $R(s_1, C) = R(s_2, C)$,
- $Post(s_1, \tau, C) \neq \varnothing \Leftrightarrow Post(s_2, \tau, C) \neq \varnothing$.

States s_1 and s_2 are strongly bisimilar, denoted $s_1 \sim s_2$, if $(s_1, s_2) \in \mathcal{R}$ for some strong bisimulation[5] $\mathcal{R}$.

Strong bisimulation is rigid as it requires that each individual step should be mimicked.

Example 9. Consider the closed IMC shown in Fig. 2 (left). Here s_5 and s_7 are bisimilar, i.e., $s_5 \sim s_7$.

Proposition 2. $\sim \not\Rightarrow \equiv$ *and* $\equiv \not\Rightarrow \sim$.

This proposition says that bisimulation and IME are incomparable.

Example 10. Consider the equivalence class $C = \{s_5, s_6, s_7\}$ under $\equiv$ for closed IMC shown in Fig. 2 (left). Here $s_5 \nsim s_6$ since s_5 can reach a state labeled with atomic proposition r while s_6 cannot.

4   Weak Interactive Markovian Equivalence

In this section we define weak interactive Markovian equivalence (WIME). WIME is a variant of IME that abstracts from stutter steps, also referred to as internal or non-observable steps. Note that weak equivalence relations are important for system synthesis as well as system analysis. To compare IMCs that model a given system at different abstraction levels, it is often too demanding to require a statewise equivalence. Instead, a state in an IMC at a high level of abstraction can be modeled by a sequence of states in the more concrete IMC. Secondly, by abstracting from internal steps, quotient IMCs are obtained that may be significantly smaller than the quotient under corresponding strong equivalence relation. We first define some auxiliary concepts followed by the definition of WIME.

Definition 14 ($*$ reachability). *Let $s, s' \in S$. Then $s \xrightarrow{\tau^*} s'$ denote an alternating sequence of states and τ transitions, i.e., $\pi = s_{01} \xrightarrow{\tau} s_{02} \xrightarrow{\tau} s_{03} \ldots s_{0n}$, where $n \geq 1$, $s_{01} = s$, $s_{0n} = s'$ and $L(s_{0i}) = L(s_{0i+1})$, $1 \leq i < n$.*

Remark 3. Note that if $n = 2$ then $s \xrightarrow{\tau^*} s'$ equals $s \xrightarrow{\tau} s'$ where $L(s) = L(s')$. Similarly, if $n = 1$ then we just have s without any outgoing τ labeled transitions.

Definition 15 ($+$ reachability). *Let $s, s' \in S$. Then $s \xrightarrow{\tau^+} s'$ denote an alternating sequence of states and τ transitions, i.e., $\pi = s \xrightarrow{\tau} \underbrace{s_1 \xrightarrow{\tau} s_2 \ldots s_n}_{n} \xrightarrow{\tau} s'$,*

where $n \geq 0$ and $L(s) = L(s_i), i = 1, \ldots, n$.

Remark 4. Note that if $n = 0$ then $s \xrightarrow{\tau^+} s'$ equals $s \xrightarrow{\tau} s'$, i.e., one step reachability in IMC. For $s \xrightarrow{\tau^+} s'$, the labeling of s and s' need not be the same but s and all the intermediate states should be equally labeled.

---

[5] Note that the definition of strong bisimulation has been slightly modified to take into account the state labels.

Example 11. Consider the closed IMC shown in Fig. 4 (left). Here $s_7 \xrightarrow{\tau^+} s_8$ since s_7 can reach s_8 in two steps such that s_7 and all the intermediate states are equally labeled, i.e., $L(s_7) = L(s_6)$.

Definition 16 (Weak predecessor based reachability[6]). *For $s \in S$ and $C, D \subseteq S$, the function $WPbr : S \times 2^S \times 2^S \to \{0, 1\}$ is defined as:*

$$WPbr(s, C, D) = \begin{cases} 1 \text{ if } \exists s' \in Post(s, \tau, C), s'' \in D \text{ s.t. } s' \xrightarrow{\tau^+} s'' \\ 0 \text{ otherwise.} \end{cases}$$

Definition 17 (WIME). *Equivalence $\mathcal{R}$ on S is a weak interactive Markovian equivalence (WIME) if we have:*

1. *$\forall (s_1, s_2) \in \mathcal{R}$ it holds: $L(s_1) = L(s_2)$,*
2. *$\forall C \in S/\mathcal{R}$ s.t. $C \in I(S)$, $\forall D \in S/\mathcal{R}$ and $\forall s', s'' \in Pred(C)$ it holds: $WPbr(s', C, D) = WPbr(s'', C, D)$,*
3. *$\forall C \in S/\mathcal{R}$ s.t. $C \in M(S)$, $\forall D \in S/\mathcal{R}$ and $\forall s', s'' \in Pred(C)$ it holds: $wr(s', C, D) = wr(s'', C, D)$ and $\forall s_1, s_2 \in C : E(s_1) = E(s_2)$,*
4. *$\forall C \in S/\mathcal{R}$ s.t. $C \notin I(S) \wedge C \notin M(S)$, we have $|C| = 1$.*

States s_1, s_2 are WIM related, denoted by $s_1 \cong s_2$, if $(s_1, s_2) \in \mathcal{R}$ for some WIME $\mathcal{R}$.

The first condition asserts that s_1 and s_2 are equally labeled. The second condition asserts that for any interactive closed equivalence class C, the weak predecessor based reachability of going from any two predecessors of C to D must be equal. Similarly, the third condition requires that for any Markovian closed equivalence class C, the weighted rate of going from any two predecessors of C to D via any state in C must be equal and all the states in C need to have identical exit rates. The last condition says that if C is neither Markovian closed nor interactive closed then the number of states in set C is 1.

Example 12. For the closed IMC shown in Fig. 4 (left), the equivalence relation induced by the partitioning $\{\{s_0\}, \{s_1, s_2\}, \{s_3\}, \{s_4\}, \{s_5, s_6, s_7\}, \{s_8\}, \{s_9\}\}$ is a WIME relation.

4.1 Quotient IMC

Definition 18. *For WIME relation $\mathcal{R}$ on $\mathcal{I}$, the quotient IMC $\mathcal{I}/\mathcal{R}$ is defined by $\mathcal{I}/\mathcal{R} = (S/\mathcal{R}, s'_0, Act, AP, \to', \Rightarrow', L')$ where:*

- *$S/\mathcal{R}$ is the set of all equivalence classes under $\mathcal{R}$,*
- *$s'_0 = C$ where $s_0 \in C = [s_0]_\mathcal{R}$,*
- *$\to' \subseteq S/\mathcal{R} \times Act \times S/\mathcal{R}$ is defined as follows:*

$$\frac{C \in I(S) \wedge WPbr(s', C, D) = 1, \; s' \in Pred(C)}{C \xrightarrow{\tau} D} \text{ and } \frac{C \notin I(S) \wedge \exists s \in C, s' \in D : s \xrightarrow{\tau} s'}{C \xrightarrow{\tau} D},$$

---

[6] Note that $WPbr$ is not really a probability, it is a Boolean indicator of whether certain states are reached in a certain way or not.

- $\Rightarrow' \subseteq S/_\mathcal{R} \times \mathbb{R}_{\geq 0} \times S/_\mathcal{R}$ *is defined as follows:*

$$\frac{C \in M(S) \wedge \lambda = wr(s',C,D),\ s' \in Pred(C)}{C \xrightarrow{\lambda} D} \quad and \quad \frac{C \notin M(S) \wedge \lambda = R(s,D),\ s \in C}{C \xrightarrow{\lambda} D},$$

- $L'(C) = L(s)$, *where* $s \in C$.

Example 13. The quotient IMC for the Fig. 4 (left) under the WIME relation with partition $\{\{s_0\}, \{s_1, s_2\}, \{s_3\}, \{s_4\}, \{s_5, s_6, s_7\}, \{s_8\}, \{s_9\}\}$ is shown in Fig. 4 (right).

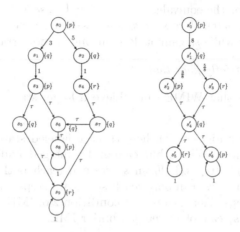

Fig. 4. An IMC $\mathcal{I}$ (left) and its quotient under a WIME $\mathcal{R}$ (right)

Definition 19. *Any IMC $\mathcal{I}$ and its quotient $\mathcal{I}/_\mathcal{R}$ under WIME $\mathcal{R}$ are $\cong$-related, denoted by $\mathcal{I} \cong \mathcal{I}/_\mathcal{R}$, if and only if there exists a WIME relation $\mathcal{R}^*$ defined on the disjoint union of state space $S \uplus S/_\mathcal{R}$ such that*

$$\forall C \in S/_\mathcal{R}, \forall s \in C \implies (s, C) \in \mathcal{R}^*.$$

Theorem 2. *Let $\mathcal{I}$ be a closed IMC and $\mathcal{R}$ be a WIME on $\mathcal{I}$. Then $\mathcal{I} \cong \mathcal{I}/_\mathcal{R}$.*

Proposition 3. *Union of WIMEs is not necessarily a WIME.*

Remark 5. WIMEs can be used for repeated minimization of a closed IMC.

4.2   WIME vs. Weak Bisimulation

Definition 20 *(Weak bisimulation [16]). Let $\mathcal{I} = (S, s_0, Act, AP, \rightarrow, \Rightarrow, L)$ be a closed IMC. An equivalence relation $\mathcal{R} \subseteq S \times S$ is a weak bisimulation on $\mathcal{I}$ if for any $(s_1, s_2) \in \mathcal{R}$ and equivalence class $C \in S/_\mathcal{R}$ the following holds:*

- $L(s_1) = L(s_2)$,
- $\exists s' \in C : s_1 \xrightarrow{\tau^+} s' \Leftrightarrow \exists s'' \in C : s_2 \xrightarrow{\tau^+} s''$,

- $s_1 \xrightarrow{\tau^*} s' \wedge s' \in MS \Rightarrow s_2 \xrightarrow{\tau^*} s'' \wedge s'' \in MS \wedge R(s', C) = R(s'', C)$ *for some* $s'' \in S$.

States s_1 *and* s_2 *are weakly bisimilar, denoted* $s_1 \approx s_2$, *if* $(s_1, s_2) \in \mathcal{R}$ *for some weak bisimulation*[7] $\mathcal{R}$.

Proposition 4. $\approx \not\Rightarrow \cong$ *and* $\cong \not\Rightarrow \approx$.

This proposition says that bisimulation and WIME are incomparable.

Example 14. Consider the equivalence class $C = \{s_5, s_6, s_7\}$ under $\cong$ for closed IMC shown in Fig. 4 (left). Here $s_5 \not\approx s_6$ since s_5 can reach a state labeled with atomic proposition r while s_6 cannot do it in one or more transitions.

Theorem 3. $\equiv$ *is strictly finer than* $\cong$.

This theorem asserts that WIME can achieve a larger state space reduction as compared to IME.

Example 15. Consider Fig. 4 (left), here three q-labeled states, i.e., s_5, s_6 and s_7 can be merged under WIME but cannot be merged under IME. For $C = \{s_5, s_6, s_7\}$, $Pred(C) = \{s_3, s_4\}$. From s_3 we can reach both s_8 and s_9 in two steps via C but from s_4 we can only reach s_9 in two steps via C but s_8 cannot be reached in two steps. This means that condition 2 of IME is not satisfied and therefore s_5, s_6 and s_7 cannot be merged under IME.

5   Preservation of Time Bounded Reachability

Let $\mathcal{I}$ be a closed IMC with state space S, initial state s_0, and let $G \subseteq S$ be a set of goal states and $I \subseteq \mathbb{R}_{\geq 0}$ a time interval with rational bounds. The time-bounded reachability event $\diamondsuit^I G$ is defined as:

$$\diamondsuit^I G = \{\pi \in Paths_\omega^\mathcal{I} | \exists t \in I.\exists s' \in \pi@t.s' \in G\}$$

This set contains all infinite paths starting in state s_0 that hit a state in G at some time point that lies in the interval I. In other words we are interested in the probability of the event $\diamondsuit^I G$. Since IMC supports non-determinism, we need to consider the probability of $\diamondsuit^I G$ relative to the specific resolution of the non-determinism in the closed IMC. More formally, we are interested in obtaining the maximum (resp. minimum) probability of $\diamondsuit^I G$ over all possible total-time dependent positional schedulers of a closed IMC $\mathcal{I}$.

$$p_\mathcal{I}^{max}(I, G) = sup_{D \in Adv_{TTDP}(\mathcal{I})} Pr_D(\diamondsuit^I G)$$

Minimum time-bounded reachability properties are defined in an analogous manner.

---

[7] Note that the definition of weak bisimulation has been slightly modified to take into account the state labels.

Next, we show that maximum (resp. minimum) time-bounded reachability probabilities are preserved under IME quotienting. In principle, this result allows performing model checking on the quotient IMC structure provided that we can obtain it in an algorithmic manner.

Theorem 4. *Let $\mathcal{I}$ be a closed IMC with state space S and $\mathcal{R}$ be an IME on $\mathcal{I}$. Then for any set of goal states $G \subseteq S$ and time interval with rational bounds $I \subseteq \mathbb{R}_{\geq 0}$:*

$$p_{\mathcal{I}}^{max}(I, G) = p_{\mathcal{I}/_{\mathcal{R}}}^{max}(I, G)$$

$$p_{\mathcal{I}}^{min}(I, G) = p_{\mathcal{I}/_{\mathcal{R}}}^{min}(I, G)$$

Corollary 1. *IME preserves transient state probabilities.*

Next, we show that time-bounded reachability probabilities are also preserved under WIME quotienting. Intuitively, this says that hiding of stutter steps does not have any affect on reachability probabilities.

Theorem 5. *Let $\mathcal{I}$ be a closed IMC with state space S and $\mathcal{R}$ be a WIME on $\mathcal{I}$. Then for any set of goal states $G \subseteq S$ and time interval with rational bounds $I \subseteq \mathbb{R}_{\geq 0}$:*

$$p_{\mathcal{I}}^{max}(I, G) = p_{\mathcal{I}/_{\mathcal{R}}}^{max}(I, G)$$

$$p_{\mathcal{I}}^{min}(I, G) = p_{\mathcal{I}/_{\mathcal{R}}}^{min}(I, G)$$

Corollary 2. *WIME preserves transient state probabilities.*

6   Conclusions and Future Work

This paper presented two equivalence relations for closed IMC models. We have shown that smaller models obtained under these equivalences can be used for verification as they preserve time-bounded reachability properties. Our work can be extended in several directions. We plan to investigate the preservation of deterministic timed automata (DTA) [1] and metric temporal logic (MTL) [19, 23] properties under IME (resp. WIME) quotienting. We also plan to investigate the relationship between IME and trace semantics for IMCs [28]. It would be interesting to study and characterize the class of systems where IME can provide better state space reductions compared to bisimulation. Another direction is to develop and implement an efficient quotienting algorithm and validate it on some case studies. Finally, it would be interesting to check if a similar technique can be used for state space reduction of Markov automata (MA) [13]. Markov automata (MA) constitute a compositional behavioral model for continuous-time stochastic and non-deterministic systems.

References

1. Alur, R., Dill, D.L.: A theory of timed automata. Theor. Comput. Sci. **126**(2), 183–235 (1994)
2. Ash, R.B., Doleans-Dade, C.A.: Probability and Measure Theory. Academic Press, San Diego (2000)
3. Baier, C., Haverkort, B.R., Hermanns, H., Katoen, J.-P.: Model-checking algorithms for continuous-time Markov chains. IEEE Trans. Software Eng. **29**(6), 524–541 (2003)
4. Baier, C., Katoen, J.-P., Hermanns, H., Wolf, V.: Comparative branching-time semantics for Markov chains. Inf. Comput. **200**(2), 149–214 (2005)
5. Bernardo, M.: Non-bisimulation-based Markovian behavioral equivalences. J. Log. Algebr. Program. **72**(1), 3–49 (2007)
6. Bernardo, M.: Towards state space reduction based on T-lumpability-consistent relations. In: Thomas, N., Juiz, C. (eds.) EPEW 2008. LNCS, vol. 5261, pp. 64–78. Springer, Heidelberg (2008). doi:10.1007/978-3-540-87412-6_6
7. Böde, E., Herbstritt, M., Hermanns, H., Johr, S., Peikenkamp, T., Pulungan, R., Rakow, J., Wimmer, R., Becker, B.: Compositional dependability evaluation for STATEMATE. IEEE Trans. Software Eng. **35**(2), 274–292 (2009)
8. Boudali, H., Crouzen, P., Haverkort, B.R., Kuntz, M., Stoelinga, M.: Architectural dependability evaluation with arcade. In: DSN, pp. 512–521. IEEE Computer Society (2008)
9. Boudali, H., Crouzen, P., Stoelinga, M.: A compositional semantics for dynamic fault trees in terms of interactive Markov chains. In: Namjoshi, K.S., Yoneda, T., Higashino, T., Okamura, Y. (eds.) ATVA 2007. LNCS, vol. 4762, pp. 441–456. Springer, Heidelberg (2007). doi:10.1007/978-3-540-75596-8_31
10. Boudali, H., Crouzen, P., Stoelinga, M.: Dynamic fault tree analysis using input/output interactive Markov chains. In: DSN, pp. 708–717. IEEE Computer Society (2007)
11. Bozzano, M., Cimatti, A., Katoen, J., Nguyen, V.Y., Noll, T., Roveri, M.: Safety, dependability and performance analysis of extended AADL models. Comput. J. **54**(5), 754–775 (2011)
12. Coste, N., Hermanns, H., Lantreibecq, E., Serwe, W.: Towards performance prediction of compositional models in industrial GALS designs. In: Bouajjani, A., Maler, O. (eds.) CAV 2009. LNCS, vol. 5643, pp. 204–218. Springer, Heidelberg (2009). doi:10.1007/978-3-642-02658-4_18
13. Eisentraut, C., Hermanns, H., Zhang, L.: On probabilistic automata in continuous time. In: LICS, pp. 342–351 (2010)
14. Guck, D., Han, T., Katoen, J.-P., Neuhäußer, M.R.: Quantitative timed analysis of interactive Markov chains. In: Goodloe, A.E., Person, S. (eds.) NFM 2012. LNCS, vol. 7226, pp. 8–23. Springer, Heidelberg (2012). doi:10.1007/978-3-642-28891-3_4
15. Hermanns, H.: Interactive Markov Chains: The Quest for Quantified Quality. Lecture Notes in Computer Science, vol. 2428. Springer, Heidelberg (2002)
16. Hermanns, H., Katoen, J.-P.: The how and why of interactive Markov chains. In: Boer, F.S., Bonsangue, M.M., Hallerstede, S., Leuschel, M. (eds.) FMCO 2009. LNCS, vol. 6286, pp. 311–337. Springer, Heidelberg (2010). doi:10.1007/978-3-642-17071-3_16
17. Hermanns, H., Katoen, J., Neuhäußer, M.R., Zhang, L.: GSPN model checking despite confusion. Technical report, RWTH Aachen University (2010)

18. Johr, S.: Model checking compositional Markov systems, Ph.D. thesis, Saarland University (2008)
19. Koymans, R.: Specifying real-time properties with metric temporal logic. Real Time Syst. **2**(4), 255–299 (1990)
20. Mateescu, R., Serwe, W.: A study of shared-memory mutual exclusion protocols using CADP. In: Kowalewski, S., Roveri, M. (eds.) FMICS 2010. LNCS, vol. 6371, pp. 180–197. Springer, Heidelberg (2010). doi:10.1007/978-3-642-15898-8_12
21. Neuhäußer, M.R.: Model checking non-deterministic and randomly timed systems, Ph.D. thesis, RWTH Aachen University (2010)
22. Neuhäußer, M.R., Katoen, J.-P.: Bisimulation and logical preservation for continuous-time Markov decision processes. In: Caires, L., Vasconcelos, V.T. (eds.) CONCUR 2007. LNCS, vol. 4703, pp. 412–427. Springer, Heidelberg (2007). doi:10. 1007/978-3-540-74407-8_28
23. Ouaknine, J., Worrell, J.: Some recent results in metric temporal logic. In: Cassez, F., Jard, C. (eds.) FORMATS 2008. LNCS, vol. 5215, pp. 1–13. Springer, Heidelberg (2008). doi:10.1007/978-3-540-85778-5_1
24. Sharma, A.: Reduction techniques for non-deterministic and probabilistic systems, Ph.D. thesis, RWTH Aachen university (2015)
25. Sharma, A.: A two step perspective for Kripke structure reduction. CoRR, abs/1210.0408 (2012)
26. Sharma, A., Katoen, J.-P.: Weighted lumpability on Markov chains. In: Clarke, E., Virbitskaite, I., Voronkov, A. (eds.) PSI 2011. LNCS, vol. 7162, pp. 322–339. Springer, Heidelberg (2012). doi:10.1007/978-3-642-29709-0_28
27. Wolf, V., Baier, C., Majster-Cederbaum, M.E.: Trace machines for observing continuous-time Markov chains. ENTCS **153**(2), 259–277 (2006)
28. Wolf, V., Baier, C., Majster-Cederbaum, M.E.: Trace semantics for stochastic systems with nondeterminism. Electr. Notes Theor. Comput. Sci. **164**(3), 187–204 (2006)
29. Zhang, L., Neuhäußer, M.R.: Model checking interactive Markov chains. In: Esparza, J., Majumdar, R. (eds.) TACAS 2010. LNCS, vol. 6015, pp. 53–68. Springer, Heidelberg (2010). doi:10.1007/978-3-642-12002-2_5

Advances in Quantitative Analysis

Delay Analysis of Resequencing Buffer in Markov Environment with HOQ-FIFO-LIFO Policy

Rostislav Razumchik[1,2] and Miklós Telek[3,4(✉)]

[1] Institute of Informatics Problems, Federal Research Center
"Computer Science and Control" of the Russian Academy of Sciences,
Moscow, Russia
[2] Peoples Friendship University of Russia (RUDN University), Moscow, Russia
rrazumchik@ipiran.ru, razumchik_rv@rudn.university
[3] Department of Telecommunications,
Technical University of Budapest, Budapest, Hungary
[4] MTA-BME Information Systems Research Group, Budapest, Hungary
telek@hit.hme.hu

Abstract. Resequencing of customers during the service process results in hard to analyze delay distributions. A set of models with various service and resequencing policies have been analyzed already for memoryless arrival, service and resequencing processes with an intensive use of transform domain descriptions. In case of Markov modulated arrival, service and resequencing processes those methods are not applicable any more. In a previous work we analyzed the Markov modulated case with HOQ-FIFO-FIFO policy (head of queue customer of the higher priority FIFO queue is moved to resequencing FIFO queue). In this work we investigate if the approach remains applicable for different service discipline for the HOQ-FIFO-LIFO policy.

It turns out that the analysis of the new service policy requires the solution of a coupled quadratic matrix equations which were separated in the HOQ-FIFO-FIFO case.

Keywords: Resequencing buffer · Delay analysis · Markov modulated arrival · Service process

1 Introduction

In models with resequencing delay distributions are of primary interest. Usually resequencing is due to some disruptive events but it also may be one of the features, which are inherent to the system (for models in the context of queueing theory see, for example, the reviews [2,3]). With the evolution and the

This work is partially supported by the Russian Foundation for Basic Research (grants 15-07-03007 and 15-07-03406) and by grant K123914 of Hungarian Scientific Research Fund (OTKA).

© Springer International Publishing AG 2017
P. Reinecke and A. Di Marco (Eds.): EPEW 2017, LNCS 10497, pp. 53–68, 2017.
DOI: 10.1007/978-3-319-66583-2_4

widespread use of matrix analytic methods [4–7], there is a belief that the more and more Markov chain based analysis of stochastic models with memoryless components can be extended for the same problem with modulating Markov environment. The transform domain delay analysis of the resequencing buffer models in [9] was an example of notoriously hard extension with modulating Markov environment. For the HOQ-FIFO-FIFO policy, which is one of the policies studied in [9], the analysis with modulating Markov environment is presented in [10].

This work is essentially a methodological study to understand if the methodology developed in [10] is general enough for applying in other queueing models, particularly for the same resequencing buffer model **but** with HOQ-FIFO-LIFO policy.

The rest of the paper is structured as follows. In Sect. 2 the system description is provided. In the next section we summarize the results concerning the joint stationary distribution, which, in fact, coincides with the one for the system from [10]. Section 4 provides the new contribution of the paper, which is the waiting time distribution **for the** HOQ-FIFO-LIFO policy. Some numerical experiments are provided in Sect. 5 and the paper is concluded **with** Sect. 6.

2   Model Description

The system under consideration is a single server queueing system with two infinite buffers: the regular buffer (or, simply, buffer) and the resequencing buffer. **Regular customers (or, simply, customers)** arrive at the system and occupy one place in the regular buffer. Resequencing signals arrive at the system according to a resequencing process. If the buffer is not empty, then, upon arrival, each resequencing signal moves one customer from the regular buffer to the resequencing buffer and itself leaves the system, otherwise it leaves the system without having any effect on it. A single server serves customers from both queues. Upon service completion one customer from the regular buffer goes to the server and only if there are no regular customers in the buffer, one customer from resequencing buffer enters the server. No service interruption is allowed. The HOQ-FIFO-LIFO policy means that the resequencing signal moves the oldest waiting regular customer to the resequencing buffer (Head Of Queue, HOQ), the service policy of the regular buffer is FIFO and of the resequencing buffer is LIFO.

Since the customers from the resequencing buffer are served if and only if the regular buffer is empty, the considered system is a variant of a priority queue with regular buffer customers as high priority customers and resequencing buffer customers as low priority customers.

We assume that regular customers arrive according to a MAP process with generator matrices $(\mathbf{A_0}, \mathbf{A_1})$ and resequencing signals arrive according to a MAP with $(\mathbf{H_0}, \mathbf{H_1})$. The service process is a MAP with $(\mathbf{S_0}, \mathbf{S_1})$. Let $\mathbf{A_J} = \mathbf{A_0} + \mathbf{A_1}$, $\mathbf{S_J} = \mathbf{S_0} + \mathbf{S_1}$, and $\mathbf{H_J} = \mathbf{H_0} + \mathbf{H_1}$ denote the phase processes of the associated MAPs (see e.g. [5] for details). The block structure of the Markov chain representing the number of high and low priority customers in the system is depicted in Fig. 1. The block represents the set of states with the same number of high and low priority customers and with different phases of the MAPs. The letters on the figures describe

– arrival of a customer: $\mathcal{A} = \mathbf{A_1} \otimes I \otimes I$,
– service of a customer: $\mathcal{S} = I \otimes \mathbf{S_1} \otimes I$,
– resequencing of a customer: $\mathcal{H} = I \otimes I \otimes \mathbf{H_1}$,
– phase change when resequencing is possible: $\mathcal{L} = \mathbf{A_0} \oplus \mathbf{S_0} \oplus \mathbf{H_0}$,
– phase change when resequencing is not possible: $\mathcal{L}' = \mathbf{A_0} \oplus \mathbf{S_0} \oplus \mathbf{H_J}$,
– phase change when resequencing is not possible and the service process is stopped: $\mathcal{L}_0 = \mathbf{A_0} \otimes I \oplus \mathbf{H_J} = \mathbf{A_0} \otimes I \otimes I + I \otimes I \otimes \mathbf{H_J}$,

where $\otimes$ ($\oplus$) denotes the Kronecker product (sum) and I the identity matrix of appropriate size. The phase of the service process is frozen (does not change) when the system is empty.

The main goal of the analysis is to evaluate the stationary waiting time distribution of a regular customer arriving at the system.

3   Joint Stationary Distribution of the Number of Customers

Before deriving the expressions for the stationary waiting time distribution one has to obtain expressions for joint stationary distribution of number of customers in regular buffer, resequencing buffer and phases of regular and resequencing arrivals and service process. Since the service order does not affect the number of customers in the system, the joint stationary distribution in the HOQ-FIFO-LIFO system is identical with the one of the HOQ-FIFO-FIFO system studied in [10]. In this section we introduce the notation and repeat results from [10], which will be used later on.

3.1   Censored Process

To simplify the analysis and obtain a Markov chain with a regular structure we censor the Markov chain in Fig. 1 for the cases when the server is busy. The structure of the censored Markov chain is depicted in Fig. 2. The transitions of upper left block of the censored chain is obtained as

$$\mathcal{L}'' = \mathcal{L}' - \mathcal{S}\mathcal{L}_0^{-1}\mathcal{A} = (\mathbf{A_0} \oplus \mathbf{S_0} \oplus \mathbf{H_J}) - (I \otimes \mathbf{S_1} \otimes I)(\mathbf{A_0} \otimes I \oplus \mathbf{H_J})^{-1}(\mathbf{A_1} \otimes I \otimes I).$$

3.2   QBD Representation of the Censored Process

Following, for example, the discussion of Sect. 13.1 in [5] we can represent the censored Markov chain as QBD process where the levels are composed by the set of states where the number of regular customers is the same (these states form the columns of blocks in Fig. 2). The generator $\mathbb{Q}$ of the censored process can be

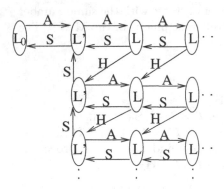

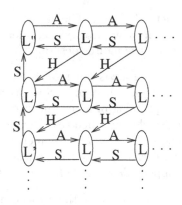

Fig. 1. Block structure of the Markov chain representing the number of regular (high priority) and resequenced (low priority) customers

Fig. 2. Block structure of the censored Markov chain representing the number of regular (high priority) and resequenced (low priority) customers

represented in hyper-block tridiagonal form, where the hyper-block refers to the set of (infinitely many) states on the same level.

$$
Q = \begin{pmatrix}
\mathbb{L}' & \mathbb{F} & 0 & 0 & 0 & \cdots \\
\mathbb{B} & \mathbb{L} & \mathbb{F} & 0 & 0 & \cdots \\
0 & \mathbb{B} & \mathbb{L} & \mathbb{F} & 0 & \cdots \\
0 & 0 & \mathbb{B} & \mathbb{L} & \mathbb{F} & \cdots \\
\vdots & \vdots & \vdots & \vdots & \vdots & \ddots
\end{pmatrix},
$$

and, due to the fact that the number of states within each level is infinite, matrices $\mathbb{L}'$, $\mathbb{L}$, $\mathbb{B}$, $\mathbb{F}$ have infinite rows and columns which are associated with the blocks in Fig. 2.

$$
\mathbb{L}' = \begin{pmatrix}
\mathcal{L}'' & 0 & 0 & \cdots \\
\mathcal{S} & \mathcal{L}' & 0 & \cdots \\
0 & \mathcal{S} & \mathcal{L}' & \cdots \\
\vdots & \vdots & \vdots & \ddots
\end{pmatrix}, \quad
\mathbb{L} = \begin{pmatrix}
\mathcal{L} & 0 & 0 & \cdots \\
0 & \mathcal{L} & 0 & \cdots \\
0 & 0 & \mathcal{L} & \cdots \\
\vdots & \vdots & \vdots & \ddots
\end{pmatrix}, \quad
\mathbb{F} = \begin{pmatrix}
\mathcal{A} & 0 & 0 & \cdots \\
0 & \mathcal{A} & 0 & \cdots \\
0 & 0 & \mathcal{A} & \cdots \\
\vdots & \vdots & \vdots & \ddots
\end{pmatrix}, \quad
\mathbb{B} = \begin{pmatrix}
\mathcal{S} & \mathcal{H} & 0 & 0 & \cdots \\
0 & \mathcal{S} & \mathcal{H} & 0 & \cdots \\
0 & 0 & \mathcal{S} & \mathcal{H} & \cdots \\
\vdots & \vdots & \vdots & \vdots & \ddots
\end{pmatrix}.
$$

In the censored Markov chain we denote the stationary probability vector of the set of states with i regular and j delayed customers by π_{ij} $(i, j \geq 0)$ and compose the following row vectors

$$
\mathbf{p}_i = (\pi_{i,0}, \pi_{i,1}, \pi_{i,2}, \pi_{i,3}, \dots), \ i \geq 0,
$$
$$
\mathbf{p} = (\mathbf{p}_0, \mathbf{p}_1, \mathbf{p}_2, \mathbf{p}_3, \dots).
$$

Henceforth we consider the distribution $\mathbf{p}$, which is the solution of the linear infinite system of equations $\mathbf{p}\,Q = \mathbf{0}$, $\mathbf{p}\,\mathbf{1} = 1$, to be known.

3.3 Distribution Right After Customer Arrival

Notice that as MAP arrivals do not see time averages (that is PASTA property does not hold) one has to calculate stationary probabilities $\tilde{\pi}_{ij}$ that after a customer arrival there are i $(i \geq 1)$ customer in the regular buffer and j $(j \geq 0)$ in the resequencing buffer. Following the same argument as in [8], we can write

$$\tilde{\pi}_{ij} = \frac{1}{\lambda}\pi_{i-1,j}\mathcal{A}, \ i \geq 1, \ j \geq 0, \text{ and } \tilde{\pi}_{00} = \frac{1}{\lambda}\pi_{idle}\mathcal{A}.$$

Here π_{idle} is the stationary distribution of the block of states representing idle server (the left most block in Fig. 1). It is found (see the details in [10, Sect. 3.6]) from the system of linear equations $\pi_{idle}(\mathcal{L}_0 - \mathcal{A}\mathbf{T}_0^{-1}\mathcal{S}) = 0$, $\pi_{idle}\mathbf{1} = 1 - \lambda/\mu$. As usual, λ denotes the average arrival rate and μ denotes the average service rate.

4 Stationary Waiting Time Distribution

The waiting time (W) is understood here, as usual, as the time lapse, starting from the instant when regular customer arrives at the system up to the instant when it enters server. Its stationary distribution will be evaluated in terms of Laplace–Stieltjes transform $\omega(s) = E(e^{-sW})$. Regular customer may enter the server either from the regular buffer or from the resequencing buffer and thus its stationary waiting time distribution can be computed as

$$\omega(s) = E(e^{-sW}) = \omega_{\mathrm{H}}(s) + \omega_{\mathrm{L}}(s)$$
$$= E(e^{-sW}I_{\{\text{served from regular buffer}\}}) + E(e^{-sW}I_{\{\text{served from resequencing buffer}\}})$$

where $I_{\{a\}}$ is the indicator of event a.

It is clear that under HOQ-FIFO-LIFO policy the stationary waiting time distribution of the regular customer that receives service from regular buffer coincides with that under the HOQ-FIFO-FIFO policy. Thus we will not repeat these derivations here and refer the reader for the details to the [10, Sect. 4.1]. Henceforth we consider $\omega_{\mathrm{H}}(s)$ to be known.

4.1 Stationary Waiting Time Distribution of the Customer that Receives Service from Resequencing Buffer

For $i \geq j \geq 0$ and $k > 0$ let $\mathbb{F}(t, i, j, k)$ be the matrix (according to the initial and final phases of the MAPs $(\mathbf{A_0}, \mathbf{A_1})$, $(\mathbf{S_0}, \mathbf{S_1})$ and $(\mathbf{H_0}, \mathbf{H_1})$) of the probabilities that k customers arrive, $i - j$ customers are served and j are moved to the resequencing buffer in time t, when the initial number of customers in the buffer is larger than i. For the Laplace transform $\tilde{\mathbb{F}}(s, i, j, k) = \int_t e^{-st}\mathbb{F}(t, i, j, k)dt$ we have

$$\tilde{\mathbb{F}}(s, 0, 0, 0) = (sI - \mathcal{L})^{-1} = \mathcal{L}(s), \tag{1}$$

and otherwise

$$\tilde{\mathbb{F}}(s,i,j,k) = I_{\{i>j\}}\mathcal{L}(s)\mathcal{S}\tilde{\mathbb{F}}(s,i-1,j,k) + I_{\{j>0\}}\mathcal{L}(s)\mathcal{H}\tilde{\mathbb{F}}(s,i-1,j-1,k) \quad (2)$$
$$+ I_{\{k>0\}}\mathcal{L}(s)\mathcal{A}\tilde{\mathbb{F}}(s,i,j,k-1),$$

where $\mathcal{L}(s)$ is defined in (1). An intuitive explanation of the first term of (2) is as follows. There is no arrival, service and resequencing up to time τ ($\mathcal{L}(s)$) than an service occurs ($\mathcal{S}$) and than $i-1$ services, j resequencing and k arrival occur in (τ, t) ($\tilde{\mathbb{F}}(s, i-1, j, k)$). The other terms follow the same pattern. The cases that the tagged customer moves to the resequencing buffer is described by $\tilde{\mathbb{F}}(s,i,j,k)\mathcal{H}$.

Similarly, let $\tilde{\mathbb{W}}(s,i,j)$ be the matrix (according to the initial and final phases of the MAPs $(\mathbf{A_0}, \mathbf{A_1})$, $(\mathbf{S_0}, \mathbf{S_1})$ and $(\mathbf{H_0}, \mathbf{H_1})$) Laplace–Stieltjes transform of the waiting time of a customer which starts its life in the resequencing buffer in LIFO position j, when the number of customers in the regular buffer is i. The LIFO position is $j = 1$ for the customer which arrived most recently to the resequencing buffer and all existing LIFO positions are increased by one when a new customer arrives to the resequencing buffer. For $i \geq 0, j \geq 1$, we have

$$\tilde{\mathbb{W}}(s,i,j) = I_{\{i>0\}}\mathcal{L}(s)\mathcal{S}\tilde{\mathbb{W}}(s,i-1,j) + I_{\{i=0\}}\mathcal{L}(s)\mathcal{S}\tilde{\mathbb{W}}(s,0,j-1) \quad (3)$$
$$+ I_{\{i>0\}}\mathcal{L}(s)\mathcal{H}\tilde{\mathbb{W}}(s,i-1,j+1) + I_{\{i=0\}}\mathcal{L}(s)\mathcal{H}\tilde{\mathbb{W}}(s,0,j) + \mathcal{L}(s)\mathcal{A}\tilde{\mathbb{W}}(s,i+1,j),$$

where $\tilde{\mathbb{W}}(s,0,0) = I$. The solution of $\tilde{\mathbb{W}}(s,i,j)$ is not trivial. We search for the solution in product form $\tilde{\mathbb{W}}(s,i,j) = \widehat{\mathbf{G}}(s)^i \widehat{\mathbf{G}}(s)^j$. The product from solution satisfies (3) for $i \geq 0, j \geq 1$ if

$$s\widehat{\mathbf{G}}(s) - \mathcal{L}\widehat{\mathbf{G}}(s) = \mathcal{S} + \mathcal{H}\widehat{\mathbf{G}}(s) + \mathcal{A}\tilde{\mathbf{G}}(s)\widehat{\mathbf{G}}(s), \quad (4)$$
$$s\tilde{\mathbf{G}}(s) - \mathcal{L}\tilde{\mathbf{G}}(s) = \mathcal{S} + \mathcal{H}\widehat{\mathbf{G}}(s) + \mathcal{A}\tilde{\mathbf{G}}^2(s), \quad (5)$$

which are obtained from (3) by substituting the product form at $i + 1 = j = 1$ and $i = j + 1 = 1$. The Eqs. (4) and (5) form a pair of coupled matrix quadratic equations whose minimal non-negative solution can be computed by efficient iterative numerical methods, but do not exhibit closed form result. A simple linearly convergent iterative method is as follows.

4.2   Iterative Solution of the Coupled Matrix Equations

The system of Eqs. (4) and (5) can be re-written as

$$\widehat{\mathbf{G}}(s) = \left(sI - \mathcal{L} - \mathcal{H} - \mathcal{A}\tilde{\mathbf{G}}(s) \right)^{-1} \mathcal{S}, \quad (6)$$
$$\tilde{\mathbf{G}}(s) = \left(sI - \mathcal{L} - \mathcal{A}\tilde{\mathbf{G}}(s) \right)^{-1} \left(\mathcal{S} + \mathcal{H}\widehat{\mathbf{G}}(s) \right). \quad (7)$$

In order to find $\widehat{\mathbf{G}}(s)$ and $\widetilde{\mathbf{G}}(s)$ for the given value of s, we start with $\widetilde{\mathbf{G}}_0(s) = 0$. Then for $i = 1, 2, \ldots$ the next two iterative steps are performed until the convergence is reached

$$\widehat{\mathbf{G}}_i(s) = \left(sI - \mathcal{L} - \mathcal{H} - \mathcal{A}\widetilde{\mathbf{G}}_{i-1}(s)\right)^{-1} \mathcal{S}, \tag{8}$$

$$\widetilde{\mathbf{G}}_i(s) = \left(sI - \mathcal{L} - \mathcal{A}\widetilde{\mathbf{G}}_{i-1}(s)\right)^{-1} \left(\mathcal{S} + \mathcal{H}\widehat{\mathbf{G}}_i(s)\right). \tag{9}$$

4.3 Delay Analysis of Customer Served from the Resequencing Buffer

Based on the previously computed matrix Laplace–Stieltjes transforms, the waiting time of the customer which enters server from the resequencing buffer can be computed as

$$\omega_{\mathrm{L}}(s) = E(e^{-sW} I_{\{\text{served from resequencing buffer}\}})$$

$$= \sum_{i=1}^{\infty} \sum_{j=0}^{\infty} \tilde{\pi}_{ij} \sum_{\ell=0}^{i-1} \sum_{k=0}^{\infty} \mathbb{F}(s, i-1, \ell, k) \mathcal{H}\widetilde{\mathbf{G}}(s)^k \widehat{\mathbf{G}}(s)\mathbf{1}$$

$$= \frac{1}{\lambda} \sum_{i=0}^{\infty} \sum_{j=0}^{\infty} \pi_{i,j} \mathcal{A} \sum_{\ell=0}^{i} \sum_{k=0}^{\infty} \mathbb{F}(s, i, \ell, k) \mathcal{H}\widetilde{\mathbf{G}}(s)^k \widehat{\mathbf{G}}(s)\mathbf{1}. \tag{10}$$

The main part of the analysis of $\omega_{\mathrm{L}}(s)$ is deferred to the next section. But in the course of the subsequent derivations we will make use of several quantities which are better introduced by considering terms of $\omega_{\mathrm{L}}(s)$ with $i = 0$. We represent $\omega_{\mathrm{L}}(s)$ as

$$\omega_{\mathrm{L}}(s) = \omega_{\mathrm{L}}^{i>0}(s) + \omega_{\mathrm{L}}^{i=0}(s)$$

$$= \frac{1}{\lambda} \sum_{i=1}^{\infty} \sum_{j=0}^{\infty} \pi_{i,j} \mathcal{A} \sum_{\ell=0}^{i} \sum_{k=0}^{\infty} \mathbb{F}(s, i, \ell, k) \mathcal{H}\widetilde{\mathbf{G}}(s)^k \widehat{\mathbf{G}}(s)\mathbf{1}$$

$$+ \frac{1}{\lambda} \sum_{j=0}^{\infty} \pi_{0,j} \mathcal{A} \sum_{k=0}^{\infty} \underbrace{\tilde{\mathbb{F}}(s, 0, 0, k)}_{(\mathcal{L}(s)\mathcal{A})^k \mathcal{L}(s)} \mathcal{H}\widetilde{\mathbf{G}}(s)^k \widehat{\mathbf{G}}(s)\mathbf{1}. \tag{11}$$

In what follows we will need the expressions for probability generating functions $\hat{\pi}_0(z) = \sum_{m=0}^{\infty} \pi_{0,m} z^m$ and $\hat{\pi}_i(z) = \sum_{j=0}^{\infty} \pi_{ij} z^j$, $i \geq 1$, which were obtained in [10]:

$$\hat{\pi}_0(z) = \pi_{0,0}(\mathcal{L}' - \mathcal{L}'' + \frac{1}{z}\mathcal{S})(\mathcal{A}\overline{\mathbf{G}}(z) + \mathcal{L}' + \frac{1}{z}\mathcal{S})^{-1}, \tag{12}$$

$$\hat{\pi}_i(z) = \hat{\pi}_{i-1}(z)\overline{\mathbf{R}}(z), \quad i \geq 1, \tag{13}$$

where $\overline{\mathbf{R}}(z)$ is the minimal non-negative solution of the quadratic matrix equation

$$\mathcal{A} + \overline{\mathbf{R}}(z)\mathcal{L} + \overline{\mathbf{R}}^2(z)(z\mathcal{H} + \mathcal{S}) = \mathbf{0}. \tag{14}$$

Derivation of $\omega_L^{i=0}(s)$. The methodology from [10], which we apply here in order to obtain the stationary waiting time distribution, is based on the technique which can be referred to as the Kronecker expansion (see [1,11]). It is based on the identity $vec(ABC) = (C^T \otimes A)vec(B)$. In this identity vec denotes the column stacking vector operator, which transforms a matrix of size $n \times m$ into a vector of size $nm \times 1$. In all further derivations we will make extensive use of the Kronecker expansion, which will appear in seemingly different but, in fact, equal forms (for example, $vec(AB) = (I^T \otimes A)vec(B) = (B^T \otimes A)vec(I) = (B^T \otimes I)vec(A)$).

Coming back to $\omega_L^{i=0}(s)$ and using the identity $vec(ABC) = (C^T \otimes A)vec(B)$, one obtains

$$\omega_L^{i=0}(s) = \frac{1}{\lambda} \sum_{j=0}^{\infty} \sum_{k=0}^{\infty} \pi_{0,j} \mathcal{A}(\mathcal{L}(s)\mathcal{A})^k \mathcal{L}(s)\mathcal{H}\widetilde{\mathbf{G}}(s)^k \widehat{\mathbf{G}}(s)\mathbf{1}$$

$$= \frac{1}{\lambda} \sum_{j=0}^{\infty} \sum_{k=0}^{\infty} \left(\mathbf{1}^T \widehat{\mathbf{G}}(s)^T \widetilde{\mathbf{G}}(s)^{k^T} \otimes \pi_{0,j}\mathcal{A}(\mathcal{L}(s)\mathcal{A})^k\right) vec(\mathcal{L}(s)\mathcal{H})$$

and

$$\omega_L^{i=0}(s) = \frac{1}{\lambda}\left(\mathbf{1}^T\widehat{\mathbf{G}}(s)^T \otimes \mathbf{1}\right)$$

$$\cdot \underbrace{\sum_{j=0}^{\infty}(I \otimes \pi_{0,j})}_{I \otimes \hat{\pi}_0(1)}(I \otimes \mathcal{A})\underbrace{\sum_{k=0}^{\infty}\left(\widetilde{\mathbf{G}}(s)^{k^T} \otimes (\mathcal{L}(s)\mathcal{A})^k\right)}_{(I-\widetilde{\mathbf{G}}(s)^T \otimes \mathcal{L}(s)\mathcal{A})^{-1}} vec(\mathcal{L}(s)\mathcal{H})$$

$$= \frac{1}{\lambda}\left(\mathbf{1}^T\widehat{\mathbf{G}}(s)^T \otimes \mathbf{1}\right)(I \otimes \hat{\pi}_0(1))(I \otimes \mathcal{A})\left(I - \widetilde{\mathbf{G}}(s)^T \otimes \mathcal{L}(s)\mathcal{A}\right)^{-1}$$
$$\cdot vec(\mathcal{L}(s)\mathcal{H})$$

$$= \frac{1}{\lambda}\left(\mathbf{1}^T\widehat{\mathbf{G}}(s)^T \otimes \hat{\pi}_0(1)\mathcal{A}\right)\left(I - \widetilde{\mathbf{G}}(s)^T \otimes \mathcal{L}(s)\mathcal{A}\right)^{-1} vec(\mathcal{L}(s)\mathcal{H}).$$

Derivation of $\omega_L^{i>0}(s)$. Having found the expression for $\omega_L^{i=0}(s)$ the last unknown quantity in $\omega_L(s)$ is $\omega_L^{i>0}(s)$. In the following we split expression (10) for $\omega_L^{i>0}(s)$ into the following two terms:

$$\omega_L^{i>0}(s) = \omega_L^{k=0}(s) + \omega_L^{k>0}(s),$$

where $\omega_L^{k=0}(s)$ includes only terms of $\omega_L^{i>0}(s)$ with $k = 0$ and $\omega_L^{k>0}(s)$ all other terms. Further we obtain the expressions for each of them individually.

Derivation of $\omega_L^{k=0}(s)$. In order to compute $\omega_L^{k=0}(s)$ we perform the Kronecker expansion and apply the relation $vec(ABC) = (C^T \otimes A)vec(B)$ two times. We have

$$\omega_L^{k=0}(s) = \frac{1}{\lambda}\sum_{i=1}^{\infty}\sum_{j=0}^{\infty}\pi_{i,j}\mathcal{A}\underbrace{\sum_{\ell=0}^{i}\tilde{\mathbb{F}}(s,i,\ell,0)\mathcal{H}\,\widehat{\mathbf{G}}(s)\mathbf{1}}_{\widehat{\mathcal{F}}_{k=0}(s,i)}$$

$$= \frac{1}{\lambda}\sum_{i=1}^{\infty}\sum_{j=0}^{\infty}\pi_{i,j}\mathcal{A}\widehat{\mathcal{F}}_{k=0}(s,i)\widehat{\mathbf{G}}(s)\mathbf{1} = \frac{1}{\lambda}\sum_{i=1}^{\infty}\sum_{j=0}^{\infty}\left(\mathbf{1}^T\widehat{\mathbf{G}}(s)^T\otimes\pi_{i,j}\mathcal{A}\right)vec(\widehat{\mathcal{F}}_{k=0}(s,i))$$

$$= \frac{1}{\lambda}\left(\mathbf{1}^T\widehat{\mathbf{G}}(s)^T\otimes 1\right)\sum_{i=1}^{\infty}\sum_{j=0}^{\infty}\left(I\otimes\pi_{i,j}\right)\left(I\otimes\mathcal{A}\right)vec(\widehat{\mathcal{F}}_{k=0}(s,i))$$

$$= \frac{1}{\lambda}\left(\mathbf{1}^T\widehat{\mathbf{G}}(s)^T\otimes 1\right)\underbrace{\sum_{i=1}^{\infty}\sum_{j=0}^{\infty}\left[vec(\widehat{\mathcal{F}}_{k=0}(s,i))^T\otimes\left(I\otimes\pi_{i,j}\right)\right]}_{\mathbf{M}(s)}vec\left(I\otimes\mathcal{A}\right).$$

Here the only unknown quantity is $\mathbf{M}(s)$. We will show now that the matrix $\mathbf{M}(s)$ can be expressed in the form $\mathbf{M}(s) = M_1(s) + \mathbf{M}(s)M_2(s)$, where $M_1(s)$ and $M_2(s)$ are known matrices. Thus for any given s it can be computed as $\mathbf{M}(s) = (I - M_2(s))^{-1}M_1(s)$. Summing over $j \geq 0$ (remembering (13)) and extracting the term with $i = 1$, one can write

$$\mathbf{M}(s) = \sum_{i=1}^{\infty}\sum_{j=0}^{\infty}\left[vec(\widehat{\mathcal{F}}_{k=0}(s,i))^T\otimes\left(I\otimes\pi_{i,j}\right)\right]$$

$$= \sum_{i=1}^{\infty}\left[vec(\widehat{\mathcal{F}}_{k=0}(s,i))^T\otimes\left(I\otimes\hat{\pi}_i(1)\right)\right]$$

$$= vec(\widehat{\mathcal{F}}_{k=0}(s,1))^T\otimes\left(I\otimes\hat{\pi}_1(1)\right)$$

$$+ \sum_{i=2}^{\infty}vec(\widehat{\mathcal{F}}_{k=0}(s,i))^T\otimes\left(I\otimes\hat{\pi}_i(1)\right). \tag{15}$$

In order to obtain the expression for the only unknown quantity $vec(\widehat{\mathcal{F}}_{k=0}(s,i))^T$ we revisit the definition of $\widehat{\mathcal{F}}_{k=0}(s,i)$. By applying (2) when $i > 0$, we obtain

$$\widehat{\mathcal{F}}_{k=0}(s,i) = \sum_{\ell=0}^{i} \widetilde{\mathbb{F}}(s,i,\ell,0)\mathcal{H}$$

$$= \sum_{\ell=1}^{i-1} \widetilde{\mathbb{F}}(s,i,\ell,0)\mathcal{H} + \widetilde{\mathbb{F}}(s,i,0,0)\mathcal{H} + \widetilde{\mathbb{F}}(s,i,i,0)\mathcal{H}$$

$$= \sum_{\ell=1}^{i-1} \mathcal{L}(s)\mathcal{S}\widetilde{\mathbb{F}}(s,i-1,\ell,0)\mathcal{H} + \sum_{\ell=1}^{i-1} \mathcal{L}(s)\mathcal{H}\widetilde{\mathbb{F}}(s,i-1,\ell-1,0)\mathcal{H}$$

$$+ \mathcal{L}(s)\mathcal{S}\widetilde{\mathbb{F}}(s,i-1,0,0)\mathcal{H} + \mathcal{L}(s)\mathcal{H}\widetilde{\mathbb{F}}(s,i-1,i-1,0)\mathcal{H}$$

$$= \mathcal{L}(s)\mathcal{S}\sum_{\ell=0}^{i-1} \widetilde{\mathbb{F}}(s,i-1,\ell,0)\mathcal{H} + \mathcal{L}(s)\mathcal{H}\sum_{\ell=0}^{i-1} \widetilde{\mathbb{F}}(s,i-1,\ell,0)\mathcal{H}$$

$$= \mathcal{L}(s)\,(\mathcal{S}+\mathcal{H})\sum_{\ell=0}^{i-1} \widetilde{\mathbb{F}}(s,i-1,\ell,0)\mathcal{H}\ ,$$

or, equivalently, in terms of $\widehat{\mathcal{F}}_{k=0}(s,i)$:

$$\widehat{\mathcal{F}}_{k=0}(s,i) = \mathcal{L}(s)\,(\mathcal{S}+\mathcal{H})\,\widehat{\mathcal{F}}_{k=0}(s,i-1)\ ,\ i \geq 1. \tag{16}$$

By applying vec operator to (16) one finds the following expression for $vec(\widehat{\mathcal{F}}_{k=0}(s,i))^T$, $i \geq 1$:

$$vec(\widehat{\mathcal{F}}_{k=0}(s,i))^T = vec(\widehat{\mathcal{F}}_{k=0}(s,i-1))^T \left[I \otimes \mathcal{L}(s)\,(\mathcal{S}+\mathcal{H})\right]^T, i \geq 1. \tag{17}$$

By substituting the (17) into (15) and remembering that according to (13) $\hat{\pi}_i(1) = \hat{\pi}_{i-1}(1)\overline{\mathbf{R}}(1)$, we find the sought-for representation for $\mathbf{M}(s)$:

$$\mathbf{M}(s) = vec(\widehat{\mathcal{F}}_{k=0}(s,1))^T \otimes \left(I \otimes \hat{\pi}_1(1)\right) + \sum_{i=1}^{\infty} \left[vec(\widehat{\mathcal{F}}_{k=0}(s,i))^T \otimes \left(I \otimes \hat{\pi}_i(1)\right)\right]$$

$$= \left(vec(\underbrace{\widehat{\mathcal{F}}_{k=0}(s,0)}_{\mathcal{L}(s)\mathcal{H}})^T \left[I \otimes \mathcal{L}(s)\,(\mathcal{S}+\mathcal{H})\right]^T\right) \otimes \left(I \otimes \hat{\pi}_0(1)\overline{\mathbf{R}}(1)\right)$$

$$+ \sum_{i=2}^{\infty} \left[vec(\widehat{\mathcal{F}}_{k=0}(s,i-1))^T \left[I \otimes \mathcal{L}(s)\,(\mathcal{S}+\mathcal{H})\right]^T \otimes \left(I \otimes \hat{\pi}_i(1)\right)\right]$$

$$= \left(vec(\mathcal{L}(s)\mathcal{H})^T \left[I \otimes \mathcal{L}(s) \, (\mathcal{S} + \mathcal{H}) \right]^T \right) \otimes \left(I \otimes \hat{\pi}_0(1)\overline{\mathbf{R}}(1) \right)$$

$$+ \sum_{i=1}^{\infty} \left[vec(\widehat{\mathcal{F}}_{k=0}(s, i))^T \left[I \otimes \mathcal{L}(s) \, (\mathcal{S} + \mathcal{H}) \right]^T \otimes \left(I \otimes \hat{\pi}_i(1) \right) \left(I \otimes \overline{\mathbf{R}}(1) \right) \right]$$

$$= \underbrace{ \left(vec(\mathcal{L}(s)\mathcal{H})^T \left[I \otimes \mathcal{L}(s) \, (\mathcal{S} + \mathcal{H}) \right]^T \right) \otimes \left(I \otimes \hat{\pi}_0(1)\overline{\mathbf{R}}(1) \right) }_{M_1(s)}$$

$$+ \underbrace{ \mathbf{M}(s) \left[\left(I \otimes \mathcal{L}(s) \, (\mathcal{S} + \mathcal{H}) \right)^T \otimes \left(I \otimes \overline{\mathbf{R}}(1) \right) \right] }_{M_2(s)}.$$

Derivation of $\omega_{\mathrm{L}}^{k>0}(s)$. Now we tackle the most complex case – the analysis of $\omega_{\mathrm{L}}^{k>0}(s)$. For $\omega_{\mathrm{L}}^{k>0}(s)$ the Kronecker expansion has to be applied multiple times. At first we recall that the definition of $\omega_{\mathrm{L}}^{k>0}(s)$ is

$$\omega_{\mathrm{L}}^{k>0}(s) = \frac{1}{\lambda} \sum_{i=1}^{\infty} \sum_{j=0}^{\infty} \pi_{i,j} \mathcal{A} \underbrace{ \sum_{\ell=0}^{i} \sum_{k=1}^{\infty} \widetilde{\mathbb{F}}(s, i, \ell, k) \mathcal{H} \widetilde{\mathbf{G}}(s)^k \, \widehat{\mathbf{G}}(s) \mathbf{1} }_{\mathcal{F}(s,i)}.$$

Let us now consider term $\mathcal{F}(s, i)$. Applying vec operator to $\mathcal{F}(s, i)$ according to the following Kronecker expansion

$$vec(ABCD) = (D^T \otimes A)vec(BC) = (vec(BC)^T \otimes (D^T \otimes A))vec(I)$$
$$= (vec(I)^T \otimes I \otimes I)(C \otimes B^T \otimes D^T \otimes A)vec(I),$$

one gets

$$vec(\mathcal{F}(s, i))$$
$$= (vec(I)^T \otimes I \otimes I) \underbrace{ \sum_{\ell=0}^{i} \sum_{k=1}^{\infty} \left(\widetilde{\mathbf{G}}(s)^k \otimes \mathcal{H}^T \otimes I^T \otimes \widetilde{\mathbb{F}}(s, i, \ell, k) \right) }_{\mathcal{F}^{\otimes}(s,i)} vec(I)$$
$$= (vec(I)^T \otimes I \otimes I)\mathcal{F}^{\otimes}(s, i)vec(I).$$

By considering the expression for $\mathcal{F}(s,i)$ and using (2), when $i>0$ and $k>0$, we obtain

$$\mathcal{F}(s,i) = \sum_{\ell=0}^{i} \sum_{k=1}^{\infty} \tilde{\mathbb{F}}(s,i,\ell,k)\mathcal{H}\tilde{\mathbf{G}}(s)^k$$

$$= \sum_{\ell=0}^{i-1} \sum_{k=1}^{\infty} \mathcal{L}(s)\mathcal{S}\tilde{\mathbb{F}}(s,i-1,\ell,k)\mathcal{H}\tilde{\mathbf{G}}(s)^k$$

$$+ \sum_{\ell=0}^{i-1} \sum_{k=1}^{\infty} \mathcal{L}(s)\mathcal{H}\tilde{\mathbb{F}}(s,i-1,\ell,k)\mathcal{H}\tilde{\mathbf{G}}(s)^k$$

$$+ \sum_{\ell=0}^{i} \sum_{k=0}^{\infty} \mathcal{L}(s)\mathcal{A}\tilde{\mathbb{F}}(s,i,\ell,k)\mathcal{H}\tilde{\mathbf{G}}(s)^{k+1}. \tag{18}$$

Having such expression for $\mathcal{F}(s,i)$ one can write out relation for the term $\mathcal{F}^{\otimes}(s,i)$ in the following form:

$$\mathcal{F}^{\otimes}(s,i)$$

$$= \underbrace{\left[\left(I\otimes I\otimes I\otimes \mathcal{L}(s)\mathcal{S}\right) + \left(I\otimes I\otimes I^T\otimes \mathcal{L}(s)\mathcal{H}\right)\right]}_{\mathbf{L}(s)} \mathcal{F}^{\otimes}(s,i-1)$$

$$+ \underbrace{\left(\tilde{\mathbf{G}}(s)\otimes I\otimes I\otimes \mathcal{L}(s)\mathcal{A}\right)}_{\mathbf{K}(s)} \left(\mathcal{F}^{\otimes}(s,i) + \widehat{\mathcal{F}}^{\otimes}_{k=0}(s,i)\right)$$

$$= [I - \mathbf{K}(s)]^{-1}[\mathbf{L}(s)\mathcal{F}^{\otimes}(s,i-1) + \mathbf{K}(s)\widehat{\mathcal{F}}^{\otimes}_{k=0}(s,i)], \tag{19}$$

where we have introduced the notation

$$\widehat{\mathcal{F}}^{\otimes}_{k=0}(s,i) = \sum_{\ell=0}^{i}\left(I\otimes \mathcal{H}^T\otimes I^T\otimes \tilde{\mathbb{F}}(s,i,\ell,0)\right), \quad i\geq 0.$$

From (2) it follows that

$$\mathcal{F}^{\otimes}(s,0) = \sum_{k=1}^{\infty}\left(\tilde{\mathbf{G}}(s)^k\otimes \mathcal{H}^T\otimes I\otimes (\mathcal{L}(s)\mathcal{A})^k\mathcal{L}(s)\right)$$

$$= \left[I - \left(\tilde{\mathbf{G}}(s)\otimes I\otimes I\otimes \mathcal{L}(s)\mathcal{A}\right)\right]^{-1}\left(\tilde{\mathbf{G}}(s)\otimes \mathcal{H}^T\otimes I\otimes \mathcal{L}(s)\mathcal{A}\mathcal{L}(s)\right),$$

and $\widehat{\mathcal{F}}^{\otimes}_{k=0}(s,0) = I\otimes \mathcal{H}^T\otimes I^T\otimes \mathcal{L}(s)$. For $i\geq 1$ from (16) we have

$$\widehat{\mathcal{F}}^{\otimes}_{k=0}(s,i) = \mathbf{L}(s)\,\widehat{\mathcal{F}}^{\otimes}_{k=0}(s,i-1), \quad i\geq 1.$$

Now we go back to $\omega_{\mathrm{L}}^{k>0}(s)$ and apply *vec* operator multiple times in the following way:

$$\omega_{\mathrm{L}}^{k>0}(s) = \frac{1}{\lambda}\sum_{i=1}^{\infty}\sum_{j=0}^{\infty}\pi_{i,j}\mathcal{A}\mathcal{F}(s,i)\widehat{\mathbf{G}}(s)\mathbf{1}$$

$$= \frac{1}{\lambda}\sum_{i=1}^{\infty}\sum_{j=0}^{\infty}\left(\mathbf{1}^T\widehat{\mathbf{G}}(s)^T \otimes \pi_{i,j}\mathcal{A}\right)vec\left(\mathcal{F}(s,i)\right)$$

$$= \frac{1}{\lambda}\left(\mathbf{1}^T\widehat{\mathbf{G}}(s)^T \otimes 1\right)\sum_{i=1}^{\infty}\sum_{j=0}^{\infty}\left(I\otimes\pi_{i,j}\right)\left(I\otimes\mathcal{A}\right)vec\left(\mathcal{F}(s,i)\right)$$

$$= \frac{1}{\lambda}\left(\mathbf{1}^T\widehat{\mathbf{G}}(s)^T \otimes 1\right)\sum_{i=1}^{\infty}\sum_{j=0}^{\infty}\left[vec\left(\mathcal{F}(s,i)\right)^T \otimes \left(I\otimes\pi_{i,j}\right)\right]vec\left(I\otimes\mathcal{A}\right)$$

$$= \frac{1}{\lambda}\left(\mathbf{1}^T\widehat{\mathbf{G}}(s)^T \otimes 1\right)\sum_{i=1}^{\infty}\sum_{j=0}^{\infty}\left[vec(I)^T\mathcal{F}^{\otimes}(s,i)^T(vec(I)^T\otimes I\otimes I)^T\right.$$
$$\left.\otimes\left(I\otimes\pi_{i,j}\right)\right]vec\left(I\otimes\mathcal{A}\right)$$

$$= \frac{1}{\lambda}\left(\mathbf{1}^T\widehat{\mathbf{G}}(s)^T \otimes 1\right)\left[vec(I)^T\otimes I\underbrace{\sum_{i=1}^{\infty}\sum_{j=0}^{\infty}\left[\mathcal{F}^{\otimes}(s,i)^T\otimes\left(I\otimes\pi_{i,j}\right)\right]}_{\mathbf{N}(s)}\right.$$

$$\cdot\left[(vec(I)^T\otimes I\otimes I)^T\otimes I\right]vec\left(I\otimes\mathcal{A}\right).$$

The only unknown quantity in the expression for $\omega_{\mathrm{L}}^{k>0}(s)$ is $\mathbf{N}(s)$. It can be found from (19) in the manner similar to $\mathbf{M}(s)$. We have

$$\mathbf{N}(s) = \left[\mathcal{F}^{\otimes}(s,1)^T \otimes \underbrace{\sum_{j=0}^{\infty}\left(I\otimes\pi_{1,j}\right)}_{I\otimes\hat{\pi}_0(1)\overline{\mathbf{R}}(1)}\right] + \sum_{i=2}^{\infty}\sum_{j=0}^{\infty}\left[\mathcal{F}^{\otimes}(s,i)^T \otimes \left(I\otimes\pi_{i,j}\right)\right]$$

$$= \left[\mathcal{F}^{\otimes}(s,1)^T \otimes \left(I\otimes\hat{\pi}_0(1)\overline{\mathbf{R}}(1)\right)\right]$$

$$+ \underbrace{\sum_{i=2}^{\infty}\left[\widehat{\mathcal{F}}_{k=0}^{\otimes}(s,i)^T \otimes \left(I\otimes\hat{\pi}_i(1)\right)\right]}_{\mathbf{Z}(s)}\left(\mathbf{K}(s)^T[I-\mathbf{K}(s)]^{-1^T}\otimes I\right)$$

$$+ \underbrace{\sum_{i=2}^{\infty}\left[\mathcal{F}^{\otimes}(s,i-1)^T\mathbf{L}(s)^T[I-\mathbf{K}(s)]^{-1^T}\otimes\left(I\otimes\hat{\pi}_{i-1}(1)\right)\left(I\otimes\overline{\mathbf{R}}(1)\right)\right]}_{\mathbf{N}(s)\left(\mathbf{L}(s)^T[I-\mathbf{K}(s)]^{-1^T}\otimes[I\otimes\overline{\mathbf{R}}(1)]\right)}.$$

For $\mathbf{Z}(s)$, using properties of the Kronecker product, one obtains the following relation:

$$
\begin{aligned}
\mathbf{Z}(s) &= \sum_{i=2}^{\infty} \left[\widehat{\mathcal{F}}_{k=0}^{\otimes}(s,i)^T \otimes \left(I \otimes \hat{\pi}_i(1) \right) \right] \\
&= \sum_{i=2}^{\infty} \left[\mathcal{F}_{k=0}^{\otimes}(s,i-1)^T \mathbf{L}(s)^T \otimes \left(I \otimes \hat{\pi}_i(1) \right) \right] \\
&= \sum_{i=1}^{\infty} \left[\mathcal{F}_{k=0}^{\otimes}(s,i)^T \mathbf{L}(s)^T \otimes \left(I \otimes \hat{\pi}_i(1)\overline{\mathbf{R}}(1) \right) \right] \\
&= \sum_{i=1}^{\infty} \left[\widehat{\mathcal{F}}_{k=0}^{\otimes}(s,i)^T \mathbf{L}(s)^T \otimes \left(I \otimes \hat{\pi}_i(1) \right) \left(I \otimes \overline{\mathbf{R}}(1) \right) \right] \\
&= \sum_{i=1}^{\infty} \left[\widehat{\mathcal{F}}_{k=0}^{\otimes}(s,i)^T \otimes \left(I \otimes \hat{\pi}_i(1) \right) \right] \left(\mathbf{L}(s)^T \otimes \left(I \otimes \overline{\mathbf{R}}(1) \right) \right) \\
&= \left[\left(\widehat{\mathcal{F}}_{k=0}^{\otimes}(s,1)^T \otimes \left(I \otimes \hat{\pi}_0(1)\overline{\mathbf{R}}(1) \right) \right) + \mathbf{Z}(s) \right] \left(\mathbf{L}(s)^T \otimes \left(I \otimes \overline{\mathbf{R}}(1) \right) \right).
\end{aligned}
$$

The latter relation allows computation of $\mathbf{Z}(s)$ and subsequently $\mathbf{N}(s)$ and $\omega_L^{k>0}(s)$. Thus the expression for $\omega_L(s)$ is obtained.

5   Numerical Example

In order to give a more complete picture of how the service and the resequencing policies influence the waiting time of an arbitrary customer, we present a simple numerical example. Due to the Little's law the mean waiting times of arbitrary customer under the HOQ-FIFO-FIFO and HOQ-FIFO-LIFO policies coincide. Thus we dwell on comparison of the standard deviation of the waiting time.

Two use cases are considered. The first one is taken from [10], where the regular customers and resequencing signals arrive according to Poisson processes with rates λ and γ, respectively. The service process has the phase-type distribution with the representation:

$$
\beta = (0.5, 0.5) \quad \mathbf{B} = \begin{pmatrix} -4 & 2 \\ 1 & -4 \end{pmatrix}, \text{ from which } \mathbf{S_0} = \begin{pmatrix} -4 & 2 \\ 1 & -4 \end{pmatrix}, \quad \mathbf{S_1} = \begin{pmatrix} 1 & 1 \\ 1.5 & 1.5 \end{pmatrix}.
$$

The service rate is $\mu = -1/(\beta\mathbf{B}^{-1}\mathbf{1}) = 2.5$ and consequently $\lambda = 2.5\rho$, where ρ and γ are the parameters of the example. As the second use case we take the same service process $(\mathbf{S_0}, \mathbf{S_1})$, but the arrival process of regular and resequencing customers are characterized by

$$
\mathbf{A_0} = \begin{pmatrix} -5 & 1.5 \\ 2 & -3 \end{pmatrix}, \mathbf{A_1} = \begin{pmatrix} 3.5p & 3.5(1-p) \\ p & (1-p) \end{pmatrix}, \mathbf{H_0} = \begin{pmatrix} -7 & 0 \\ 0 & -7q \end{pmatrix}, \mathbf{H_1} = \begin{pmatrix} 7q & 7(1-q) \\ 7q^2 & 7q(1-q) \end{pmatrix}.
$$

Indeed they mean order 2 phase-type renewal processes with mean intensity $\lambda = \frac{120}{70-25p}$ $(\rho = \frac{240}{350-125p})$ and $\gamma = \frac{7q}{1-q+q^2}$. By tuning the values of p and q

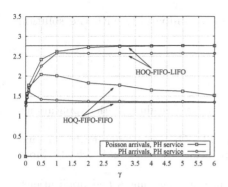

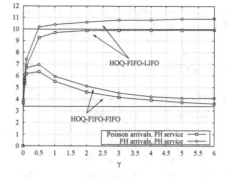

Fig. 3. $\rho = 0.72$ **Fig. 4.** $\rho = 0.88$

we can set the load and the resequencing rate. In Figs. 3 and 4[1] one can see the graphs of the standard deviation of the waiting times as function of resequencing rate γ for two arbitrary values of load $\rho = 0.72$ and $\rho = 0.88$ and both use cases.

When γ is low the second order characteristics of the waiting time are almost the same. As the resequencing rate γ grows, the difference in the behaviour of the both curves becomes more significant. This difference comes from the following fact. As the resequencing rate γ grows almost all customers get resequenced. Thus under the HOQ-FIFO-FIFO policy they are served according to FIFO and under the HOQ-FIFO-LIFO policy – according to LIFO. Intuitively in the latter case the variance of the waiting time is bigger because LIFO policy can generate some extremely high response times. Indeed we may have to wait for a very long time in order to take care of the first arrival to the resequencing buffer.

Finally, as γ grows the standard deviations of waiting time under the HOQ-FIFO-FIFO and HOQ-FIFO-LIFO policies tend to the standard deviations of the waiting time (horizontal lines in the figures for the Poisson arrival case) in the standard $M/PH/1$ FIFO and $M/PH/1$ LIFO queues respectively. At $\gamma = 0$ we also have the case of pure FIFO queue.

6   Conclusion

The delay analysis of the HOQ-FIFO-LIFO policy shows that the majority of the analysis steps (recursive evolution equation like description of properly chosen performance measures, Kronecker expansion based treatment of non-commuting matrices, describing the relation of infinite summations from 0 to ∞ with the one from 1 to ∞) remain applicable, but also new analysis elements are required. In particular, the analysis of the HOQ-FIFO-LIFO service policy requires the solution of a coupled quadratic matrix equation, which was separated in the HOQ-FIFO-FIFO case. In spite, the computational complexity of the HOQ-FIFO-LIFO case is not higher than the one of the HOQ-FIFO-FIFO case,

---

[1] Standard deviation of customer's waiting time as function of resequencing intensity (γ) for two different load (ρ) levels, two different policies and two use cases.

because the solution of the coupled equation is comparable with the solution of two separate ones.

References

1. Alexander, G.: Kronecker Products and Matrix Calculus: With Applications. Wiley, New York (1982)
2. Dimitrov, B., Green, D., Rykov, V., Stanchev, P.: On performance evaluation and optimization problems in queues with resequencing. Advances in Stochastic Modelling, pp. 55–72 (2002)
3. Van Do, T.: Bibliography on g-networks, negative customers and applications. Math. Comput. Modell. **53**(1), 205–212 (2011)
4. He, Q.-M.: Fundamentals of Matrix-Analytic Methods. Springer, New York (2013). doi:10.1007/978-1-4614-7330-5
5. Latouche, G., Ramaswami, V.: Introduction to matrix analytic methods in stochastic modeling. In: Society for Industrial and Applied Mathematics (1999)
6. Neuts, M.F.: Matrix Geometric Solutions in Stochastic Models. Johns Hopkins University Press, Baltimore (1981)
7. Neuts, M.F.: Structured stochastic matrices of M/G/1 type and their applications. Marcel Dekker, New York (1989)
8. Ozawa, T.: Sojourn time distributions in the queue defined by a general QBD process. Queueing Syst. Theory Appl. **53**(4), 203–211 (2006)
9. Pechinkin, A.V., Razumchik, R.V.: On temporal characteristics in an exponential queueing system with negative claims and a bunker for ousted claims. Autom. Remote Control **72**(12), 2492–2504 (2011)
10. Razumchik, R., Telek, M.: Delay analysis of a queue with re-sequencing buffer and markov environment. Queueing Syst. **82**(1–2), 7–28 (2016)
11. Steeb, W.H., Hardy, Y.: Matrix Calculus and Kronecker Product: A Practical Approach to Linear and Multilinear Algebra. World Scientific, River Edge (2011)

Analysis of Timed Properties Using the Jump-Diffusion Approximation

Paolo Ballarini[1], Marco Beccuti[2(✉)], Enrico Bibbona[3], Andras Horvath[2(✉)],
Roberta Sirovich[4], and Jeremy Sproston[2]

[1] Laboratoire MICS, CentraleSupèlec, Université Paris Saclay, Paris, France
[2] Dipartimento di Informatica, Università di Torino, Turin, Italy
{beccuti,horvath}@di.unito.it
[3] Dipartimento di Scienze Matematiche,
"G. L. Lagrange", Politecnico di Torino, Turin, Italy
[4] Dipartimento di Matematica, Università di Torino, Turin, Italy

Abstract. Density dependent Markov chains (DDMCs) describe the interaction of groups of identical objects. In case of large numbers of objects a DDMC can be approximated efficiently by means of either a set of ordinary differential equations (ODEs) or by a set of stochastic differential equations (SDEs). While with the ODE approximation the chain stochasticity is not maintained, the SDE approximation, also known as the diffusion approximation, can capture specific stochastic phenomena (e.g., bi-modality) and has also better convergence characteristics. In this paper we introduce a method for assessing temporal properties, specified in terms of a timed automaton, of a DDMC through a jump diffusion approximation. The added value is in terms of runtime: the costly simulation of a very large DDMC model can be replaced through much faster simulation of the corresponding jump diffusion model. We show the efficacy of the framework through the analysis of a biological oscillator.

Keywords: Diffusion approximation · Stochastic differential equations with jumps · Statistical model checking

1 Introduction

Context. Advances in modelling lead to increasingly complex models of concurrent systems whose analysis, consequently, has become a critical issue. In particular the analysis of quantitative aspects of these systems by means of stochastic models (e.g., Markov chains) may be impaired by the combinatorial explosion of their state space. To cope with this problem several approaches have been proposed in the literature including, e.g., decomposition and aggregation, bounding techniques, compact representations of the state space. However, when the model accounts for large groups of individuals (e.g., Internet users, molecule populations) these techniques may turn out to be insufficient, meaning that *discrete event simulation* (DES) is the most practical option for analysing the system's performance. Indeed, DES based approaches do not require the explicit

© Springer International Publishing AG 2017
P. Reinecke and A. Di Marco (Eds.): EPEW 2017, LNCS 10497, pp. 69–84, 2017.
DOI: 10.1007/978-3-319-66583-2_5

storage of the state space, but instead exploit a set of sample executions (i.e., traces) in order to devise arbitrarily accurate (statistical) *estimates* of relevant indicators of a model's behaviour.

Fluid approximations. In the case of large interacting populations, an alternative to simulation is to use a *deterministic approximation* in which the behavior is represented by a set of ordinary differential equations (ODE) [17]. However, this approach is not suitable for the study of models where stochasticity (bimodality, high variance) plays an important role even for large population counts. To analyze a model's stochastic nature, in [18] a *diffusion approximation* was proposed, based on a set of stochastic differential equations (SDE), that can be applied up to the first visit of the boundary of the state space. Both the deterministic and the diffusion approximation are such that every state variable is approximated by a continuous variable, i.e., the state variables are made "fluid". Since in real systems the boundaries of the state space often can be visited many times, in [7] we proposed an extension, namely, a *jump diffusion approximation*, to properly approximate the original model at the boundaries as well. A further extension was made in [3] that uses partial fluidification of the state space, which results in a *switching jump diffusion approximation*, allowing us to mimic better the original process in the case of low population counts.

Contribution. Starting from [7], in this paper, we propose a new statistical model checking method based on jump diffusion approximation. This method takes as input a DDMC and, following [9–11,16], a formal description of a property, in this paper described as a (deterministic) timed automaton [1]. The jump diffusion approximation of the DDMC is used to generate trajectories of the system and the deterministic timed automaton is used to accept or reject each trace. Based on the proportion of the accepted traces confidence intervals are derived for the probability that the system exhibits the property in question.

Applicability. The applied theoretical framework requires a sequence of DDMCs indexed by a parameter N [17]. The sequence is such that the state space, the transition intensities and also the vector describing the initial state increase as N increases. Four possibly overlapping ranges of values can be identified for N. N can be so small that the corresponding DDMC can be analyzed by analytical approaches. As N grows, analytical analysis becomes unfeasible, but the DDMC can still be evaluated efficiently by simulation. By further increasing N, even simulation of the DDMC becomes impractical but the model can still exhibit important stochastic behavior. This is the range in which the approach we propose is convenient to use: in this range the diffusion approximation provides results with reasonable precision in much shorter time than dealing with the original DDMC. For even larger values of N, the stochastic behavior disappears and the model can be analyzed with a deterministic approximation.

Organization. The paper is organized as follows. Section 2 introduces DDMCs and Sect. 3 discusses their approximations. In Sect. 4 we provide the definition of the applied timed automata. We discuss the issues related to assessing properties through a diffusion approximation in Sect. 5. In Sect. 6 a case study is presented. Conclusions are drawn in Sect. 7.

2 Nearly Density Dependent Markov Chains

Continuous time Markov chains (CTMC) are often used to describe the interaction of groups of identical objects. Informally, such CTMCs are called *density dependent* if the intensities of the interactions can be expressed as a function of the *density of the objects present in the area* (or volume) described by the model (as opposed to being expressed as a function of the number of objects itself).

Definition 1. *Consider a sequence of CTMCs, denoted by $X^{[N]}(t)$, indexed by $N \in \mathbb{N} \setminus \{0\}$ and with state space $\mathcal{S}^{[N]} \subseteq \mathbb{Z}^k$ (i.e., every state is identified by a vector of k integers), that describe the interaction of k groups of identical objects. The sequence $X^{[N]}(t)$ is called* density dependent *if the associated transition intensities, given any two states $r \in \mathcal{S}^{[N]}$ and $r + m \in \mathcal{S}^{[N]}$ that are connected by a transition, can be written in the form*

$$q^{[N]}_{r,r+m} = N f \left(\frac{r}{N}, m \right) \tag{1}$$

where $f : \mathbb{R}^k \times \mathbb{Z}^k \to \mathbb{R}_{\geq 0}$ is a bivariate function whose first argument is a vector that provides the density for each group of objects in state r and its second argument is the change in the state due to the transition from state r to state $r + m$ ($\mathbb{R}_{\geq 0}$ is the set of non-negative real numbers).

The indexing parameter N can represent the size of the considered area or volume, or the total number of objects in the model (in this case the vector r/N is a vector of proportions). Note that in Definition 1 a single function, namely f, provides the intensity of every transition of every CTMC of the sequence of CTMCs. This implies that in every CTMC the transitions have the same effect on the state.

The above definition can be relaxed by substituting (1) with

$$q^{[N]}_{r,r+m} = N f \left(\frac{r}{N}, m \right) + N g \left(\frac{r}{N}, m, N \right) \tag{2}$$

where $g : \mathbb{R}^k \times \mathbb{Z}^k \times \mathbb{N} \to \mathbb{R}_{\geq 0}$ is a trivariate function and $g(r/N, m, N) \in O(1/N)$. Sequences of CTMCs in which the transition intensities are in the form given in (2) are referred to as *nearly density dependent*. The rationale behind the definition is the following. As N grows, thanks to $g(r/N, m, N) \in O(1/N)$, the term $Ng(r/N, m, N)$, which is *not density dependent*, remains in the order of a constant. The other term instead grows proportionally to N. Accordingly, as N grows the density dependent nature of the process prevails. Indeed, *density dependent* and *nearly density dependent* processes can be studied with the same approximations.

As for notation, the set of possible changes in the state due to a transition will be denoted by C. Formally, a vector $m \in \mathbb{Z}^k$ is in C if and only if there exist two states $r \in \mathcal{S}^{[N]}$ and $r + m \in \mathcal{S}^{[N]}$ such that there is a transition from r to $r + m$. Note that, like the function f, also the set C is shared by every member of a given sequence of DDMC.

Example 1. As an example we consider a simple epidemic model in which two groups are involved, namely, susceptible and infected individuals. Accordingly, each state is described by a pair (i, j) providing the number of susceptible and infected people, respectively. We assume that the modelled individuals are uniformly distributed over an area split into N equally sized cells and that three kinds of events are possible. The number of susceptible individuals grows with an intensity proportional to the number of cells: $q^{[N]}_{(i,j),(i+1,j)} = N\lambda_1$. Due to the contact of two infected and one susceptible person in one of the cells, one susceptible individual becomes infected; this happens with intensity $q^{[N]}_{(i,j),(i-1,j+1)} = \frac{ij(j-1)}{2} \frac{1}{N^3} N\lambda_2$, where the first term is the number of ways the three individuals can be selected, the second term is the probability that the three selected individuals are together in a given cell, and the multiplication by N is due to the fact that the contact can occur in any cell. Infected individuals can become immune independently of each other and independently of the number of cells; the associated intensity is $q^{[N]}_{(i,j),(i,j-1)} = j\lambda_3$. The intensity of the first type of event is independent of the actual state and proportional to N and thus it is a special form of (1). The intensity of the other two kinds of events can be rewritten as

$$q^{[N]}_{(i,j),(i-1,j+1)} = N\left(\frac{\lambda_2}{2}\frac{i}{N}\left(\frac{j}{N}\right)^2\right) - N\left(\frac{1}{N}\frac{\lambda_2}{2}\frac{i}{N}\frac{j}{N}\right), \quad q^{[N]}_{(i,j),(i,j-1)} = N\lambda_3\frac{j}{N}$$

where the first intensity is nearly density dependent while the second is density dependent. The set of possible state changes is $C = \{(1, 0), (-1, 1), (0, -1)\}$.

3 Approximations of Nearly Density Dependent CTMCs

All approximations we describe in the following use a process with a continuous state space and thus are considered "fluid" approximations. In order to proceed we need to introduce the sequence of normalized CTMCs given by $Z^{[N]}(t) = X^{[N]}(t)/N$, called also the density process. The reason to use $Z^{[N]}$ instead of the original process is that normalization brings all CTMCs of a given density dependent sequence to the same scale, making them comparable.

The first approximation we consider uses a set of ODEs in which there is one equation per group. Accordingly, the original stochastic behavior is approximated by a deterministic process. The set of ODEs used in the approximation is provided by the following result of Kurtz [17]. Given a nearly density dependent sequence of CTMCs $X^{[N]}(t)$ with initial state that tends to z_0 as N tends to infinity, i.e., $\lim_{N\to\infty} Z^{[N]}(0) = \lim_{N\to\infty} X^{[N]}(0)/N = z_0$, if the function $\sum_{l \in C} lf(y, l)$ satisfies some relatively mild conditions, then the density process

$Z^{[N]}(t)$ converges to a deterministic function $z(t)$. The function $z(t)$ is the solution of the following set of ODEs

$$dz(t) = \sum_{l \in C} lf\left(z(t), l\right) dt, \quad z(0) = z_0. \tag{3}$$

We note that (3) is equivalent to the more familiar form $\frac{dz(t)}{dt} = \sum_{l \in C} lf\left(z(t), l\right)$; however we prefer the form in (3) because it has more in common with the other approximations that we introduce later.

The approximation given by $z(t)$ has the following property:

$$\lim_{N \to \infty} \mathbb{P}\left\{\sup_{t \leq T} \left|Z^{[N]}(t) - z(t)\right| > \delta\right\} = 0, \tag{4}$$

for every $\delta > 0$ and where T is the upper limit of the considered finite time horizon. Moreover, it was shown in [17] that the difference between the deterministic approximation and the original stochastic behavior is characterized by

$$\sup_{t \leq T} \left|Z^{[N]}(t) - z(t)\right| = O\left(1/\sqrt{N}\right) \tag{5}$$

The practical meaning of (5) is that the error of the deterministic approximation decreases as $1/\sqrt{N}$.

Another approximation of a density dependent sequence $X^{[N]}$, which is based on stochastic differential equations and thus it preserves the stochastic nature of the original process, was proposed in [18,19]. This approximation, denoted by $Y^{[N]}(t)$, is obtained by the following set of SDEs:

$$dY^{[N]}(t) = \sum_{l \in C} lf\left(Y^{[N]}(t), l\right) dt + \sum_{l \in C} \frac{l}{\sqrt{N}} \sqrt{f\left(Y^{[N]}(t), l\right)} dW_l(t) \tag{6}$$

where the $W_l(t)$ with $l \in C$ are independent standard one-dimensional Brownian motions. The approximation holds up to the first time $Y^{[N]}(t)$ reaches a boundary of the state space. In (6) the first term is the same used by the deterministic approximation in (3), while the second term is a noise that mimics the stochasticity of the original CTMCs.

For what concerns the relation of the diffusion approximation and the original density process, in [18] it has been proven that, for any finite N, we have

$$\sup_{t \leq T} \left|Z^{[N]}(t) - Y^{[N]}(t)\right| = O\left(\log N / N\right) \tag{7}$$

In practice, one uses $N \cdot z(t)$ or $N \cdot Y^{[N]}(t)$ to approximate the original CTMC $X^{[N]}(t)$. The difference between $N \cdot z(t)$ and $X^{[N]}(t)$ according to (5) is in the order of $N(1/\sqrt{N}) = \sqrt{N}$. Between $N \cdot Y^{[N]}(t)$ and $X^{[N]}(t)$ according to (7) it is instead $N(\log N / N) = \log N$ which is much lower than $\sqrt{N}$.

A limitation of the previous approach based on SDEs is that it can be applied only to models where the probability of reaching a boundary of the area of the process is negligible. In order to overcome this limitation, in [7] we introduced a jump diffusion process in which the jumps are used to capture the behavior of the process at the boundaries. We provide here a brief description of the jump diffusion process, denoted by $J^{[N]}(t)$; for a detailed treatment, see [3,7].

The main idea is to split the transitions of the model into two sets depending on the current state. In particular, we denote by $C^{\circ}(y)$ the set of transitions that change one or more components of the state which are at the boundary in state y. The jump diffusion process is defined then by

$$dJ^{[N]}(t) = \sum_{l \in C - C^{\circ}(J^{[N]}(t))} lf\left(J^{[N]}(t), l\right) dt+ \tag{8}$$

$$\sum_{l \in C - C^{\circ}(J^{[N]}(t))} \frac{l}{\sqrt{N}} \sqrt{f\left(J^{[N]}(t), l\right)} dW_l(t) + \sum_{l \in C^{\circ}(J^{[N]}(t))} \frac{l}{N} dM_l^{[N]}(t)$$

where the first two terms are analogous to those in (6) but are restricted to those transitions that change components away from the boundaries. If none of the components are at the boundary of the state space then $J^{[N]}(t)$ behaves exactly as $Y^{[N]}(t)$. The term $M_l^{[N]}(t)$ corresponds to Poisson counting processes that gives rise to jumps that mimic the behavior of the original CTMC at the boundaries. In other words, when the process reaches a boundary then discrete jumps regulated by a Poisson process make it jump back eventually to the inner part of the state space. The intensity associated with $dM_l^{[N]}(t)$ is $\mu_l(t) = Nf\left(J^{[N]}(t), j\right)$, i.e., it is taken directly from the original CTMC (note that $J^{[N]}(t)$ provides directly a vector of densities as required by f). Then $dM_l^{[N]}(t)$ is multiplied by l/N because that is the effect of the transition in the normalized state space.

Recent studies [8] have shown that the jump diffusion approximation has similar characteristics to those of the pure diffusion approximation and, in particular, that the approximation it introduces is as good as that of the "pure" diffusion process, that is:

$$\sup_{t \leq T} \left| Z^{[N]}(t) - J^{[N]}(t) \right| = O\left(\log N/N\right) \tag{9}$$

Numerical evaluation of the goodness of the jump diffusion approximation has been illustrated instead in [3,7].

4 Timed Automata

In this section, we introduce a timed automata-based formalism for the specification of timed properties of CTMCs. As is standard when using timed automata for the specification of properties of stochastic systems (e.g., [10–13,20]), we use *deterministic* timed automata (DTA): that is, each input sequence of the timed automaton (which in our context is a trajectory, i.e., a function from time to the state space of the CTMC, representing a particular behavior of the CTMC or of its diffusion approximation) corresponds to a single run of the timed automaton. In order to provide a uniform framework for timed properties interpreted on CTMC and on jump diffusion approximations, our DTA are labeled with constraints both on clocks and on variables characterizing the state space, but are not labeled with actions corresponding to individual CTMC transitions (which have no meaning in the jump-diffusion diffusion approximation setting). Edges of our variant of DTA are *urgent*: they are taken as soon as they are enabled. Urgency of edges allows for a natural interpretation of our DTA not only on behaviors of CTMC, but also on trajectories of their diffusion approximations.

We denote by $\mathcal{S} \subseteq \mathbb{Z}^k$ the state space and by $\mathcal{V} = \{\vartheta_1, ..., \vartheta_k\}$ a set of k variables, where we interpret ϑ_i as a variable corresponding to the i-th element of the vector representing a state. Let $\mathcal{C}$ be a finite set of variables called *clocks*.

Definition 2. *A constraint is defined by the following grammar:*

$$\Phi ::= \varphi \leq \varphi \mid \mathsf{c} \leq \lambda \mid \mathsf{c} \geq \lambda \mid \Phi \wedge \Phi,$$
$$\varphi ::= \varphi + \varphi \mid \varphi - \varphi \mid \varphi * \varphi \mid \varphi/\varphi \mid \vartheta_i \mid \lambda,$$

where $\mathsf{c} \in \mathcal{C}$ *is a clock,* $\vartheta_i \in \mathcal{V}$ *and* $\lambda \in \mathbb{Q}$ *is a rational constant. A guard constraint is a constraint* Φ *such that, for each* $a \in \mathcal{V} \cup \mathcal{C}$, *there is at most one subformula of* Φ *featuring* a. *An* invariant constraint *is a guard constraint* Φ *in which there is no subformula of the form* $\mathsf{c} \geq \lambda$. *We write* Guards$(\mathcal{V}, \mathcal{C})$ *and* Invariants$(\mathcal{V}, \mathcal{C})$ *to denote the set of guard constraints and invariant constraints, respectively, over* $\mathcal{V}$ *and* $\mathcal{C}$.

Examples of invariant constraints include $\vartheta_1 \leq 10 \wedge \vartheta_2 \geq \vartheta_3$ and $\vartheta_1 \geq 3 \wedge \mathsf{c}_1 \leq 15$, whereas $\mathsf{c}_1 \geq 3 \wedge \mathsf{c}_2 \leq 10 \wedge \vartheta_1 \geq 3$ is an example of a guard constraint that is not an invariant constraint (due to the conjunct $\mathsf{c}_1 \geq 3$).

A function $v : \mathcal{C} \to \mathbb{R}_{\geq 0}$ is referred to as a *clock valuation*, and the set of all clock valuations is denoted by Val$(\mathcal{C})$. For any $v \in$ Val$(\mathcal{C})$, $\gamma \in \mathbb{R}_{\geq 0}$ and $\mathsf{C} \subseteq \mathcal{C}$, we use $v + \gamma$ to denote the clock valuation that increments all clock values in v by γ (that is, $(v+\gamma)(\mathsf{c}) = v(\mathsf{c}) + \gamma$ for all $\mathsf{c} \in \mathcal{C}$), and $v[\mathsf{C}{:=}0]$ to denote the clock valuation in which clocks in C are reset to 0 (that is, $v[\mathsf{C}{:=}0](\mathsf{c}) = 0$ for $\mathsf{c} \in \mathsf{C}$, and $v[\mathsf{C}{:=}0](\mathsf{c}) = v(\mathsf{c})$ for $\mathsf{c} \in \mathcal{C} \setminus \mathsf{C}$). The clock valuation that assigns 0 to all clocks in $\mathcal{C}$ is denoted by $\mathbf{0}$. Let Φ be a constraint, let $y \in \mathcal{S}$ be a state and let $v \in$ Val$(\mathcal{C})$ be a clock valuation. Then we write $(y, v) \models \Phi$ if and only if substituting ϑ_i by y_i (where y_i is the i-th element of the vector y) and c by $v(\mathsf{c})$ in Φ results in Φ resolving to true. For example, for y such that $y_1 = 4$ and v such that $v(\mathsf{c}_1) = 12.1$, we write $(y, v) \models \vartheta_1 \geq 3 \wedge \mathsf{c}_1 \leq 15$.

Definition 3. *A timed automaton is a tuple* $(\mathcal{L}, \ell_{\mathrm{init}}, \mathcal{F}, \mathcal{C}, \mathrm{Inv}, \mathcal{E})$ *comprising: (1) a finite set* $\mathcal{L}$ *of locations, with an initial location* $\ell_{\mathrm{init}} \in \mathcal{L}$ *and a set* $\mathcal{F} \subseteq \mathcal{L}$ *of final locations; (2) a finite set* $\mathcal{C}$ *of clocks; (3) an invariant condition* $\mathrm{Inv} : \mathcal{L} \to$ Invariants$(\mathcal{V}, \mathcal{C})$; *(4) a set* $\mathcal{E} \subseteq \mathcal{L} \times$ Guards$(\mathcal{V}, \mathcal{C}) \times 2^{\mathcal{C}} \times \mathcal{L}$ *of edges, where each edge* $(\ell, \Phi, \mathsf{C}, \ell') \in \mathcal{E}$ *comprises a source location* ℓ, *an enabling condition* Φ, *a set* C *of clocks to be reset to 0, and a target location* ℓ'. *A timed automaton is* deterministic *if, for any location* $\ell \in \mathcal{L}$ *and for any pair* $(\ell, \Phi_1, \mathsf{C}_1, \ell_1), (\ell, \Phi_2, \mathsf{C}_2, \ell_2) \in \mathcal{E}$, *we have that* $\Phi_1 \wedge \Phi_2$ *is unsatisfiable.*

We use DTA to determine whether a trajectory $X : \mathbb{R}_{\geq 0} \to \mathcal{S}$ satisfies a timed property. More precisely, the DTA reads the trajectory X and traverses edges between locations on the basis of (1) the states visited by the trajectory as time passes and (2) the current values of the clocks. The values of the clocks increase at the same rate as real-time. The DTA *must* leave its current location ℓ without letting time pass if there exists an edge $(\ell, \Phi, \mathsf{C}, \ell') \in \mathcal{E}$ such that the enabling condition Φ is currently satisfied (hence, the DTA can be regarded as having an "urgent" semantics in which an enabled edge must be taken as soon as possible): this satisfaction of the enabling condition of the guard may occur, for example, because the value of a state variable falls below some threshold, or the value of a clock reaches a particular value. Furthermore, an additional constraint on the trajectory is imposed by the invariant conditions: during a period in which the DTA is in a particular location ℓ, the invariant condition $\mathrm{Inv}(\ell)$ must be satisfied by the states visited by the trajectory and by the current value of the clocks during that period, otherwise the trajectory will be regarded as not satisfying the timed property. A set of clocks can be reset to 0 when an edge is taken. If the DTA, starting from the initial location, reaches a final location when reading the trajectory X, then we say that the trajectory is accepted by the DTA (which, intuitively, corresponds to the trajectory X satisfying the timed property represented by the DTA), otherwise it is rejected.

In the following, we describe formally the acceptance of trajectories by a DTA. Let $(\ell, \Phi, \mathsf{C}, \ell') \in \mathcal{E}$ be an edge of a DTA $\mathcal{A}$. Then we write $\mathsf{source}(\ell, \Phi, \mathsf{C}, \ell') = \ell$, $\mathsf{guard}(\ell, \Phi, \mathsf{C}, \ell') = \Phi$, $\mathsf{reset}(\ell, \Phi, \mathsf{C}, \ell') = \mathsf{C}$, and $\mathsf{target}(\ell, \Phi, \mathsf{C}, \ell') = \ell'$. Let $\ell \in \mathcal{L}$ be a location of $\mathcal{A}$, and let $y \in \mathcal{S}$ be state and $v \in \mathrm{Val}(\mathcal{C})$. We write $(y, v) \not\models \mathsf{Guards}(\ell)$ if and only if $(y, v) \not\models \mathsf{guard}(e)$ for all $e \in \mathcal{E}$ such that $\mathsf{source}(e) = \ell$. A pair $(\ell, v) \in \mathcal{L} \times \mathrm{Val}(\mathcal{C})$ is called a *configuration*. We write $(\ell, v) \xrightarrow{\gamma, e} (\ell', v')$ to denote the DTA-transition from configuration (ℓ, v) to configuration (ℓ', v') after $\gamma > 0$ time units have elapsed and by taking the edge e. The transition $(\ell, v) \xrightarrow{\gamma, e} (\ell', v')$ exists if (1) $\mathsf{source}(e) = \ell$, (2) $v' = (v + \gamma)[\mathsf{reset}(e) {:=} 0]$, and (3) $\mathsf{target}(e) = \ell'$. A path of $\mathcal{A}$ is a finite sequence of DTA-transitions $\pi = (\ell_0, v_0) \xrightarrow{\gamma_0, e_0} (\ell_1, v_1) \xrightarrow{\gamma_1, e_1} \cdots \xrightarrow{\gamma_{m-1}, e_{m-1}} (\ell_m, v_m)$. Let $\Lambda^\pi = \{\lambda_0^\pi, \lambda_1^\pi, \ldots, \lambda_m^\pi\}$ be the set of constants such that $\lambda_0^\pi = 0$ and $\lambda_i^\pi = \sum_{k=0}^{i-1} \gamma_k$ for all i such that $1 \leq i \leq m$.

Definition 4. *Let $\mathcal{A}$ be a DTA. We say that $X : \mathbb{R}_{\geq 0} \to \mathcal{S}$ is accepted by the DTA if there exists a path $\pi = (\ell_0, v_0) \xrightarrow{\gamma_0, e_0} (\ell_1, v_1) \xrightarrow{\gamma_1, e_1} \cdots \xrightarrow{\gamma_{m-1}, e_{m-1}} (\ell_m, v_m)$ of $\mathcal{A}$ such that $\ell_0 = \ell_{\mathrm{init}}$, $v_0 = \mathbf{0}$, $\ell_m \in \mathcal{F}$ and, for all $0 \leq i < m$, the following conditions are satisfied:*

- *for all $0 \leq \gamma' < \gamma_i$, we have $(X(\lambda_i^\pi + \gamma'), v_i + \gamma') \models \mathrm{Inv}(\ell_i)$ and $(X(\lambda_i^\pi + \gamma'), v_i + \gamma') \not\models \mathsf{Guards}(\ell_i)$;*
- *$(X(\lambda_i^\pi + \gamma_i), v_i + \gamma_i) \models \mathsf{guard}(e_i)$.*

5 Assessing Timed Automata Based Properties by Diffusion Approximations

In this paper we limit our attention to illustrating the practical applicability of the approach. According to (9), there is a correspondence between the trajectories of the CTMC and those of the approximating jump diffusion process. Moreover, the larger N is, the tighter the relation gets. It is natural hence to expect that over a certain threshold for N, which depends on the considered model, one can safely use trajectories of the diffusion process instead of trajectories of the CTMC to assess DTA-based temporal properties.

There is, however, a fundamental difference between a CTMC and a diffusion process. A diffusion process exhibits extreme oscillatory nature along its drift in any infinitesimal interval. This means that if a diffusion exceeds a given limit **L** for the first time then it goes below **L** with probability 1 afterwords in any infinitesimal interval. Consider now a diffusion process $X(t)$ and a DTA with three locations. The initial location is with invariant $X(t) \leq$ **L** and has a transition enabled if $X(t) >$ **L**. The second location is with invariant $X(t) \geq$ **L**, it does not have an enabled transition associated with the situation $X(t) \leq$ **L** and it has a transition enabled when $X(t) \geq 2$**L** that leads to the third location which is a final one. Due to the oscillatory nature of the diffusion process, with probability 0 a trajectory is accepted by the DTA. Note however that such a situation is coherent with what happens in a CTMC as N grows large. For large values of N, the trajectories of a CTMC are more and more similar to those of a diffusion. Consequently, the probability of the set of those trajectories of the CTMC that are accepted by the above described DTA tends to 0 as N tends to infinity. The characteristics of the diffusion process and that of the CTMCs with large N must be taken into account during the definition of the DTA in order to avoid results that are consequences of these characteristics and not the properties of the studied phenomenon. In practice, the problem is alleviated by using piecewise constant abstractions of the trajectories of the diffusion process.

We consider now the three kinds of approximations errors that occur during the analysis of CTMCs based on diffusion processes. First, the diffusion process is an approximation of the original CTMC. The goodness of this approximation was discussed in Sect. 3. Second, the analysis is carried out based on traces generated by approximate simulation. Indeed, exact simulation can be carried out only in special cases of diffusion (for example, in case of a Wiener process without drift) but in general the process is multidimensional and it includes a state dependent drift and a state dependent noise that cannot be simulated exactly in general. Third, a diffusion process fluctuates in any infinitesimal interval which means that it is not possible to obtain a

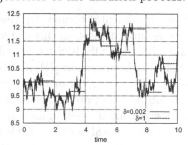

Fig. 1. Two versions of the same trajectory of a pure Wiener process with different time steps.

complete representation of a trace. Indeed, the temporal properties are assessed based on a constant piecewise approximation of an infinitely fluctuating trace.

Let us illustrate this third source of approximation error in some detail. In Fig. 1 we plotted two versions of the same trajectory with two different time steps (δ). Consider now a DTA which accepts only those traces along which the process never exceeds level 12 in the time interval $[0, 10]$. Clearly, using the piecewise constant abstraction of the trace, the trace with $\delta = 1$ is accepted while the other is rejected.

The previous example indicates that the choice of the time step during the generation of the traces is of fundamental importance to achieve good approximation of the original behavior. The same problem, i.e., not knowing the fluctuation between two consecutive time points, is present to a somewhat lesser extent also when jump diffusion processes are used to obtain approximations of more classical measures, like transient probabilities. In that case, it is of crucial importance to find with sufficient precision the time instants when the process reaches the boundary, i.e., the time instants when the change from pure diffusion process to jump diffusion process has to be made. When assessing temporal properties described by DTA, the problem appears also inside the state space around the thresholds present in the automaton. The choice of the time step was discussed to some extent in [3, 7]. In theory, it is possible to add intermediate points given a trace but this can be done only in very special cases, like the one used before, i.e., the pure Wiener process.

All the three kinds of approximation error decrease as the indexing parameter N increases. Numerical experiments suggest that in the situation when it is reasonable to use the diffusion approximation, i.e., when the CTMC is too large for the analysis but there are still important stochastic behaviors in the system, the approximation errors are in an acceptable range.

6   Experimental Results

The experimental results described in this section were carried out using a prototype implementation integrated in the GreatSPN suite [2], for the SDE part, and with the COSMOS statistical model checker [6] (which uses a generalisation of the DTA formalism [5]), for the CTMC part.

Case study: A model of the Wnt pathway. We consider a model of the Wnt/β-catenin pathway, an intracellular signalling pathway involved in neuroinflammation, a key mechanism in numerous brain diseases [14]. Such model [15] accounts for 8 biochemical species (Table 1) regulated through 12 reactions (Table 2). It consists of three main actors: the β-catenin (denoted B) and Axin2 proteins (A), forming a negative feedback loop, and the Wnt protein (here subsumed by the LRP5-6 membrane receptor, i.e., L), representing the extracellular signal. The behavior can be summarised as follows. With scarcity of extracellular Wnt molecules (low L), a degradation complex (C, a trimer resulting by β-catenin binding to previously formed GSK3-Axin2 dimer, i.e., GA) causes the phosphorylation and subsequent destruction of β-catenin located in the cell's cytosol. On

Table 1. Species of the Wnt pathway.

Name	Description	Init. values
A	Axin2 protein	0
A_m	Axin2 mRNA	0
G	GSK3 protein	$50 \cdot N$
L	LRP5/6 coreceptor	$20 \cdot N$
B	free β-catenin	0
AL	Axin2-LRP5/6 complex	$50 \cdot N$
GA	GSK3-Axin2 complex	0
C	GSK3-Axin-β-catenin complex	0

Table 2. Reactions of the Wnt pathway.

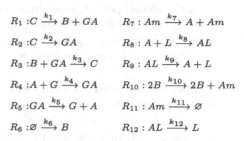

$R_1 : C \xrightarrow{k_1} B + GA$ $R_7 : Am \xrightarrow{k_7} A + Am$

$R_2 : C \xrightarrow{k_2} GA$ $R_8 : A + L \xrightarrow{k_8} AL$

$R_3 : B + GA \xrightarrow{k_3} C$ $R_9 : AL \xrightarrow{k_9} A + L$

$R_4 : A + G \xrightarrow{k_4} GA$ $R_{10} : 2B \xrightarrow{k_{10}} 2B + Am$

$R_5 : GA \xrightarrow{k_5} G + A$ $R_{11} : Am \xrightarrow{k_{11}} \varnothing$

$R_6 : \varnothing \xrightarrow{k_6} B$ $R_{12} : AL \xrightarrow{k_{12}} L$

Table 3. Kinetic rate constants and initial populations for the DDMC model of the Wnt pathway (both dependent on index N).

kinetic rates					
k_1	7	k_4	0.2/N	k_7	0.7	k_{10}	0.7/N
k_2	200	k_5	1.2	k_8	10/N	k_{11}	0.025
k_3	0.1/N	k_6	0.4·N	k_9	0.08	k_{12}	0.1

initial population	
G	$N \cdot 50$
L	$N \cdot 20$
AL	$N \cdot 50$

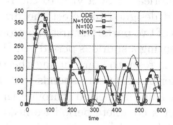

Fig. 2. Sample paths of A (Axin2 protein) for various values of N and the behavior obtained by the ODE description.

the other hand with an abundant Wnt signal (high L), the degradation complex is deactivated (as Axin is degraded through reversibly binding with receptor L, i.e., forming the AL complex) resulting in an accumulation of β-catenin which in turn activates (through transcription of the Axin2 messenger RNA, i.e., A_m) the expression Axin2, and therefore determining its own destruction (i.e., negative feedback loop).

In [15] the model is given in ODE form and it is shown to exhibit sustained oscillations (Fig. 2) for specific parameter settings. Here we consider a sequence of DDMCs indexed by N (here proportional to the volume) and of the parameters of the ODE model in [15][1]. Table 3 depicts the kinetic rate constants and the initial populations of the Wnt-pathway DDMC[2] whereas Fig. 2 compares species A's projection of a sample path of the CTMC for various values of N with the deterministic trajectory of the corresponding ODEs (notice that for readability

---

[1] I.e., for $N = 1$ we assumed the *discrete* initial populations and reaction intensities being equal to the *continuous* ones as given in [15], note that this is in agreement with a cell volume $V = 10^9/n_A$ where n_A is the Avogadro number given that species concentrations and kinetic rate constants of the ODE model are expressed in nM.

[2] Notice that zero-order and second-order reactions' rates are dependent on N because for these conversion from continuous to discrete rates depends on cell's volume.

the CTMC paths have been normalised, i.e., the molecule count of A is divided by N). Furthermore observe that ODEs exhibit sustained oscillations, which, after the second period, have almost constant amplitude, and that, for increasing values of N, CTMC trajectories approximated quite accurately the ODE's. The choice of N when analyzing a real scenario depends on the considered volume; in wet-lab experiments typically molecular value is usually greater than 500.

Automaton. Inspired by Mikeev *et al.* [20] we propose a DTA (Fig. 3) to measure the duration of the period exhibited by the population of the Axin2 protein. The rationale for *noisy period* detection [20] is to split the domain of the observed species, i.e., A, in three subintervals: **low** (i.e., $A \leq \mathbf{L}$), **mid** (i.e., $L < A \leq \mathbf{H}$) and **high** (i.e., $A \geq \mathbf{H}$)[3]. A *noisy period realisation* [4] corresponds to the time interval occurring between two successive entries to the **low** region of the state-space interleaved by a visit to the **high** region. The single clock DTA in Fig. 3 is indeed designed to detect the first noisy period realisation of species A. It consists of two parts: the first one processes the initial **low₀-high₀-low₁** traversal (representing a spurious period), at the end of which (**low₁-mid₂** transition) the clock x is reset to start timing the realisation of the first non-spurious period whose termination corresponds with the **low₃-mid_{end}** edge. Note that ignoring the first spurious period (through the first part of the DTA) is necessary since to detect a complete period we need to identify the actual starting point (i.e., the first **low-mid** crossing that follows a visit to **high**) which we cannot do from the initial state because the system starts at $A = 0$. Furthermore note that trajectories are accepted on condition that the observed duration of the first period is within $T_{min} \leq x \leq T_{max}$ which, by choosing different values for T_{min} and T_{max}, allows us to assess the probability density of the period duration (Fig. 4). Observe that for any element $N > 1$ of the Wnt-DDMC sequence the probability of non-sustainably oscillating paths (i.e., paths non-perpetually traversing the **low-mid-high** regions) is negligible, therefore, given that L and H are properly chosen (so to be above, resp. below, the average height of minimal, resp. maximal, peaks of oscillations), the DTA accepts all trajectories of the model.

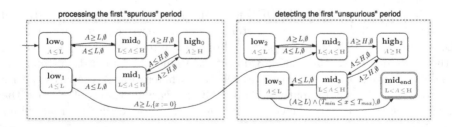

Fig. 3. DTA to study the oscillation period of Axin2 proteins.

---

[3] **L** and **H**, where $\mathbf{L} < \mathbf{H}$, are two thresholds chosen so that the minimal, resp. maximal, peaks of oscillation are most likely to fall below **L**, resp. above **H**.

Experiments. We compare the proposed approach by assessing of DTA-based oscillation-period properties on both a few CTMCs (of the Wnt-DDMC sequence) and on their SDE approximation. The experiments were executed on a server with 48 core AMD Opteron(tm) Processor 6176 by considering five CTMCs of the Wnt-DDMC sequence corresponding to the following values of N, i.e., $N \in \{100, 500, 1000, 2000, 5000\}$, and while the SDE results were computed with the GSPN prototype the CTMCs results were computed with COSMOS [6] which, to the best of our knowledge, is one of the most efficient statistical model checkers. We have run two families of experiments. The first one is devoted to assessing the density function of the oscillation period (Fig. 4) and employs the DTA of Fig. 3. The second one is devoted to comparing both runtime and accuracy of the two approaches w.r.t. estimating the duration of the oscillation period (Table 4) and employs a slightly modified DTA[4].

Table 4 compares, as a function of N, the execution times and the confidence intervals for the mean duration of the first non-spurious oscillation period with confidence level set to 0.99 in case of generating 10000 traces. Columns two and three depict the value of the **L**, resp. **H**, parameter of the DTA; the fourth and fifth (resp. seventh and eighth), columns show the runtime and estimated confidence-interval computed through our SDE prototype (resp. COSMOS); the sixth column shows the proportion of the number of jumps occurred because of hitting the border during the simulation of the SDE; finally the ninth column gives the speed up obtained by our SDE approach. In Fig. 4 the probability density functions (pdf) of the length of the first non-spurious period are plotted for N equal to 100, 500, 1000 and 2000. The pdf for N equal to 5000 (not shown for the lack of space) confirms the trend toward a closer correspondence between the SDE and COSMOS results.

Table 4. Comparing SDE and COSMOS results considering 10000 traces.

N	**L**	**H**	SDE Time	SDE Average period	$\frac{Jump}{Tot.}$	COSMOS Time	COSMOS Average period	Speedup
100	50	1×10^4	180 h	[103.174, 103.344]	0.90	30 h	[112.628, 113.133]	0.17
500	50	5×10^4	88 h	[111.021, 111.115]	0.76	167 h	[112.654, 112.784]	1.9
1,000	500	1×10^5	54 h	[111.977, 112.089]	0.60	344 h	[112.381, 112.499]	6,37
2,000	500	1.5×10^5	39 h	[112.408, 112.474]	0.45	705 h	[112.504, 112.578]	18.08
5,000	500	5×10^5	28 h	[112.693, 112.725]	0.26	1763 h	[112.708, 112.749]	62.96

---

[4] I.e., we use the DTA in Fig. 3 but without the $T_{min} \leq x \leq T_{max}$ conjunct on the edge from **low₃-mid**ₑₙ𝒹, which allows us to obtain the value of the clock x at the moment of reaching the final location of this modified DTA: this value gives the length of the first non-spurious oscillation period.

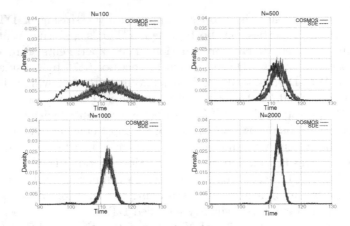

Fig. 4. Probability mass functions of the length of the first non-spurious oscillation period with bin length equal to 0.1 and with N equal to 100 (top left), 500 (top right), 1000 (bottom left), 2000 (bottom right).

Discussion. As expected, the SDE approach becomes more convenient, in terms of runtime and precision, as N increases.[5] In particular, for $N = 100$ the precision is strongly affected by the SDE approximation error, moreover the SDE execution time is greater than COSMOS since for each SDE trace the process hits the boundaries, on average, along 90% of a trace length (col. 6, Table 4). With $N = 1000$ the SDE based analysis is about twice faster than that based on the CTMC (col. 9, Table 4) while the precision is acceptable: indeed the SDE approach is able to reproduce the multimodal behavior of the pdf generated by COSMOS (Fig. 4 bottom-left plot, i.e., $N = 1000$). Such trend is confirmed by experiments with $N = 2000$, (speedup $\sim$6x, even closer approximation) and with $N = 5000$ (exhibiting a 63x speedup obtained with the SDE approach).

7   Conclusions

In this paper we presented a framework that allows for assessing temporal properties, described in terms of DTA, of DDMCs through their jump diffusion approximation. The applicability of the approach was illustrated through a case study regarding a biological oscillator. As future work we aim to study the theoretical limits of assessing DTA-based temporal properties of diffusion processes. Furthermore, the approach can be extended to hybrid jump diffusion processes, which are obtained by partial fluidification of DDMCs, that are useful to study systems in which not all population counts are high and thus fluidification of all state variables would lead to large approximation errors.

---

[5] Observe that the dimension of integration step is dynamically computed through a heuristic function which provides a good trade-off between speed-up and precision of the solution.

References

1. Alur, R., Dill, D.L.: A theory of timed automata. Theor. Comput. Sci. **126**(2), 183–235 (1994)
2. Amparore, E.G., Beccuti, M., Donatelli, S.: (Stochastic) model checking in Great-SPN. In: Ciardo, G., Kindler, E. (eds.) PETRI NETS 2014. LNCS, vol. 8489, pp. 354–363. Springer, Cham (2014). doi:10.1007/978-3-319-07734-5_19
3. Angius, A., Balbo, G., Beccuti, M., Bibbona, E., Horvath, A., Sirovich, R.: Approximate analysis of biological systems by hybrid switching jump diffusion. Theor. Comput. Sci. **587**, 49–72 (2015)
4. Ballarini, P.: Analysing oscillatory trends of discrete-state stochastic processes through HASL statistical model checking. STTT **17**(4), 505–526 (2015)
5. Ballarini, P., Barbot, B., Duflot, M., Haddad, S., Pekergin, N.: HASL: a new approach for performance evaluation and model checking from concepts to experimentation. Perform. Eval. **90**, 53–77 (2015)
6. Ballarini, P., Djafri, H., Duflot, M., Haddad, S., Pekergin, N.: COSMOS: a statistical model checker for the hybrid automata stochastic logic. In: Proceedings of the QEST 2011, pp. 143–144. IEEE Computer Society (2011)
7. Beccuti, M., Bibbona, E., Horvath, A., Sirovich, R., Angius, A., Balbo, G.: Analysis of petri net models through stochastic differential equations. In: Ciardo, G., Kindler, E. (eds.) PETRI NETS 2014. LNCS, vol. 8489, pp. 273–293. Springer, Cham (2014). doi:10.1007/978-3-319-07734-5_15
8. Bibbona, E., Sirovich, R.: Strong approximation of density dependent Markov chains on bounded domains by jump diffusion processes. Technical report, Università di Torino (2017)
9. Bortolussi, L., Hillston, J.: Model checking single agent behaviours by fluid approximation. Inf. Comput. **242**, 183–226 (2015)
10. Bortolussi, L., Lanciani, R.: Model checking markov population models by central limit approximation. In: Joshi, K., Siegle, M., Stoelinga, M., D'Argenio, P.R. (eds.) QEST 2013. LNCS, vol. 8054, pp. 123–138. Springer, Heidelberg (2013). doi:10.1007/978-3-642-40196-1_9
11. Bortolussi, L., Lanciani, R.: Fluid model checking of timed properties. In: Sankaranarayanan, S., Vicario, E. (eds.) FORMATS 2015. LNCS, vol. 9268, pp. 172–188. Springer, Cham (2015). doi:10.1007/978-3-319-22975-1_12
12. Chen, T., Han, T., Katoen, J.-P., Mereacre, A.: Model checking of continuous-time Markov chains against timed automata specifications. Log. Meth. Comput. Sci. **7**(1), 1–34 (2011)
13. Donatelli, S., Haddad, S., Sproston, J.: Model checking timed and stochastic properties with CSL^{TA}. IEEE T. Software Eng. **35**(2), 224–240 (2009)
14. Gressens, P., Steenwinckel, J.V., Schang, A., Sigaut, S., Degos, V., Lebon, S., Schwendimann, L., Le Charpentier, T., Hagberg, H., Soussi, N., Fleiss, B.: Microglial Wnt signaling inhibition promotes microglia activation and oligodendrocyte maturation blockade. J. Neurochem. **134**, 122 (2015)
15. Jensen, P.B., Pedersen, L., Krishna, S., Jensen, M.H.: A Wnt oscillator model for somitogenesis. Biophys. J. **98**(6), 943–950 (2010)
16. Kolesnichenko, A., de Boer, P., Remke, A., Haverkort B.R.: A logic for model-checking mean-field models. In: Proceedings of the DSN 2013, pp. 1–12. IEEE Computer Society (2013)
17. Kurtz, T.G.: Solutions of ordinary differential equations as limits of pure jump Markov processes. J. Appl. Probab. **1**(7), 49–58 (1970)

18. Kurtz, T.G.: Limit theorems and diffusion approximations for density dependent Markov chains. In: Wets, R.J.B. (ed.) Stochastic Systems: Modeling, Identification and Optimization, I, pp. 67–78. Springer, Heidelberg (1976)
19. Kurtz, T.G.: Strong approximation theorems for density dependent Markov chains. Stoc. Proc. Appl. **6**(3), 223–240 (1978)
20. Mikeev, L., Neuhäußer, M.R., Spieler, D., Wolf, V.: On-the-fly verification and optimization of DTA-properties for large Markov chains. Form. Method. Syst. Des. **43**(2), 313–337 (2013)

Stability Analysis of a Multiclass Retrial System with Coupled Orbit Queues

Evsey Morozov[1] and Ioannis Dimitriou[2(✉)]

[1] Karelian Research Centre RAS, Institute of Applied Mathematical Research,
Petrozavodsk State University, Petrozavodsk, Russian Federation
emorozov@karelia.ru
[2] Department of Mathematics, University of Patras, 26500 Patras, Greece
idimit@math.upatras.gr

Abstract. In this work we consider a single-server system accepting N types of retrial customers, which arrive according to independent Poisson streams. In case of blocking, type-i customer, $i = 1, 2, ..., N$ is routed to a separate type-i orbit queue of infinite capacity. Customers from the orbit queues try to access the server according to the constant retrial policy. We consider coupled orbit queues. More precisely, the orbit queue i retransmits a blocked customer of type-i to the main service station after an exponentially distributed time with rate μ_i, when at least one other orbit queue is non-empty. Otherwise, if all other orbit queues are empty, the orbit queue i changes its retransmission rate from μ_i to μ_i^*. Such a scheme arises in the modeling of cooperative cognitive wireless networks, in which a node is aware of the status of other nodes, and accordingly, adjusts its retransmission parameters in order to exploit the idle periods of the other nodes. Using the regenerative approach we obtain the necessary conditions of the ergodicity of our system, and show that these conditions have a clear probabilistic interpretation. We also suggest a sufficient stability condition. Simulation experiments show that the obtained conditions delimit the stability domain with remarkable accuracy.

Keywords: Multiclass retrial queues · Stability · Constant retrial rates · Coupled orbit queues · Cooperative cognitive network

1 Introduction

We consider a fairly general single server retrial system with multiple classes of retrial customers (i.e., multiclass retrial systems), fed by independent arrival streams. If upon arrival, a customer of either type finds all servers unavailable joins an infinite capacity orbit queue[1] from where re-attempts to connect with the server after some random time, according to a constant retrial policy; e.g., [4,6].

In this work, using the regenerative approach, we obtain the necessary stability conditions of this multiclass system with an arbitrary number of coupled

---

[1] In this work the terms "orbit" and "orbit queue" are identical.

© Springer International Publishing AG 2017
P. Reinecke and A. Di Marco (Eds.): EPEW 2017, LNCS 10497, pp. 85–98, 2017.
DOI: 10.1007/978-3-319-66583-2_6

orbit queues, and general service times. To the best of our knowledge, these results are completely new, and it is the main contribution of this work. Some stationary performance measures are obtained as well, as a by-product of the stability analysis. We also suggest a sufficient stability condition, which is verified by simulation experiments.

Retrial queues have been extensively studied in the literature, and for further reading we refer to the books [3,11] and the survey papers [2,16]. Clearly, the analysis, and in particular the stability problem of a multiclass system, is much more challenging than that of the single-class variant. Due to their mathematical difficulty, there are only few works regarding the analysis of multi-class retrial systems.

We mention the seminal papers in [12,17], where the authors derived the expected number of customers at each orbit queue as a solution of a linear system of equations; see also [18,26]. In [5,7] the authors derived necessary and sufficient stability conditions, respectively, for the case of a single-server, multi-class retrial queue with constant retrial rates. Recently, stability conditions for the multiclass system with classical retrial policy along with some generalizations were given in [21]. In all above mentioned works the authors used the regenerative approach [22–25], which has become an elegant methodology to study the stability conditions of such systems.

Contrary to other works in the related literature, the major contribution in this work is that we assume that the service rate at each orbit queue (i.e., the re-transmission rate) depends on the number of customers in the other orbit queues (i.e., coupled orbit queues). More precisely, we assume that an orbit queue is aware of the status of the other orbit queues, and accordingly, reconfigures its retransmission parameters. Recently, the stability conditions for a two class retrial system with coupled orbit queues were obtained in [13–15] by using results from the theory of random walks in the quarter plane. In this work, we investigate the stability conditions of the model with an arbitrary number of orbit queues, and show that the regenerative approach is an adequate method to handle it.

Retrial systems with coupled orbit queues [13–15] have potential applications in the modeling of wireless multiple access systems. In particular, they are natural for the modeling of relay-assisted cognitive cooperative wireless systems [27–29]. Such a system operates as follows: There is a finite number of source users that transmit packets to a common destination node, and a finite number of relay nodes (i.e., orbit queues) that assist source users by retransmitting their blocked packets; e.g., [27,28]. More precisely, when a direct source user transmission is blocked, (i.e., the destination node is unavailable), it forwards its blocked packet at a relay node (i.e., a relay overhears the transmission, and stores the blocked packet), which in turn retransmits the blocked packet after some random time.

It is evidently proved that the current trend towards dense networks and the spatial reuse of resources potentially increase the impact of wireless interference, and thus it is essential to take it into account in the network planning. Moreover, although nowadays there is an increasing demand for variety of wireless applications, the usable radio spectrum is of limited physical extent. Recent studies on the spectrum usage have revealed that substantial portion of the licensed spectrum is underutilized, and thus there is an imperative need for developing the cognitive radio communication, which is a promising solution to the spectrum underutilization problem [20, 28].

In the full cognitive radio [20] a wireless node is capable to obtain knowledge of its operational environment (i.e., it is "smart"), and to dynamically adjusts its operational parameters accordingly. Thus, in order to achieve full spectrum utilization of the shared channel, it adjusts its retransmission parameters according to the state of the other relay node (i.e., coupled relay nodes); e.g., [8, 10, 13]. Moreover, in other applications in cellular networks, the available transmission rate for users in a particular cell is decreasing as the number of users in the neighboring cells increase [8]. Another important category of related applications deals with processor sharing models, where several customer classes simultaneously use one or more servers, whose rate allocations and total processing rates depend on the number of customers in each of the classes [9, 19].

The paper is organized as follows: in Sect. 2 we briefly describe the mathematical model and obtain some basic results. In Sect. 3, we develop the stability analysis of a multi-class retrial system with coupled orbit queues, and deduce the necessary stability condition, while in Sect. 4 we propose a sufficient stability condition. Finally, in Sect. 5, we present simulation results, which demonstrate a remarkable consistency with the theoretical results.

2 Description of the Model

We consider a single-server, multiclass retrial queueing system with no buffer, which operates as follows. Let $\{t_n\}$ be the arrival instants of primary customers with the iid exponential interarrival times $\tau_n = t_{n+1} - t_n$, $n \geq 1$, with the rate $\lambda = 1/\mathsf{E}\tau \in (0, \infty)$. (Here and in what follows, we omit serial index to denote a generic element of an iid sequence.) There are N classes of arrivals, and we denote $\{S_n^{(i)}, n \geq 1\}$ the iid service times of class-i customers with the rate $\gamma_i = 1/\mathsf{E}S^{(i)} \in (0, \infty)$, $i = 1, \ldots, N$. It is assumed that a new arrival is class-i with the probability p_i, and thus the arrival rate of class-i customers is $\lambda_i := \lambda p_i$, and $\lambda = \sum_{i=1}^{N} \lambda_i$.

Our *main assumption* is that the retrial rate is *state-dependent* (i.e., orbit queues are coupled). More precisely, we assume that orbit queue i retransmits after an exponentially distributed time period with rate μ_i, if at least one other orbit queue is non-empty. Otherwise, that is, if all other orbit queues are empty, it changes its retransmission rate to μ_i^*, $i = 1, \ldots, N$.

Let $S(t)$ be the remaining service time at instant t^- ($S(t) = 0$, if the server is free). Denote the total idle time of the server in $[0, t]$ by,

$$I(t) = \int_0^t 1(S(u) = 0) du,$$

where 1 is an indicator function. Denote also by $A_i(t)$ the number of class-i arrivals in $[0, t]$. Then, $V_i(t) := \sum_{n=1}^{A_i(t)} S_n^{(i)}$ is the work generated by class-i customers, and $V(t) := \sum_i V_i(t)$ is the total amount of work arrived in $[0, t]$. Thus,

$$V(t) = \sum_{i=1}^{N} \sum_{n=1}^{A_i(t)} S_n^{(i)}, \ t \geq 0.$$

Denote by $N_i(t)$ the number of class-i orbital customers and $W_i(t)$ the work-load (remaining work) in orbit queue i, at instant t^-, $i = 1, \ldots, N$. Let also $S(t)$ be the remaining work (i.e., the remaining service time) in the server at instant t^-. Define the forward interarrival time $\tau(t) = \inf_n(t_n - t : t_n - t > 0)$ at instant t.

We consider the one-dimensional non-Markovian process $X(t) := N(t) + Q(t), t \geq 0$, where $Q(t) \in \{0, 1\}$ is the number of customers in the primary system at instant t^-, and $N(t) := \sum_i N_i(t)$ is the *summary orbit size* (at instant t^-). In order to study the process $X := \{X(t), t \geq 0\}$, let $X(t_n) = X_n$, put $T_0 = 0$, and define recursively,

$$T_{n+1} = \inf\left(t_k > T_n : X_k = 0\right), n \geq 0.$$

It is easy to see that $\{T_n\}$ are classical regenerations of the basic process X. Note that $T_{n+1} - T_n$ are iid *regeneration periods*, and let T denote the generic period. We call the process X (and the basic system) *positive recurrent* if the first regeneration period is finite, $T_1 < \infty$ with probability 1 (w.p.1), and the mean generic period is finite, $ET < \infty$ [23,30]. Throughout the paper we assume *zero initial state*, in which case customer 1 arrives at the empty system at instant $t_1 = 0$. In this case the instant $T_0 = 0$ is indeed the 1st regeneration epoch, and the first regeneration period $T_1 =_{st} T$ (stochastically)[2]. Thus the regenerations constitute a *zero-delayed* renewal process [1]. In what follows we use the following result of the renewal theory: in the zero-delayed renewal process, the remaining renewal (regeneration) time at instant t,

$$T(t) = \min_k\left(T_k - t : T_k - t > 0\right) \not\Rightarrow \infty, \ t \to \infty, \tag{1}$$

if and only if $ET < \infty$. (Here symbol $\Rightarrow$ stands for convergence in probability.)

Hence to establish positive recurrence in the zero-delayed case, it suffices to show (1). In this work our aim is to show that, under predefined conditions, (1) holds true. Before proceeding further, we note that the positive recurrence is the key ingredient of the regenerative stability analysis, see [1, 22–25]. Denote traffic intensity for each class, $\rho_i = \lambda_i/\gamma_i$, $i = 1, \ldots, N$.

---

[2] We mention first the general case of non-zero initial conditions, in which case $T_1 \neq T$. For such a case, positive recurrence means both $ET < \infty$ and $T_1 < \infty$ w.p. 1. For zero initial state $T_1 =_{st} T$, and thus in order to prove the positive recurrence, it only remains to show that $ET < \infty$. See also some comments in Remark 1.

3 Necessary Stability Conditions

Recall that our *main assumption* is that retrial rates are *state-dependent*. More precisely, we assume that all retrial times are exponentially distributed, and, for orbit queue i, the retrial rate is μ_i, if at least one other orbit queue is non-empty. Otherwise, this rate changes to μ_i^*, $i = 1, \ldots, N$.

To obtain necessary stability condition, we assume that the model is *stable*, i.e., the basic regenerative process $\{X(t)\}$ is positive recurrent. In the following we provide some preparatory results that are important for the proof of our main result given in Theorem 1.

Denote by P_b the stationary busy probability of the server, i.e. $\mathsf{P}_0 = 1 - \mathsf{P}_b$ is the stationary idle probability. More exactly, $\mathsf{P}(S(t) = 0) \to \mathsf{P}_0$ as $t \to \infty$. This limit exists since the interarrival time τ is *spread-out* [1]. Denote by $B(t)$ the busy time of the server in $[0, t]$. It is evident that $B(t)$ equals the departed work and can be expressed via idle time as,

$$B(t) = t - I(t).$$

We start with the following balance equation:

$$
\begin{aligned}
V(t) &= \sum\nolimits_{i=1}^{N} W_i(t) + S(t) + B(t) \\
&= \sum\nolimits_{i=1}^{N} W_i(t) + S(t) + t - I(t), \ t \geq 0.
\end{aligned}
\tag{2}
$$

Under positive recurrence, see [31],

$$\sum\nolimits_{i=1}^{N} W_i(t) + S(t) = o(t), \ t \to \infty,$$

where $o(t)$ is any quantity such that $\lim_{t \to \infty} \frac{o(t)}{t} = 0$; [31]. By the Strong Law of Large Numbers, w.p.1,

$$\frac{V(t)}{t} = \sum\nolimits_{i=1}^{N} \frac{\sum_{n=1}^{A_i(t)} S_n^{(i)}}{A_i(t)} \frac{A_i(t)}{t} \to \sum\nolimits_{i=1}^{N} \rho_i, \ t \to \infty.
\tag{3}$$

Since the busy time process, $B(t)$, $t \geq 0$, is a *cumulative* process with the positive recurrent process of regenerations $\{T_n\}$, there exists (w.p.1) the limit [31]

$$\lim_{t \to \infty} \frac{B(t)}{t} = \mathsf{P}_b.
\tag{4}$$

Moreover, since the input process is Poisson, then, the weak limit $S(t) \Rightarrow S_e$, $t \to \infty$ exists as well, and $\mathsf{P}_b = \mathsf{P}(S_e > 0)$ is the *stationary busy probability* of the server. It follows from (2)–(4) that $\sum_{i=1}^{N} \rho_i \leq 1$, and we prove the basic (expected) strict inequality

$$\mathsf{P}_b = \sum\nolimits_{i=1}^{N} \rho_i < 1.
\tag{5}$$

Theorem 1. *Assume that the N-class retrial system with N coupled orbit queues is positive recurrent. Then,*

$$P_b = \sum_{i=1}^{N} \rho_i \le \min_{1 \le i \le N} \left[\frac{\max(\mu_i, \mu_i^*)}{\lambda_i + \max(\mu_i, \mu_i^*)} \right] < 1. \tag{6}$$

Proof: Denote by I_0 the duration of an empty period within a regeneration cycle. Let also B be a generic busy period, i.e., the time the server is busy within a regeneration cycle. Then, the regeneration period can be represented as $T =_{st} B + I_0$. By the positive recurrence of the cumulative process $I(t)$, $t \ge 0$, there exists the limit (w.p.1)

$$\lim_{t \to \infty} \frac{I(t)}{t} = \frac{EI_0}{ET} = P_0 = 1 - P_b.$$

Since,

$$V(t) = o(t) + B(t) = o(t) + t - I(t), \ t \to 0,$$

then,

$$\lim_{t \to \infty} \frac{V(t)}{t} = \sum_{i=1}^{N} \rho_i = 1 - \lim_{t \to \infty} \frac{I(t)}{t} \le 1 - \frac{EI_0}{ET}. \tag{7}$$

We show that $EI_0 > 0$. Since τ is exponential, then there exist constants $\delta_0 > 0$, $\varepsilon_0 > 0$, such that for each class-i, $i = 1, ..., N$,

$$P(\tau > S^{(i)} + \delta_0) \ge \varepsilon_0 > 0. \tag{8}$$

Denote the indicator function,

$$1_i = 1(\text{a class} - i \text{ customer starts a new regeneration cycle}).$$

Note that $E1_i = p_i > 0$, and that regeneration cycle may contain *one class-i customer only*. Thus,

$$I_0 \ge I_0 \, 1_i \, 1(\tau > S^{(i)} + \delta_0) \ge \delta_0 \, 1_i \, 1(\tau > S^{(i)} + \delta_0).$$

Note that (5) follows from (7) and from the inequality

$$EI_0 \ge \delta_0 \, p_i \, P(\tau > S^{(i)} + \delta_0) \ge \delta_0 \, p_i \, \varepsilon_0 > 0.$$

However the inequality (5) can be strengthened if we apply a balance between the input to each orbit queue, and the output from the same orbit queue.

Denote $A_i^{(0)}(t)$ the number of class-i customers joining orbit queue i in interval $[0, t]$, $i = 1, \dots, N$. Denote, in interval $[0, t]$, $T_{0b}^{(i)}(t)$ the time when server is free, orbit queue i is busy and at least one orbit queue $j \ne i$ is busy. Note that in this case retrial rate from orbit queue i is μ_i. Let also $T_{00}^{(i)}(t)$ be the time when orbit queue i is busy, the server is free, and all other orbit queues are empty. In this case the retrial rate from orbit queue i is μ_i^*, $i = 1 \dots, N$.

Then $A_i^{(0)}(t) = \sum_{k=1}^{A_i(t)} I_k$, where indicator $I_k = 1$ if the kth class-i customer joins orbit queue i. Since $A_i(t)/t \to \lambda_i$, then,

$$\lim_{t \to \infty} \frac{A_i^{(0)}(t)}{t} = \lim_{t \to \infty} \frac{1}{A_i(t)} \sum_{k=1}^{A_i(t)} I_k \frac{A_i(t)}{t} = \lambda_i P_b,$$

where we use the equality

$$\lim_{t\to\infty} \tfrac{1}{A_i(t)} \sum_{k=1}^{A_i(t)} I_k = \mathsf{P}_b, \tag{9}$$

since, by PASTA [32], the limiting fraction of all class-i customers, which see the server busy, and join orbit queue i, equals the fraction of time the server is busy. Note that the limit (9) is independent of i. Note also that the sum,

$$T_{0b}^{(i)}(t) + T_{00}^{(i)}(t) =: T_0^{(i)}(t),$$

is the time period, in interval $[0, t]$, where the server is free and orbit queue i is busy.

It is easy to see that, within $[0, t]$, the (successful) output from orbit queue i is possible during summary time $T_0^{(i)}(t)$. Denote by $\hat{D}_i(t)$ the number of class-i customers leaving the orbit queue i in $[0, t]$. Due to the fact that when orbit queue i is busy and server is free, the process of (successful) retrial attempts is Poisson, either with rate μ_i, or with rate μ_i^*, then we have the stochastic equality,

$$A_i^{(0)}(t) = \hat{D}_i(t) =_{st} D_i(T_{0b}^{(i)}(t)) + D_i(T_{00}^{(i)}(t)), \tag{10}$$

where $D_i(t)$ denotes the Poisson process (with the corresponding rate). Using results from renewal theory, and the Strong Law of Large numbers, the limits (w.p.1) exist and

$$\lim_{t\to\infty} \tfrac{1}{t} D_i(T_{0b}^{(i)}(t)) = \lim_{t\to\infty} \tfrac{D_i(T_{0b}^{(i)}(t))}{T_{0b}^{(i)}(t)} \tfrac{T_{0b}^{(i)}(t)}{t} = \mu_i \mathsf{P}_{0b}^{(i)}, \tag{11}$$

where $\mathsf{P}_{0b}^{(i)}$ is the stationary probability that server is free, orbit queue i, and at least one other orbit queue are busy. Analogously,

$$\lim_{t\to\infty} \tfrac{1}{t} D_i(T_{00}^{(i)}(t)) = \lim_{t\to\infty} \tfrac{D_i(T_{00}^{(i)}(t))}{T_{00}^{(i)}(t)} \tfrac{T_{00}^{(i)}(t)}{t} = \mu_i^* \mathsf{P}_{00}^{(i)}, \tag{12}$$

where $\mathsf{P}_{00}^{(i)}$ is the stationary probability that server and all orbit queues $j \neq i$ are idle, while orbit queue i is busy. Due to the fact that under positive recurrence, orbit queue i size is $N_i(t) = o(t)$, then, from (10)–(12) and the *local* balance

$$A_i^{(0)}(t) = \hat{D}_i(t) + o(t),$$

we obtain,

$$\lambda_i \mathsf{P}_b = \mu_i^* \mathsf{P}_{00}^{(i)} + \mu_i \mathsf{P}_{0b}^{(i)}, \quad i = 1, \ldots, N. \tag{13}$$

Note that

$$\mathsf{P}_{00}^{(i)} + \mathsf{P}_{0b}^{(i)} =: \mathsf{P}_0^{(i)}, \tag{14}$$

is the stationary probability that server is free and orbit queue i is busy. Finally,

$$\mathsf{P}_0^{(i)} \leq \mathsf{P}_0 = 1 - \mathsf{P}_b. \tag{15}$$

Now, it follows from (13)–(15) that

$$\lambda_i P_b \leq \max(\mu_i, \mu_i^*)(1 - P_b), \ i = 1, \ldots, N, \tag{16}$$

implying the following relation between parameters:

$$P_b = \sum_{i=1}^{N} \rho_i \leq \min_{1 \leq i \leq N} \left[\frac{\max(\mu_i, \mu_i^*)}{\lambda_i + \max(\mu_i, \mu_i^*)}\right] < 1. \tag{17}$$

Similarly, we can deduce (less interesting) inequality

$$P_b = \sum_{i=1}^{N} \rho_i \geq \max_i \left[\frac{\min(\mu_i, \mu_i^*)}{\lambda_i} P_0^{(i)}\right].$$

$\square$

Denote by $I(t) \cap B_i(t)$ the set of instants of time (in $[0, t]$) when server is free and orbit queue i is busy. Then we have the following balance relation:

$$A_i^{(0)}(t) = o(t) + D_i(I(t) \cap B_i(t)).$$

Using arguments from renewal theory, we can easily show that,

$$\lim \sup_{t \to \infty} \frac{D_i(t)}{t} \leq \max(\mu_i, \mu_i^*).$$

Then, it follows from

$$\frac{D_i(B_i(t))}{t} = \frac{D_i(B_i(t))}{B_i(t)} \frac{B_i(t)}{t},$$

that the following inequality holds:

$$\lambda_i P_b \leq \max(\mu_i, \mu_i^*) P_b(i), \tag{18}$$

where,

$$P_b(i) = \lim_{t \to \infty} \frac{B_i(t)}{t},$$

is the stationary probability that orbit queue i is busy. Now (18) gives the following inequality (which however is not uniform in i):

$$P_b(i) \geq \frac{\lambda_i}{\max(\mu_i, \mu_i^*)} \sum_{i=1}^{N} \rho_i, \ i = 1, \ldots, N.$$

Remark 1. It is easy to extend this analysis to the m-server system (with stochastically equivalent servers), in which case, the r.h.s. in inequality (6) (and in (5)) must be replaced by m. Note also that it is straightforward to extend the analysis to an arbitrary initial state $X(0)$, in which case positive recurrence means that $\mathsf{E}T < \infty$ and the first regeneration period $T_1 < \infty$ w.p.1, for more details see [23].

Remark 2. Note that (6) has a very interesting (but expected) probabilistic interpretation. Indeed, rewrite the term in brackets as $1 - \frac{\lambda_i}{\lambda_i + \max(\mu_i, \mu_i^*)}$, and without loss of generality let $\min_{1 \leq i \leq N} [1 - \frac{\lambda_i}{\lambda_i + \max(\mu_i, \mu_i^*)}] = 1 - \frac{\lambda_k}{\lambda_k + \max(\mu_k, \mu_k^*)}$, for some $k = 1, \ldots, N$. Then, (6) is rewritten as $P_b + \frac{\lambda_k}{\lambda_k + \max(\mu_k, \mu_k^*)} < 1$, which is expected since the mean number of external arrivals between two consecutive departures during a busy period must be smaller than 1.

4   On the Sufficient Stability Conditions

Based on the results presented in [5, 13], we claim that the sufficient stability condition is

$$\sum_i \rho_i + \max_i \left[\max(\tfrac{\lambda}{\lambda+\mu_i}, \tfrac{\lambda}{\lambda+\mu_i^*}) \right] < 1, \tag{19}$$

which coincides with the corresponding stability condition on p. 30 in [13] for $N = 2$ and $\gamma_1 = \gamma_2$. The proof is based on negative drift arguments and regenerative approach, and will be presented in a separate work.

The simulation experiments in Sect. 5 enhances our claim. We also define in the following an additional metric, called Δ, in order to evaluate the "difference" between necessary (17) and sufficient condition (19). More precisely, it is easy to check that (19) can be rewritten as

$$\sum_i \rho_i \le \min_i \left(\tfrac{\min(\mu_i, \mu_i^*)}{\lambda+\min(\mu_i, \mu_i^*)} \right). \tag{20}$$

Since the function $f(x) = x/(\lambda + x)$ is monotone increasing in x, then (20) implies (17):

$$\min_i \tfrac{\min(\mu_i, \mu_i^*)}{\lambda+\min(\mu_i, \mu_i^*)} \le \min_i \tfrac{\min(\mu_i, \mu_i^*)}{\lambda_i+\min(\mu_i, \mu_i^*)} \le \max_i \tfrac{\max(\mu_i, \mu_i^*)}{\lambda_i+\max(\mu_i, \mu_i^*)}.$$

Denote the difference between the two conditions as

$$\Delta = \max_i \tfrac{\max(\mu_i, \mu_i^*)}{\lambda_i+\max(\mu_i, \mu_i^*)} - \min_i \tfrac{\min(\mu_i, \mu_i^*)}{\lambda+\min(\mu_i, \mu_i^*)} > 0. \tag{21}$$

Our aim is to study the dependence of Δ on parameters λ_i, μ_i, μ_i^*.

Remark 3. Assume that $\mu_i = \mu_i^*$ for all i (i.e., non-coupled system). Then (13) becomes $\lambda_i \mathsf{P}_b = \mu_i \mathsf{P}_0^{(i)}$, and we obtain explicit formula for the stationary probability that orbit queue i is busy and server is free:

$$\mathsf{P}_0^{(i)} = \tfrac{\lambda_i}{\mu_i} \sum_{k=1}^{N} \rho_k, \quad i = 1, \ldots, N.$$

5   Simulations

In the following, we present some numerical results which illustrate that conditions (17), (19) are necessary and sufficient, respectively, for the ergodicity of our system. This is verified for exponential service time distribution. For $N = 2$, and $\gamma_1 = \gamma_2 = \gamma$, set

$$\Gamma_1 := \min_i \left[\frac{\max(\mu_i, \mu_i^*)}{\lambda_i + \max(\mu_i, \mu_i^*)} \right] - \rho,$$

$$\Gamma_2 := 1 - \rho - \max_i [\max(\tfrac{\lambda}{\lambda + \mu_i}, \tfrac{\lambda}{\lambda + \mu_i^*})]. \tag{22}$$

According to the theoretical findings, the necessary condition is $0 \le \Gamma_1 < 1$, and the sufficient $\Gamma_2 > 0$. As simulation shows, these measures allow to delimit stability regions with a remarkable accuracy.

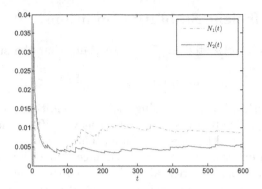

Fig. 1. Orbit queue dynamics for the exponential system with $\gamma = 40$, $\Gamma_1 = 0.8625 < 1$, $\Gamma_2 = 0.6523 > 0$, with $\mu_1 = 20$, $\mu_2 = 10$, $\mu_1^* = 30$, $\mu_2^* = 20$, $\lambda_1 = 2$, $\lambda_2 = 1$.

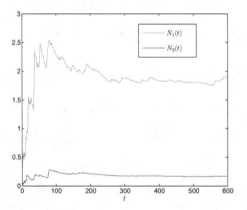

Fig. 2. Orbit queue dynamics for the exponential system with $\gamma = 20$, $\Gamma_1 = 0.2 < 1$, $\Gamma_2 = -0.0738$, with $\mu_1 = 20$, $\mu_2 = 30$, $\mu_1^* = 20$, $\mu_2^* = 40$, $\lambda_1 = 10$, $\lambda_2 = 1$.

Example 1: Orbit Queue Dynamics. In Figs. 1, 2 and 3 we study the dynamics of the orbit queues depending on the value of the above introduced measures. In Fig. 1 we observe the orbit queue dynamics when $\Gamma_1 = 0.8625 < 1$, $\Gamma_2 = 0.6523 > 0$. Therefore, under such conditions, our system is stable as expected.

Similar observations can be deduced by Fig. 2. There, the sufficient condition $\Gamma_2 > 0$ is violated. In particular, $\Gamma_2 = -0.0738$. Note that in such a scenario we have set $\lambda_1 = 10$, and reduced γ from 40 to 20. We can easily observe the effect of changing these parameters especially on the number of customers in orbit queue 1. Recall that the necessary condition $0 \leq \Gamma_1 = 0.2 < 1$ is still valid.

In Fig. 3, we have further reduced the service rate γ from 40 to 12. In such a case the server becomes very slow, which in turn has a negative effect on the system performance. Indeed, in such a case both stability conditions are violated, and we can easily observe that the system becomes unstable. In particular, orbit queue 1 is unstable.

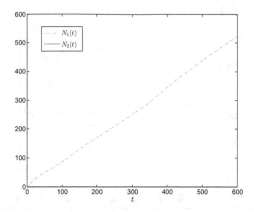

Fig. 3. Orbit queue dynamics for the exponential system with $\gamma = 12$, $\Gamma_1 = -0.1667 < 0$, $\Gamma_2 = -0.4405 < 0$, with $\mu_1 = 20$, $\mu_2 = 10$, $\mu_1^* = 20$, $\mu_2^* = 40$, $\lambda_1 = 10$, $\lambda_2 = 1$.

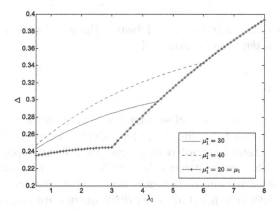

Fig. 4. Study the "difference" for $\mu_1 = 20$, $\mu_2 = 10$, $\mu_2^* = 20$, $\lambda_2 = 1$.

Example 2: Effect of Δ. In Figs. 4 and 5 we study the dependence of Δ on the system parameters. Recall that Δ is a measure of "distance" between sufficient and necessary stability conditions.

Figure 4 shows how the "difference" Δ varies for increasing values of λ_1, and for different values of μ_1^*, by setting $\mu_1 = 20$, $\mu_1^* = 30$, $\mu_2 = 10$, $\mu_2^* = 20$, $\lambda_2 = 1$. We can observe that as μ_1^* increases, Δ is relatively larger for small values of λ_1. When λ_1 passes a certain threshold value, the effect of μ_1^* vanishes. Moreover, for such a scenario we can easily observe that for increasing values of λ_1, the "difference" between necessary and sufficient condition increases too.

Finally, in Fig. 5 we can observe how Δ decreases for increasing values of μ_1^* by setting $\mu_1 = 20$, $\mu_2^* = 20$, $\lambda_1 = 2$, $\lambda_2 = 1$. In particular, that decrease becomes more apparent as μ_2 increases too. Therefore, we conclude that when

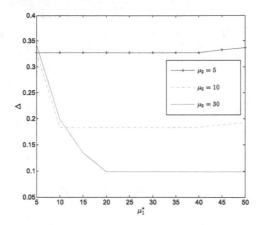

Fig. 5. Study the "difference" for $\mu_1 = 20$, $\mu_2^* = 20$, $\lambda_1 = 2$, $\lambda_2 = 1$.

the orbit queues retransmit faster and faster, the gap between necessary and sufficient condition tends to be vanished.

6 Conclusion

We considered a multi-class bufferless retrial system accepting N types of retrial customers, which arrive according to independent Poisson streams. In case of blocking, class-i customer is routed to a separate infinite capacity orbit queue i, $i = 1, 2, ..., N$. When at least one other orbit queue is non-empty, the orbit i retransmits a class-i customer to the main service station after an exponentially distributed time with rate μ_i. If all other orbit queues are empty, orbit queue i changes its retransmission rate from μ_i to μ_i^*.

Using the regenerative approach we obtained the necessary stability conditions of the system, which have a clear probabilistic interpretation. Moreover we proposed a sufficient stability condition and study numerically the "difference" between sufficient and necessary conditions. Simulation experiments shown that the obtained conditions delimit the stability domain with remarkable accuracy.

Acknowledgments. The research of EM is supported by Russian Foundation for Basic Research, projects 15-07-02341, 15-07-02354, 15-07-02360.

References

1. Asmussen, S.: Applied Probability and Queues. Wiley, New York (1987)
2. Artalejo, J.R.: Accessible bibliography on retrial queues: progress in 2000–2009. Math. Comput. Model. **51**(9–10), 1071–1081 (2010)
3. Artalejo, J.R., Gomez-Corral, A.: Retrial Queueing Systems: A Computational Approach. Springer, Heidelberg (2008). doi:10.1007/978-3-540-78725-9

4. Avrachenkov, K., Yechiali, U.: Retrial networks with finite buffers and their application to internet data traffic. Prob. Eng. Inf. Sci. **22**, 519–536 (2008)
5. Avrachenkov, K., Morozov, E., Nekrasova, R., Steyaert, B.: Stability analysis and simulation of N-class retrial system with constant retrial rates and Poisson inputs. Asia Pac. J. Oper. Res. **31**(2), 1440002 (2014). (18 pages)
6. Avrachenkov, K., Nain, P., Yechiali, U.: A retrial system with two input streams and two orbit queues. Queueing Syst. **77**(1), 1–31 (2014)
7. Avrachenkov, K., Morozov, E., Steyaert, B.: Sufficient stability conditions for multi-class constant retrial rate systems. Queueing Syst. **82**, 149–171 (2016)
8. Bonald, T., Borst, S., Hegde, N., Proutiere, A.: Wireless data performance in multi-cell scenarios. In: Proceedings of the ACM Sigmetrics/Performance 2004, pp. 378–388. ACM, New York (2004)
9. Bonald, T., Massoulié, L., Proutiére, A., Virtamo, J.: A queueing analysis of max-min fairness, proportional fairness and balanced fairness. Queueing Syst. **53**(1–2), 65–84 (2006)
10. Borst, S., Jonckheere, M., Leskela, L.: Stability of parallel queueing systems with coupled service rates. Discrete Event Dyn. Syst. **18**(4), 447–472 (2008)
11. Falin, G.I., Templeton, J.G.D.: Retrial Queues. Chapman & Hall, London (1997)
12. Falin, G.I.: On a multiclass batch arrival retrial queue. Adv. Appl. Prob. **20**, 483–487 (1988)
13. Dimitriou, I.: A two class retrial system with coupled orbit queues. Prob. Eng. Inf. Sci. **31**(2), 139–179 (2017)
14. Dimitriou, I.: A queueing system for modeling cooperative wireless networks with coupled relay nodes and synchronized packet arrivals. Perform. Eval. (2017). doi:10.1016/j.peva.2017.04.002
15. Dimitriou, Ioannis: Modeling and analysis of a relay-assisted cooperative cognitive network. In: Thomas, Nigel, Forshaw, Matthew (eds.) ASMTA 2017. LNCS, vol. 10378, pp. 47–62. Springer, Cham (2017). doi:10.1007/978-3-319-61428-1_4
16. Kim, J., Kim, B.: A survey of retrial queueing systems. Ann. Oper. Res. **247**(1), 3–36 (2016)
17. Kulkarni, V.G.: Expected waiting times in a multiclass batch arrival retrial queue. J. Appl. Prob. **23**, 144–154 (1986)
18. Langaris, C., Dimitriou, I.: A queueing system with n-phases of service and $(n1)$-types of retrial customers. Eur. J. Oper. Res. **205**, 638–649 (2010)
19. Liu, X., Chong, E., Shroff, N.: A framework for opportunistic scheduling in wireless networks. Comput. Netw. **41**, 451–474 (2003)
20. Mitola, J., Maguire, G.: Cognitive radio: making software radios more personal. IEEE Pers. Commun. **6**(4), 13–18 (1999)
21. Morozov, E., Phung-Duc, T.: Stability analysis of a multiclass retrial system with classical retrial policy. Perform. Eval. (2017). doi:10.1016/j.peva.2017.03.003
22. Morozov, E.: The tightness in the ergodic analysis of regenerative queueing processes. Queueing Syst. **27**, 179–203 (1997)
23. Morozov, E.: Weak regeneration in modeling of queueing processes. Queueing Syst. **46**, 295–315 (2004)
24. Morozov, E.: A multiserver retrial queue: regenerative stability analysis. Queueing Syst. **56**, 157–168 (2007)
25. Morozov, E., Delgado, R.: Stability analysis of regenerative queues. Autom. Remote Control **70**, 1977–1991 (2009)
26. Moutzoukis, E., Langaris, C.: Non-preemptive priorities and vacations in a multiclass retrial queueing system. Stoch. Models **12**(3), 455–472 (1996)

27. Pappas, N., Kountouris, M., Ephremides, A., Traganitis, A.: Relay-assisted multiple access with full-duplex multi-packet reception. IEEE Trans. Wirel. Commun. **14**, 3544–3558 (2015)
28. Sadek, A., Liu, K., Ephremides, A.: Cognitive multiple access via cooperation: protocol design and performance analysis. IEEE Trans. Inf. Theor. **53**(10), 3677–3696 (2007)
29. Sendonaris, A., Erkip, E., Aazhang, B.: User cooperation diversity-Part I: system description. IEEE Trans. Commun. **51**, 1927–1938 (2003)
30. Sigman, K., Wolff, R.W.: A review of regenerative processes. SIAM Rev. **35**, 269–288 (1993)
31. Smith, W.L.: Regenerative stochastic processes. Proc. R. Soc. Ser. A **232**, 6–31 (1955)
32. Wolff, R.: Work conserving priorities. J. Appl. Prob. **7**, 327–337 (1970)

Model Checking

Model Checking the STL Time-Bounded Until on Hybrid Petri Nets Using Nef Polyhedra

Adrian Godde and Anne Remke[✉]

Institute of Mathematics and Computer Science, Westfälische Wilhelms-Universität
Münster, Münster, Germany
{a.godde,anne.remke}@uni-muenster.de

Abstract. Hybrid Petri nets have been extended with so-called general transitions, which add one random variable and one dimension to the underlying state space for each firing of a general transition. We propose an algorithm for model checking the time-bounded until operator in hybrid Petri nets with two general transition firings, based on boolean-set operations on Nef polyhedra. A case study on (dis)-charging an electrical vehicle shows the feasibility of the approach. Results are validated against a simulation tool and computation times are compared.

1 Introduction

Critical infrastructures are vital for societal well-fare and industrial operations. The high requirements that are placed on their dependability can often be verified for different situations using *model checking*, where an abstract model of the system is checked against a set of properties [1]. The modeling formalism of Petri nets has been introduced in [2] for communication models and extended to Hybrid Petri nets [3], which form a subclass of Hybrid Automata [4]. Recently, also stochastic firing times have been added to the hybrid formalism in [5]. These so-called *Hybrid Petri nets with general transitions* (HPnGs) can represent systems with discrete, continuous and probabilistic features. The continuous behaviour is however restricted to a piece-wise linear evolution. Model checking properties expressed in *Stochastic Time Logic* (STL) against HPnGs has been proposed before, either limited to HPnGs with a single general transition firing [6] or excluding the time-bounded until operator [7]. For model checking HPnGs, we need to identify regions in the underlying state space representation - a so-called *Stochastic Time Diagram* (STD) - where certain properties hold. These regions have been shown to correspond to (possibly open) convex polyhedra. For model checking the until operator we need to perform among others, the *set difference* operation on these polyhedra, yielding non-convex polyhedra, which e.g., the Parma Polyhedra Library (PPL) [8] does not cover.

This paper proposes model checking algorithms for hybrid Petri nets with multiple general transitions firings based on boolean set-operations on *Nef polyhedra*[9]. These include non-convex polyhedra and are closed under the set operations $\cup$, $\neg$, $\setminus$ and $\cap$. To validate the described concepts, the algorithms are

© Springer International Publishing AG 2017
P. Reinecke and A. Di Marco (Eds.): EPEW 2017, LNCS 10497, pp. 101–116, 2017.
DOI: 10.1007/978-3-319-66583-2_7

implemented for HPnGs with *two* general transitions firings, resulting in a three dimensional state-space. We used the *Computational Geometrical Algorithms Library*, which offers a library for Nef polyhedra in only three dimensions.

Related work. Parametric reachability analysis [5] characterises the possible evolutions of a system as trees of *parametric locations*, on which certain properties can be checked. Region-based analysis [6] constructs the underlying Stochastic Time Diagram, where states with similar properties are collected in regions. This approach resembles the analysis of (probabilistic) timed automata [10,11], or systems with piecewise-constant derivatives [12]. The syntax of STL for specifying properties of HPnGs is similar to MITL [13] or the *temporal layer* of STL/PSL [14]. Model checking for HPnGs with a single general transition firing is presented in [15] and for multiple firings, however *without* the time-bounded until operator in [7]. This paper presents model checking the time-bounded Until in HPnGs with multiple general transition firings. While the approach in this paper aims for exact model checking of HPnGs, related work also considers abstraction and simulation techniques. Statistical model checking for HPnGs is presented in [16] and e.g. simulation for Fluid Stochastic Petri Nets in [5]. Abstraction is often applied Stochastic Hybrid Systems, as presented in [17,18].

2   Hybrid Petri Nets with General Transitions

Hybrid Petri nets with multiple general transition firings (HPnGs) have discrete, continuous, and probabilistic features. Figure 1 illustrates the primitives of an HPnG. In the following, we provide an informal model definition, discuss their evolution and state-space representation and refer to [5] for the formal definition.

An HPnG is a tuple $P = (\mathcal{P}, \mathcal{T}, \mathcal{A}, m_0, x_0, \Phi)$. The set $\mathcal{P}$ of places consists of the subsets $\mathcal{P}^D$ of *discrete places* and $\mathcal{P}^C$ of *continuous places*. The discrete marking m contains the number of tokens of the discrete places and the continuous marking x the amount of fluid (from $\mathbb{R}_+$) in the continuous places.

The set $\mathcal{T}$ of transitions consists of *immediate transitions* $(\mathcal{T}^I)$, *deterministic transitions* $(\mathcal{T}^D)$, *general transitions* $(\mathcal{T}^G)$, and *continuous transitions* $(\mathcal{T}^F)$, which can all fire multiple times. A transition *firing* alters the marking of a place. *Discrete arcs* $\mathcal{A}^D$, *continuous arcs* $\mathcal{A}^F$, *test arcs* $\mathcal{A}^T$ and *inhibitor arcs* form the set of arcs $\mathcal{A}$. Test arcs connect transitions and places and enable the transition iff the connected place contains at least as many tokens or fluid as determined by the arc's weight. For inhibitor arcs the opposite holds.

The various sets of parameters are aggregated in the tuple $\Phi = (\Phi_b^P, \Phi_w^T, \Phi_p^T, \Phi_d^T, \Phi_f^T, \Phi_g^T, \Phi_w^A, \Phi_s^A, \Phi_p^A)$. Continuous places have a maximum capacity Φ_b^P. Φ_w^T and Φ_p^T, assign weight and priority to transitions, which is used for conflict resolution. Firing times for deterministic transitions can be found in Φ_d^T. Φ_f^T assigns a rate to continuous transitions. General transition Φ_g^T fire according to a cumulative distribution function (CDF) g_i. All arcs have a weight Φ_w^A, and fluid arcs have a *share* Φ_s^A and *priority* Φ_p^A.

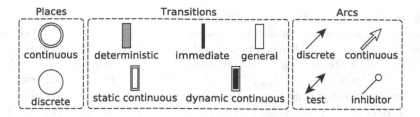

Fig. 1. Graphical representation of HPnG components.

The *evolution* of an HPnG is characterised as the change of *state* over time. We collect all firing times of general transitions in the datastructure s, which contains all realizations for each general transition that occurred before time t.

The state of an HPnG with fixed general transition firing times at time $t \in [0, t_{max}]$ is defined as $\Gamma(s, t) := (m, x, c, d, g)$. Next to the discrete and continuous marking, the state contains three more parameters. The vector c carries a clock c_i for each deterministic transition $T_i^D \in \mathcal{T}^D$, that counts the time for which the transition has already been enabled. The drifts[1] d of the continuous places is stored in the state tuple. Finally, vector g contains the time a general transition has been enabled since its last firing.

Given an HPnG P with fixed general transition firing times s and a state $\Gamma(s, t)$, the subsequent evolution of the Petri net is completely determined. Since the markings at time t are captured in the state, the current drifts as well as the enabled transitions are known, which fully characterises the behaviour of the system. Hence, the change in both markings and the future enabling or disabling of transitions can be predicted. The entire change of the state over time starting from an initial configuration $\Gamma(s, 0)$ up to $\Gamma(s, t_{max})$ only depends on the firing times of the general transitions.

The most significant changes in the state of an HPnG during its evolution are due to *events*. Events mark either the firing, enabling or disabling of any kind of transition or a rate adaptation for a continuous transition. Between two events only the markings of the continuous places and the clocks of the enabled transitions change according to their specific drifts. All the other parameters of the state are constant. For the computation of the next event we refer to [5].

While the discussion so far has conditioned the firing times of all general transitions to fixed realizations, we need to take into account all possible evolutions to evaluate the overall behaviour of an HPnG. This can be done by using a graphical representation of the Petri net's state space, which is called *Stochastic Time Diagram* (STD). This multidimensional representation features one dimension for the time and one dimension for each firing of a general transition. Furthermore, it combines states with similar characteristics over time into so-called regions. STDs for HPnGs with a single general one-shot transition [6] have already been extended to an arbitrary number of general transition firings [7].

---

[1] The change of fluid per time.

In an STD, the possible firing times s_i of the general transition T_i^G, $i \in \{1, \ldots, n-1\}$, are plotted against the system time t. A point in an STD is hence a tuple $(s_1, \ldots, s_{n-1}, t)$ of the s_i and t, thereby identifying a single state $\Gamma(s, t)$ of the HPnG. We write (s, t) as an abbreviation for such a point. An STD includes all evolutions in a subset $[0, t_{max}]^n$ of $\mathbb{R}^n$, as the individual general transition firing times s_i and the time t are limited by the observed time interval $[0, t_{max}]$. The time of occurrence of events partitions the STD into so-called *regions*. They aggregate sets of points (s, t), whose associated states $\Gamma(s, t)$ only differ in their continuous marking $\Gamma(s, t).x$ and their clocks $\Gamma(s, t).c$.

All states contained in a region share the same discrete marking and the same drifts of fluid places and clocks. All these parameters of a state can only be altered by events. Therefore, the borders of a region correlate to the time, at which an event takes place. According to Proposition 1 from [7], this time can be expressed as a linear function, i.e. a hyperplane equation, of s and t. Hence, the regions are surrounded by hyperplanes, which we call *event hyperplanes*.

3   Model Checking Time-Bounded Until

Semantics of time-bounded Until is presented in Sect. 3.1, the translation to Nef polyhedra in Sect. 3.2 and the model checking algorithm in Sect. 3.3.

3.1   Time-Bounded Until

Following the definition of STL as in [15], the *time-bounded Until* $\phi_1 \, \mathcal{U}^{[t_1, t_2]} \, \phi_2$ is a temporal modality that requires property ϕ_1 to hold until another property ϕ_2 holds in the time interval $[t_1, t_2]$. Excluding the nesting of Until operators, the formulas ϕ_1 and ϕ_2 are restricted to the negation and conjunction of atomic properties, that compare the continuous or discrete marking to a certain constant. By defining a so-called *satisfaction relation* $\models$ between a state of an HPnG and an Until formula, the STL is equipped with the means to describe the properties of a hybrid Petri net.

$$\Gamma(s, t) \models \phi_1 \mathcal{U}^{[t_1, t_2]} \phi_2 \iff \exists \tau \in [t + t_1, t + t_2] : \Gamma(s, \tau) \models \phi_2 \wedge$$
$$(\forall \tau' \in [t, \tau) : \Gamma(s, \tau') \models \phi_1).$$

Investigating the satisfaction of a time-bounded Until formula requires the examination of an evolution in the time interval $[t, t + t_2]$. The state $\Gamma(s, t)$ and its successors have to fulfil three conditions to satisfy an Until formula:

1. The property ϕ_1 must hold in all states $\Gamma(s, \tau')$, where $\tau' \in [t, t + t_1)$. If $t_1 = 0$, this condition is omitted.
2. The property ϕ_1 must hold in all states $\Gamma(s, \tau'')$, where $\tau'' \in [t + t_1, \tau)$ and $\tau \in [t + t_1, t + t_2]$.
3. The property ϕ_2 must hold eventually in some state $\Gamma(s, \tau)$.

In an HPnG with $n - 1$ general transition firings, this corresponds to identifying those combinations of firing times s, such that $\Gamma(s, t)$ fulfils $\phi_1 \, \mathcal{U}^{[t_1, t_2]} \, \phi_2$, which are combined in the so-called *satisfaction set* $Sat(\phi, t)$.

The probability operator as in Definition 1, compares the probability that from a system state $\Gamma(t)$ an evolution is taken that fulfils ϕ with a given threshold $\triangleright p$. This can be computed by an $n - 1$-dimensional integration of the density function that corresponds to each general transition over the satisfaction set.

Definition 1. *Let P be an HPnG with n general transition firings. A system state $\Gamma(t)$ satisfies a probability operator $\mathbb{P}_{\triangleright \, p}(\phi)$ for an STL formula ϕ, a probability $p \in [0, 1]$, and a comparison operator $\triangleright \in \{<, \leq, >, \geq\}$, iff:*

$$\Gamma(t) \models \mathbb{P}_{\triangleright \, p}(\phi) \iff Prob(\phi, t) \triangleright p.$$

Where $Prob(\phi, t) := \int \cdots \int_{Sat(\phi, t)} g_1(s_1) * \cdots * g_n(s_n) \, ds_n \ldots ds_1$ is the multiple integral over the domain $Sat(\phi, t)$ with the probability density functions g_i of the general transition firing times.

3.2   Translation to Nef Polyhedra

According to [9], Nef Polyhedra can be defined as intersections of a finite number of open half-spaces. Moreover, they are closed w.r.t. the operations $\cup$, $\neg$, $\setminus$ and $\cap$. Note that the requirement of open half-spaces is not a restriction, since every closed half-space can be expressed as the complement of an open half-space. Hence, convex polytopes are a subclass of Nef polyhedra, since they are defined as the intersection of closed half-spaces.

Regions in an STD correspond to convex polytopes, also in the case of multiple stochastic firings and a region in an STD is the union of the faces of a hyperplane arrangement [7]. According to [9], a point set $P \subset \mathbb{R}^d$ is a Nef polyhedron if there exists a finite family $\mathbf{H}$ of hyperplanes in $\mathbb{R}^d$, such that P is the union of certain faces of the corresponding arrangement $\mathcal{A}(\mathbf{H})$. Hence, the regions of an STD are also Nef polyhedra. We use Nef polyhedra and the fact that they are closed under boolean-set operations $\cup$, $\neg$, $\setminus$ and $\cap$ for model checking.

The satisfaction set $Sat(\phi_1 \, \mathcal{U}^{[t_1, t_2]} \, \phi_2, t)$ is computed in three steps. First, the set $M \subseteq [0, t_{max}]^{n-1}$ of general transition firing times is determined, for which:

$$\forall s \in M, t' \in [t, t + t_1] : \Gamma(s, t') \models^{s, t} \phi_1.$$

The elements s in M fulfil the first condition for satisfying an Until formula, as described in the previous subsection. Then all combinations of firing times are identified for which ϕ_1 holds until eventually the right-hand formula ϕ_2 holds within $[t + t_1, t + t_2]$. They are combined in another set $N \subseteq [0, t_{max}]^{n-1}$ and fulfil Conditions 2 and 3 from Sect. 3.1. All points s contained in both sets M and N fulfil all three conditions and satisfy the Until formula ϕ. Hence, the intersection of M and N forms the satisfaction set $Sat(\phi, t)$. To compute both M and N, we first determine the subsets of all regions in the respective time interval, in which ϕ_1 and ϕ_2 hold. The computation of these subsets is identical for M and N, except for the treatment of the upper and lower boundary.

3.3   Model Checking

We explain the details of the model checking approach in terms of Nef polyhedra, following the conditions discussed in Sect. 3.1.

Checking condition 1: Examining the first interval $[t, t+t_1]$ reveals the system evolutions for which ϕ_1 holds permanently inside the interval. These so-called *candidates* that potentially fulfil the Until formula are computed by function COMPUTECANDIDATESETS (c.f. line 1, Algorithm 1). Function CHECKINTERVAL (line 2) computes two Nef polyhedra containing all states, which satisfy ϕ_1 and ϕ_2, in the respective intervals. Figure 2 illustrates the output of the algorithm, i.e., polyhedra P_1, which contains all points that fulfill ϕ_1 and P_2 for ϕ_2. As $l = t$ and $u = t + t_1$, ϕ_1 should hold and ϕ_2 is not allowed to hold during the time interval $[l, u]$. Hence, we exclude all points which fulfil ϕ_2 too early, i.e., in $[t, t + t_1]$. Thus, P_1 is limited to the volumes in which ϕ_1 but not ϕ_2 holds by subtracting P_2 from P_1 (line 3).

A system evolution defined by s does not satisfy ϕ_1 during a given time interval iff there are points (s, t') with $t' \in [t, t + t_1]$ that lie outside the Nef polyhedron P_1. Hence, a segment, that fulfils ϕ_1 at time t and does not pierce through any facet of P, represents a system evolution, which constantly satisfies ϕ_1 during $[t, t+t_1]$. Figure 3 shows this concept for a two-dimensional STD, where evolution (a) passes through a facet of P and does hence not fulfil ϕ_1 throughout the interval $[t, t + t_1]$. However, evolution (b) does stay in the area covered by P during the time interval. Hence, (b) is considered a candidate evolution.

Algorithm 1. Determining the t-satisfying sets for ϕ_1 in the interval $[t, t + t_1]$

Require: regions of the STD $\mathcal{R}$, subformulas ϕ_1, ϕ_2, checking time t, time bound t_1
Ensure: Returns set of firing times which t-satisfies ϕ_1 in the time interval $[t, t + t_1]$
1: **function** COMPUTECANDIDATESETS($\mathcal{R}$, ϕ_1, ϕ_2, t, t_1)
2:      $P_1, P_2 \leftarrow$ CHECKINTERVAL($\mathcal{R}$, ϕ_1, ϕ_2, t, $t + t_1$)
3:      $P \leftarrow P_1 \setminus P_2$
4:      **return** IDENTIFYFULFILLINGSYSTEMEVOLUTIONS($t, t + t_1$, P)

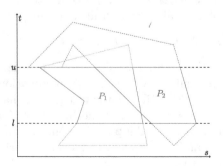

Fig. 2. Polyhedra P_1 and P_2 corresponding to formulas ϕ_1 and ϕ_2

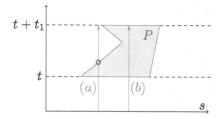

Fig. 3. An evolution that does not satisfy a formula in the complete interval $[t, t + t_1]$ (a) vs. one that satisfies the formula throughout the interval (b) in an STD in $\mathbb{R}^2$.

Algorithm 2. Identifing system evolutions that fulfil ϕ during $[l, u]$

Require: time interval $[l, u]$, Nef polyhedron P, i.e., satisfaction set of ϕ in $[l, u]$
Ensure: Returns set of firing times s that identify fulfilling system evolutions
 1: **function** IDENTIFYFULFILLINGSYSTEMEVOLUTIONS(l,u,P)
 2:     $M \leftarrow$ INTERSECTIONTOSUBSPACE(P, H_l)
 3:     **for all** facets $f \in P$ **do**
 4:         **if** $f \neq H_l \wedge f \neq H_u$ **then**
 5:             $M \leftarrow M\backslash$ PROJECTTOSUBSPACE(f)
 6:     **return** M

Based on this idea, Procedure IDENTIFYFULFILLINGSYSTEMEVOLUTIONS (c.f. Algorithm 2) determines s that specify evolutions that are completely enclosed by P in the given time interval $[t, t + t_1]$. The algorithm is defined for arbitrary intervals $[l, u]$, as it is reused later for the second time interval. By intersecting the input polyhedron with the hyperplane $H_l := t = l$ and subsequently projecting the result to $\mathbb{R}^{n-1}$, a set M is generated, that contains all firing times of the general transitions, for which ϕ holds at time l (line 2). In Fig. 4, the interval M is in $\mathbb{R}$ and results from the intersection of the Nef polyhedron P with the hyperplane H_l, which corresponds to the lower dashed line. In the next step, M is restricted using the facets of P to keep only those firing times, which correspond to evolutions that fulfil ϕ_1 in the complete interval. A facet whose hyperplane is equal to the interval borders l or u is excluded from this process (line 4). Projecting the facet f, that lies inside the interval (l, u) to $\mathbb{R}^{n-1}$ identifies those evolutions, which intersect with f. Hence, the projection of such a facet is substracted from the set M (line 5). After having processed all facets of P, M is reduced to the firing times, for which the corresponding system evolutions do not intersect with facets of P.

Figure 5 shows the result of the method applied to our running example. The facets of P between l and u have been used to restrict M to the marked interval, such that the remaining evolutions represented by M constantly fulfil ϕ in $[l, u]$. If M is empty after IDENTIFYFULFILLINGSYSTEMEVOLUTIONS returns from the call in COMPUTECANDIDATESETS, no system evolution constantly fulfils ϕ_1 in the first interval $[t, t + t_1]$. The model hence does not satisfy the Until formula and the process can be stopped at that point.

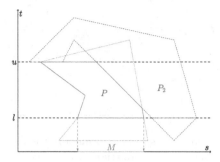

Fig. 4. Hyperplane inters. to identify evolutions satisfying ϕ_1 at a time l (Algorithm 2, line 2).

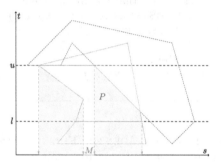

Fig. 5. Restriction of satisfying evolutions using the facets of P (Algorithm 2, line 5).

Checking condition 2 and 3: When analyzing the interval $[t + t_1, t + t_2]$ the observation of a system evolution should be cancelled, when a ϕ_2 state is reached. Algorithm 3 handles the second interval, which first identifies the time points at which ϕ_2 holds. Similar to Algorithm 1 the first step computes the subsets of the regions in the interval $[t + t_1, t + t_2]$, which satisfy ϕ_1 and ϕ_2 using the previously mentioned procedure CHECKINTERVAL (line 2). The next task is to determine the points, at which an evolution switches from P_1 to P_2, that is the facets of P_2 which intersect with P_1 have to be found. Points that are shared between P_1 and P_2 do not need to be considered, which is why they are excluded from P_1 (line 3). By intersecting the limited P_1 with the boundary of P_2 (line 4), we receive the desired point set. The interior and the border of a Nef polyhedron are again Nef polyhedra, so that the result set B of the operation is a Nef polyhedron, too.

For each point on a facet of B then both ϕ_1 and ϕ_2 hold. A system evolution with general transition firing times s, which crosses such a facet at some time $\tau \in [t + t_1, t + t_2]$, potentially fulfils the Until formula. If, in addition, all points of the evolution in the interval $[t, \tau)$ satisfy the formula ϕ_1, the firing times s belong to the set N. For this purpose, we can reuse the function IDENTIFYFUL-FILLINGSYSTEMEVOLUTIONS (Algorithm 2), which has been introduced for the computation of the set M in the first interval (line 8).

Algorithm 3. Determining t-satisfying sets for $\psi_1\,\mathcal{U}^{[t_1,t_2]}\,\psi_2$.

Require: regions of STD $\mathcal{R}$, subformulas ϕ_1, ϕ_2, checking time t, interval $[t_1,t_2]$
Ensure: Returns set of firing times which t-satisfy $\phi_1\,\mathcal{U}^{[0,t_2]}\phi_2$ at time $t+t_1$
 1: **function** COMPUTESATISFYINGSETS($\mathcal{R}$, ϕ_1, ϕ_2, t, t_1, t_2)
 2:      $P_1, P_2 \leftarrow$ CHECKINTERVAL($\mathcal{R}$, ϕ_1, ϕ_2, $t+t_1$, $t+t_2$)
 3:      $P_1 \leftarrow P_1 \setminus P_2.interior()$
 4:      $B \leftarrow P_1 \cap P_2.boundary()$
 5:      **for all** facets $f \in B$ **do**
 6:          $f_{projected} \leftarrow$ PROJECTTOSUBSPACE(f)
 7:          $Temp \leftarrow P_1 \cap \{(f_1,\dots,f_{n-1},t')|(f_1,\dots,f_{n-1}) \in f_{projected}, t' \in [t+t_1,t+t_2]\}$

 8:          $N \leftarrow N \cup$ IDENTIFYFULFILLINGSYSTEMEVOLUTIONS($t+t_1$, f, $Temp$)
 9:      **return** $N \cup$ INTERSECTIONTOSUBSPACE(P_2, H_{t+t_1})

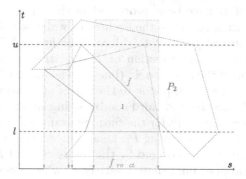

Fig. 6. Limiting P_1 with the prisms generated from facets of B (Algorithm 3, lines 6–7).

Each facet f of B is visited separately to identify the fulfilling system evolutions. The area $f_{projected}$, that results from projecting f to $\mathbb{R}^{n-1}$ (line 6), defines the subset of P_1, which is processed for the facet. We use $f_{projected}$ as a *basis*, to create a so-called *prism*, which is then intersected with P_1 (line 7). Figure 6 depicts the facets f of B which are projected to $\mathbb{R}^{n-1}$, and the prisms for each of the projected facets. Note that in $\mathbb{R}^2$ the prisms are simple rectangles.

The intersections $Temp$ of the individual prisms and P_1 are passed successively to the function IDENTIFYFULFILLINGSYSTEMEVOLUTIONS. Instead of a time point the facet f is passed as the upper boundary, which corresponds for each s to the time τ for which Φ_2 holds. The hyperplane $\mathcal{H}_u$, which is used as the upper limit inside Algorithm 2, is however just replaced by the hyperplane of the facet. Otherwise, nothing else is changed. The function call determines the projected sets in $\mathbb{R}^{n-1}$ from the intersection of H_{t+t_1} and $Temp$. Afterwards they are limited with the projection of the facets of $Temp$, which are located below f. The result is a set of firing times, which fulfil the second and third condition for the satisfaction of the Until formula.

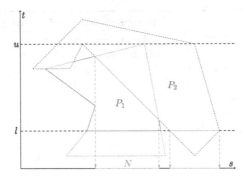

Fig. 7. Result of the checking process according to Algorithm 3 for the 2-d example.

After all facets of B have been processed in this manner, N does not yet contain all relevant firing times, as the evolutions which satisfy ϕ_2 immediately when entering the interval $[t + t_1, t + t_2]$ have been ignored. They can however be easily computed by the intersection of P_2 and the hyperplane H_{t+t_1} and subsequent projection of the result to $\mathbb{R}^2$ (line 9). The union of N with the result of this last operation now corresponds to the set of all general transition firing times, which fulfil the second and third condition in the interval $[t + t_1, t + t_2]$.

Figure 7 shows the final result for the running example. The left two facets of B can not be reached without leaving P_1, hence only the marked interval is part of N. In addition, the interval covered by the intersection of H_l with P_2 is part of N, since these evolutions immediately satisfy ϕ_2 when entering $[t + t_1, t + t_2]$. Intersecting the sets M and N as computed by Algorithms 1 and 3, yields the satisfaction set $Sat(\phi, t)$ for the time bounded Until formula. The evolutions identified by the elements in M fulfil the first condition for the satisfaction of the Until formula, while the points in N fulfil the second and third condition. Hence, checking an Until formula ϕ at time t results in $Sat(\phi, t) = M \cap N$.

4   Case Study: Electric Vehicle Charging

We examine the charging process for electric vehicles; following a *just-in-time* strategy, which aligns the charging process to the client's behaviour. By predicting the return time of the driver, the state of charge of the battery is kept at an optimal state for the battery's lifetime before charging the battery to its full capacity just before the driver returns (cf. [19]). Note that we do not consider discharging the battery for grid balancing. The developed model checking algorithms allow to investigate the impact of different parameters on the probability that the battery is fully charged when beeing picked up.

Figure 8 depicts a simplified version[2] of the HPnG model used in this case study. It models the charging process (left part of Fig. 8), which is observed for

---

[2] The model used for computation consists of 19 places and 26 transitions.

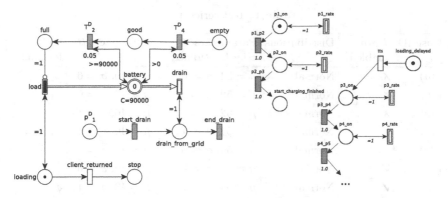

Fig. 8. HPnG model for charging a battery in an electric vehicle (based on [19]).

144 time units[3]. The central element is the place *battery* with a capacity of 90000 (in Watt-hours) and its dynamic continuous source transition *load*. The latter hides a time-dependent loading process (partly depicted in the right part of Fig. 8) that starts, when the car is plugged to the charger, corresponding to the firing of the first general transition *tts*.

As long as the place *loading* contains one token, the car is connected to the charger. The duration of the charging cycle is determined by the second general transition *client_returned*, which specifies, after how many time units the driver of the car unplugs it. The abstract state of the battery is either *empty*, *good*, or *full*, as indicated by the token that moves between those states during the charging process. The delay of 0.05 time units (30 s) for the firing of T_4^D and T_2^D models the reaction time of the charging station to begin or respectively end the process. If the *battery* is fully charged, T_D^2 fires after 0.05 time units and moves the token to the place *full*, thereby disabling the transition *load* via the inhibitor arc. The charging strategy is modelled via the dynamic continuous transition *load* and as implemented in this model, fills the battery in two consecutive steps. In a first phase, the capacity is brought quickly to the level of 40 kWh. The second phase charges the battery up to its capacity limit, such that the client can use the maximum range of his electric vehicle. In the mean time, the charging is stopped and while in the current model the power level of the battery remains unchanged, the available capacity could be offered to the grid operator to flexibly charge/discharge. Additionally, the transition *drain* allows the grid to consume battery capacity, if enabled, i.e., if the place *drain_from_grid* contains a token. A client might consider this charging strategy reasonable, if the battery is charged at least up to some threshold when the car is unplugged, specified in STL as follows:

$$\phi_1 := m_{loading} \geq 1 \, \mathcal{U}^{[0,144]} \, x_{battery} \geq c,$$

where the constant c represents the required state of charge. The place *loading* contains a token while the general transition *client_returned* has not fired.

---

[3] Corresponding to 1440 min or 24 h.

Table 1. Test series 1–3.

Case ID	drain enabled?	Distributions	Parameters client_returned	Parameters tts	STL formula
(1a)	✗	Normal	$\mu = 54, \sigma = 6$	$\mu = 40, \sigma = 6$	ϕ_1
(1b)	✗	Normal	$\mu = 52, \sigma = 6$	$\mu = 40, \sigma = 6$	ϕ_1
(1c)	✗	Normal	$\mu = 52, \sigma = 12$	$\mu = 40, \sigma = 12$	ϕ_1
(1d)	✗	Normal	$\mu = 54, \sigma = 12$	$\mu = 40, \sigma = 12$	ϕ_1
(2a)	✓	Normal	$\mu = 54, \sigma = 6$	$\mu = 40, \sigma = 6$	ϕ_1
(2b)	✓	Normal	$\mu = 54, \sigma = 6$	$\mu = 40, \sigma = 6$	ϕ_2
(2c)	✓	Normal	$\mu = 54, \sigma = 12$	$\mu = 40, \sigma = 12$	ϕ_2
(3b)	✗	Uniform	$a = 52, b = 56$	$a = 46, b = 50$	$t = 36$
(3c)	✗	Uniform	$a = 52, b = 56$	$a = 46, b = 50$	$t = 54$
(3d)	✗	Uniform	$a = 52, b = 56$	$a = 46, b = 50$	$t = 66$
(3e)	✗	Uniform	$a = 52, b = 56$	$a = 46, b = 50$	$t = 72$

A first series of tests (1a to 1d in Table 1) investigates the influence of different distribution parameters on the probability that ϕ_1 is satisfied. Here, the firing times of both general transitions client_returned and tts follow normal distributions with varying mean and standard deviation. The client unplugs the car from the charger on average after $\mu = 54$ time units (9 h) with a standard deviation between $\sigma = 6$ (1 h) and $\sigma = 12$. Then ϕ_1 is checked for a threshold of $c = 81000$ Wh, which corresponds to 90% of the battery capacity. The second test series assumes that the grid also consumes battery power. Our model however fills the place battery with a fixed total amount of 90 kWh, so that the reachable level in battery is reduced by the amount of power, which is consumed by transition drain. If drain is enabled for one time unit with e.g. a rate of 3 kWh, the battery can only be charged up to 77 kWh, and ϕ_1 will not hold. Hence, for scenarios (2b) and (2c), the constant in ϕ_1 is set to $c = 75000$ Wh, as indicated in Table 1. The last test series determines the probability that the battery is charged to 81 kWh at a specific time point. As shown in Table 1, the firing times of client_returned and tts are both uniformly distributed between $[52, 56]$ and $[46, 50]$ in Scenarios 3b–3e. This corresponds to starting the second phase of charging after 460 to 500 min and to unplugging the car after 520 to 560 min. The following formula is investigated at different times:

$$\phi_3 := x_{battery} \leq 81000.$$

All configurations, as summarized in Table 1, have been model checked using the algorithms presented in this paper. Note that, depending on the configuration of the HPnG, the corresponding STD consists of 335 to 360 regions. In addition, the results have been compared with a statistical model checking tool for HPnGs [16], which samples the firing times of all general transitions. Model checking a formula then reduces to testing the single evolution for the desired property.

Table 2. Comparison of results for all test series.

Case ID	Nef polyhedra tool probability	Simulation mean	Simulation confidence interval
(1a)	0.9026	0.9015	[0.8965, 0.9065]
(1b)	0.8556	0.8557	[0.8507, 0.8607]
(1c)	0.7016	0.7020	[0.6970, 0.7070]
(1d)	0.7411	0.7401	[0.7351, 0.7451]
(2a)	0	0	[0, 0]
(2b)	0.8643	0.8654	[0.8604, 0.8704]
(2c)	0.6667	0.6662	[0.6612, 0.6712]
(3b)	0.9985	1	[1, 1]
(3c)	0.0313	0.0289	[0.0239, 0.0339]
(3d)	0.0317	0.0334	[0.0284, 0.0384]
(3e)	0.0314	0.0287	[0.0237, 0.0337]

By repeating this process several times, the tool is able to approximate the probability for the satisfaction of the formula. After a finite number of iterations, the tool outputs a mean value with a confidence interval. Table 2 summarises the results rounded to four decimal places.

In test series 1 and 2 the analytical results lie well within the 99% confidence intervals provided by the simulation. Assuming that the chosen distribution $\mathcal{N}(54, 6^2)$ correctly models the return time distribution of the client (transition *client_returned*), the second charging phase (transition *tts*) starts early enough to satisfy the property ϕ_1 in over 90% of all cases. As expected, a higher variation in the return time distribution, leads to a lower probability of Φ_1 to hold and to larger confidence intervals in the simulation. Recall, that test series 3 model checks an atomic formula at different time points. Case (3c), for example, implies, that the battery is charged to less than 81 kWh after 54 time units in only about 3% of the possible evolutions. Only for case (3b) the analytical result of 0.9985, does not match the simulation value of 1. This deviation can be traced back to a numerical error in the integration over the satisfaction set. By increasing the number of iterations in the numerical integration, the error can be reduced, but the higher precision can significantly increase the computation time. Table 3 compares the computation times the statistical model checker and of the Nef polyhedra tool for different iteration numbers. The computation times in the highlighted column correspond to the results presented in Table 2.

In most test cases, the Nef polyhedra tool with 4096 iterations computes the probabilities faster than the simulation tool. Only in the cases where the probability amounts to 0 or 1, the simulation tool outperforms the analytical approach. Even for 8192 iterations the Nef polyhedra tool is often faster than the simulation tool. Note that the computation time of our approach consists of three parts: building the STD, computing the satisfaction set, and integrating

Table 3. Comparison of computation times for Nef polyhedra tool and simulation tool.

case ID	Nef polyhedra tool runtime			Simulation runtime	Number of simulations
	1024 iterations	4096 iterations	8192 iterations		
1a)	3.157s	3.454s	4.144s	31.122s	23573
1b)	3.213s	3.443s	4.15s	64.1s	32774
1c)	3.167s	3.397s	4.145s	84.179s	55525
1d)	3.187s	3.384s	4.119s	81.049s	51059
2a)	1.718s	1.715s	1.717s	1.523s	100
2b)	2.394s	2.626s	3.378s	41.256s	30927
2c)	2.397s	2.628s	3.334s	143.403s	59020
3b)	3.105s	6.675s	15.303s	388s	100
3c)	3.112s	6.015s	15.333s	10.588s	7447
3d)	3.043s	6.783s	16.04s	7.594s	8569
3e)	3.482s	6.063s	15.295s	10.777s	7396

over the satisfaction set. The former two steps are not affected by the number of iterations, which is why in test series 1, for example, the generation of the STD and the following computation of the satisfaction set take about 3.14 s for every configuration. The mere integration thus requires about 0.02 s with 1024 iterations, about 0.3 s with 4096 iterations, and about 1.1 s with 8192 iterations, which implies that doubling the number of iterations approximately quadruples the computation time. However, our computations have shown that the precision gain is often negligible or even non-existent. Computations have been performed on an *Intel Core i5-750 @2.67* GHz CPU.

5   Conclusion

We proposed an algorithm for model checking the time-bounded until operator in Hybrid Petri nets with multiple general transition firings based on operations on Nef polyhedra. It has been shown that an STD consists of a set of Nef polyhedra, which is a special class of polytopes, that is defined by intersections of half-spaces. The proposed algorithm has been implemented in C++ using the *Computational Geometrical Algorithms Library* CGAL [20], which offers an implementation for operations on Nef polyhedra in three dimensions. This allows to model check HPnGs with two firings of general transitions. A case study with several scenarios not only shows the feasibility of the approach, but also validates the results against a dedicated simulator for HPnGs. The analysis of HPnGs with multiple general transition firings requires an implementation of polyhedra in arbitrary dimensions and a more sophisticated technique for the integration over the multidimensional satisfaction set. Future work will investigate the use of Hypro, a toolbox for the Reachability Analysis of Hybrid Systems using Geometric Approximations [21].

References

1. Baier, C., Katoen, J.-P.: Principles of Model Checking. The MIT Press, Cambridge (2008)
2. Petri, C.A.: Kommunikation mit Automaten. Ph.D. thesis, Uni. Hamburg (1962)
3. David, R., Alla, H.: On hybrid petri nets. Discrete Event Dyn. Syst. **11**(1), 9–40 (2001)
4. Henzinger, T.A.: The theory of hybrid automata. In: 11th Annual IEEE Symposium on Logic in Computer Science, p. 278, IEEE Computer Society (1996)
5. Gribaudo, M., Remke, A.: Hybrid petri nets with general one-shot transitions. Perform. Eval. **105**, 22–50 (2016)
6. Ghasemieh, H., Remke, A., Haverkort, B., Gribaudo, M.: Region-based analysis of hybrid petri nets with a single general one-shot transition. In: Jurdziński, M., Ničković, D. (eds.) FORMATS 2012. LNCS, vol. 7595, pp. 139–154. Springer, Heidelberg (2012). doi:10.1007/978-3-642-33365-1_11
7. Ghasemieh, H., Remke, A., Haverkort, B.: Hybrid petri nets with multiple stochastic transition firings. In: 8th EAI International Conference on Performance Evaluation Methodologies and Tools (ICST), pp. 217–224 (2014)
8. Bagnara, R., Hill, P.M., Zaffanella, E.: The parma polyhedra library: toward a complete set of numerical abstractions for the analysis and verification of hardware and software systems. Sci. Comput. Program. **72**(1–2), 3–21 (2008)
9. Nef, W.: Beiträge zur Theorie der Polyeder: mit Anwendungen in der Computergraphik. Beiträge zur Mathematik, Informatik und Nachrichtentechnik (1978)
10. Alur, R., Dill, D.L.: A theory of timed automata. Theor. Comput. Sci. **126**(2), 183–235 (1994)
11. Kwiatkowska, M., Norman, G., Segala, R., Sproston, J.: Automatic verification of real-time systems with discrete probability distributions. Theor. Comput. Sci. **282**(1), 101–150 (2002)
12. Asarin, E., Maler, O., Pnueli, A.: Reachability analysis of dynamical systems having piecewise-constant derivatives. Theor. Comput. Sci. **138**(1), 35–65 (1995)
13. Alur, R., Feder, T., Henzinger, T.A.: The benefits of relaxing punctuality. J. ACM **43**, 116–146 (1996)
14. Nickovic, D., Maler, O.: AMT: a property-based monitoring tool for analog systems. In: Raskin, J.-F., Thiagarajan, P.S. (eds.) FORMATS 2007. LNCS, vol. 4763, pp. 304–319. Springer, Heidelberg (2007). doi:10.1007/978-3-540-75454-1_22
15. Ghasemieh, H., Remke, A., Haverkort, B.R.: Survivability evaluation of fluid critical infrastructures using hybrid petri nets. In: IEEE 19th Pacific Rim International Symposium on Dependable Computing, pp. 152–161. IEEE (2013)
16. Pilch, C., Remke, A.: Statistical model checking for hybrid petri nets with multiple general transitions. In: 47th IEEE/IFIP International Conference on Dependable Systems and Networks. IEEE (2016). http://go.wwu.de/pbptn
17. Abate, A., D'Innocenzo, A., Benedetto, M.D.D.: Approximate abstractions of stochastic hybrid systems. IEEE Trans. Autom. Control **56**(11), 2688–2694 (2011)
18. Fränzle, M., Hahn, E.M., Hermanns, H., Wolovick, N., Zhang, L.: Measurability and safety verification for stochastic hybrid systems (2011)
19. Hüls, J., Remke, A.: Coordinated charging strategies for plug-in electric vehicles to ensure a robust charging process. In: 10th EAI International Conference on Performance Evaluation Methodologies and Tools, ICST (2016)

20. The CGAL Project: CGAL User and Reference Manual. CGAL Editorial Board, 4.10 ed. (2017)
21. Schupp, S., Ábrahám, E., Makhlouf, I.B., Kowalewski, S.: HyPro: a C++ library of state set representations for hybrid systems reachability analysis. In: Barrett, C., Davies, M., Kahsai, T. (eds.) NFM 2017. LNCS, vol. 10227, pp. 288–294. Springer, Cham (2017). doi:10.1007/978-3-319-57288-8_20

A New Approach to Predicting Reliable Project Runtimes via Probabilistic Model Checking

Ulrich Vogl[(✉)] and Markus Siegle[(✉)]

Universität der Bundeswehr München, 85577 Neubiberg, Germany
{ulrich.vogl,markus.siegle}@unibw.de

Abstract. For more than five decades, efforts of calculating exact probabilistic quantiles for generally distributed project runtimes have not been successful due to the tremendous computation requirements, paired with hard restrictions on the available computation power. The methods established today are PERT (Program Evaluation and Review Technique) and CCPM (Critical Chain Project Management). They make simplifying assumptions by focusing on the critical path (PERT) or estimating appropriate buffers (CCPM). In view of this, and since today's machines offer an increased computation power, we have developed a new approach: For the calculation of more exact quantiles or – reversely – of the resulting buffer sizes, we combine the capabilities of classical reduction techniques for series-parallel structures with the capabilities of probabilistic model checking. In order to avoid the state space explosion problem, we propose a heuristic algorithm.

Keywords: Project planning · Stochastic graph model · Series-parallel reduction · Probabilistic model checking · PERT · CCPM

1 Introduction

In project planning, predicting the total runtime of activity chains and/or concurrent project activities, is an important task. In most situations, due to the influence of many unpredictable factors, probabilistic methods are more appropriate than deterministic ones. Two well-known representatives, PERT (Program Evaluation and Review Technique [13, pp. 303–365]) and CCPM (Critical Chain Project Management [4]), mostly only yield inaccurate results, due to methodical simplifications (taken in order to make them computable).

PERT is limited in principle by focussing on the critical path. Sub-critical paths are completely neglected – even if they appear in a high number or with a significant variance influencing the total runtime distribution. (A more detailed analysis of PERT networks, albeit under Markovian assumptions, has been described in [7].) CCPM works in a more differentiated (but also non-preemptive) way: Using this method, all non-critical paths are augmented by feeding buffers in order to lower the influence of the critical path. For determining the size of the buffer, the so-called Cut & Paste Method and the Root Square Error Method have been suggested. But these methods for buffer sizing handle the variance of the side-paths

© Springer International Publishing AG 2017
P. Reinecke and A. Di Marco (Eds.): EPEW 2017, LNCS 10497, pp. 117–132, 2017.
DOI: 10.1007/978-3-319-66583-2_8

only in an indirect, usually inaccurate manner, and thus their effectiveness is hard to quantify. This motivated us to take a review on this topic, seeking for a new, better approach. In particular, since the introduction of both methods in 1958 resp. 1996 the available computation power has increased significantly, and new calculation methods such as probabilistic model checking (pMC) [8], together with efficient tools like PRISM [12], have become available.

We propose a method for complete and accurate calculation of a project's total runtime distribution, where our key ideas are as follows:

- The nodes of a stochastic graph model (SGM), i.e. a directed acyclic graph (DAG), represent the activities of a project. Each node is equipped with a continuous probability distribution, representing its individual runtime.
- The nodes of the graph are reduced in a step-by-step manner, ultimately leading to only a single node, with a related result distribution which represents the project's total runtime.
- We seek to find subgraphs which can be reduced to a single node by serial or parallel reduction, as explained below in Sect. 2.
- When no further series-parallel reduction is possible, we identify the starting and end points of a so-called complex cluster (a generally structured subgraph). We use the concept of syncpoints (see Sect. 3 below) for defining such clusters. The cluster is then reduced by a *complex reduction* to a single node, for which step pMC is employed. In order to avoid state space explosion, it is essential to limit the size of the graph to be fed into pMC. Therefore the clusters analysed by pMC should be as small as possible.
- One faces the challenge of finding an appropriate (heuristic) fitting for the given source distributions, because pMC tools usually only accept exponential distributions. "Fitting" in this context means the approximate modelling of a given distribution by some phase type distribution, for instance by matching the first moments.
- It is an interesting side effect that already one complex reduction step can often eliminate a local complexity hotspot and thereby enable further series-parallel reduction steps.

1.1   Related Work

Melchiors and Kolisch [10,11] have presented a heuristic approach for establishing and assessing scheduling policies for dynamically arriving, concurring project activities competing for limited resources. In contrast to our approach which works on the operational planning level, their work is targeted at the "tactical" planning level (aka "Macro Process Planning"). It uses aggregated work packages as well as global estimates of runtime and precedence relationships. It assumes high variability project environments, allowing for dynamically emerging activities as well as dynamically changing dependencies between activities. Several heuristics (based on CTMCs and MDPs) are combined, to gain a computable model state space (including a preemptive modelling approach). The final valuation and performance assessment of the priority policy methods is done via simulation.

Kapici [5] presented an approach where complex projects are mapped to a newly developed stochastic model. He then derives statements on the adherance of deadlines, costs or other results. His approach is fully simulation-based.

There are further approaches, most of them based on simulation and/or complexity-reducing heuristics. The destinctive feature of our work, however, is the accurate consideration of the individual project activities and their precedence relations, combined with the power of a probabilistic model checker.

1.2   Paper Structure

The rest of this paper is structured as follows: Sect. 2 provides background information on PERT, CCPM and the analysis of stochastic graph models. The shortcomings of these classical methods motivated us to develop an innovative approach, which is presented in detail in Sect. 3. To illustrate our method, Sect. 4 presents some non-trivial examples. Finally, Sect. 5 summarizes the results and describes some ideas for continuing work.

2   State of the Art

2.1   Stochastic Graph Models

Stochastic Graph Models are a simple and intuitive formalism for modelling the structure of projects, parallel programs, collections of interdependent tasks, etc. They have been described in detail, e.g., in [6].

Definition 1. *A Stochastic Graph Model (SGM) is a directed acyclic graph $G = (V, E, exec)$ with the following properties:*

1. *V is a finite set of vertices (aka nodes), and $E \subseteq V \times V$ is the set of directed edges. G is connected and has a single source and a single sink node.*
2. *$exec : V \mapsto Distr$ is a function which assigns to each vertex its associated continuous nonnegative runtime distribution. The runtime distributions of all vertices are mutually independent.*
3. *A tuple $p = (e_1, e_2, \ldots, e_k) \in E^k$ of edges is called a directed path, if and only if $\forall_{1 < i \leq k} : start(e_i) = end(e_{i-1})$, where $start(e)/end(e)$ denotes the start/end node of edge e. The set of all possible paths in G (induced by E) is denoted by $paths(E)$.*

A vertex of a SGM starts its execution as soon as all its predecessor vertices have completed. The goal of SGM analysis is to determine the total runtime distribution, i.e. the elapsed time between the start of the source node and the finishing of the sink node.

Definition 2. *An edge $e \in E$ is called **redundant** iff there is a path $p = (e_1, e_2, \ldots, e_k)$ of edges with $e \notin p$, $start(e) = start(e_1)$ and $end(e) = end(e_k)$.*

For the rest of this paper, without any loss of information, we assume the SGM to be free of redundant edges (if a particular SGM is not so, they can be discovered and removed easily).

There is a class of SGMs, featuring a series-parallel structure, which is amenable to efficient analysis. This class is characterised by the following definition and theorem.

Definition 3. *Two vertices n_i and n_j of a SGM are said to be* serially connected *iff n_j is the only successor of n_i and n_i is the only predecessor of n_j (or vice versa). A set of vertices $P \subseteq V$ with $|P| \geq 2$ is called* parallelly connected *iff all vertices $n \in P$ have the same set of predecessors and the same set of successors.*

Theorem 1. *(from [6, p. 184]) Two serially connected vertices n_1 and n_2 may be serially reduced to a single vertex n_{12} as follows:*
$pred(n_{12}) = pred(n_1)$, $succ(n_{12}) = succ(n_2)$, $exec(n_{12}) = exec(n_1) *$
$exec(n_2)$, *where $pred(n)/succ(n)$ denotes the set of predecessor/successor nodes of n, and $*$ denotes the convolution operator on continuous distributions. Two or more parallelly connected vertices $n_1, n_2, \ldots, n_k$ may be parallelly reduced to a single vertex $n_{1\ldots k}$ as follows:*
$pred(n_{1\ldots k}) = pred(n_1)$, $succ(n_{1\ldots k}) = succ(n_1)$,
$exec(n_{1\ldots k}) = \max(exec(n_1), \ldots, exec(n_k))$,
where $\max$ denotes the maximum operator on continuous distributions.

Definition 4. *A SGM G is called* series-parallel reducible *if it can be reduced to a single node by successive serial and parallel reduction steps.*

In short, serial reduction means that two serially connected nodes are reduced to a single node whose distribution is the convolution of the two operand distributions (the convolution yields the distribution of the sum of the execution times). Parallel reduction means that two or more "parallel" nodes are reduced to a single node which is distributed according to the maximum of the operand runtimes. Series-parallel reduction is a very efficient method for analysing SGMs. However, while many graph structures are series-parallel reducible, in practice many SGMs are not of this class, see e.g. the graph shown in Fig. 1, which is no longer series-parallel reducible if the traverse edge A-D is inserted.

2.2   A Simple SGM

We consider a simple SGM consisting of four significant nodes $\{A, B, C, D\}$, enclosed by the source node S and the sink node E (shown in Fig. 1). It is assumed that nodes S and E have negligible runtime, i.e. their runtime is deterministic with value zero. The runtimes of the four significant nodes all follow an Erlang-distribution with n phases and basic rate λ (having mean $\mu = \frac{n}{\lambda}$ and variance $\sigma^2 = \frac{n}{\lambda^2}$): For A and B with $\lambda = 1, n = 10$, for C with $\lambda = 0.2, n = 2$, and for D with $\lambda = 0.201, n = 2$. The resulting μ and σ^2 values are given in Fig. 1.

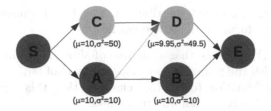

Fig. 1. Example – the traversal edge A-D is initially neglected (Color figure online)

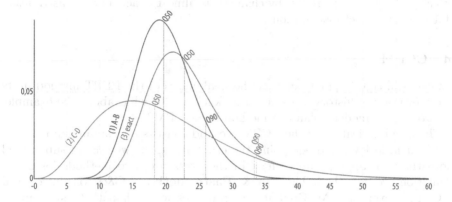

Fig. 2. Comparison of the densities; 50%- and 90%-quantiles (Color figure online)

For the sake of simplicity, we first neglect the traverse edge from A to D, such that the SGM is series-parallel reducible. Its overall runtime distribution is thus

$$exec(ABCD) = \max(exec(A) * exec(B), exec(C) * exec(D))$$

This density is depicted in Fig. 2 as curve (3) (green).

2.3   PERT

PERT focusses only on the critical path, thereby taking only the mean execution times into account. For the SGM from Fig. 1 (without traversal edge A-D), this results in the critical path (S-A-B-E) (orange coloured nodes), which has a mean value of 20 time units and a variance of 20 (standard deviation 4.47 time units). We calculated exact distributions for two variants and depicted the associated densities in Fig. 2:

(1) only the critical path (S-A-B-E) (curve (1), blue)
(2) only the sub-critical side-path (S-C-D-E) (curve (2), red)

Curves (1) and (2), representing the competing paths, possess indeed similar means (20 resp. 19.95 time units), but quite different variances (20 resp. 99.5 time units squared). The crucial conclusion is gained by comparing curves (1) and curve (3). The former represents the (seeming) PERT view, whereas the latter equals the precise overall distribution. The PERT-caused error becomes most apparent if we focus on the 90% quantiles:

- PERT (curve (1)) suggests, that a runtime of about 25.9 time units would be sufficient to finish the project at a confidence level of 90%.
- The accurate calculation (curve (3)) clarifies, that this confidence level in reality is reached only at about 34 time units.

Figure 2 also shows that for this particular SGM, the 90%-quantile of the sub-critical path (curve (2), red) – by chance – is almost exact, but its distribution is far from the exact distribution.

2.4   CCPM

Now one will rightly point out that the weaknesses of the PERT method in its almost 60 years of history are sufficiently known. But what about the example, fed into a more modern planning method like CCPM?

The deciding feature of the CCPM method consists of planning all paths at a 50%-quantile level and equipping the critical path as well as all sub-critical side-paths with appropriate buffers. In the literature, two methods for buffer calculation are described: The **Cut & Paste method (C&PM)**, introduced by CCPM inventor E.M. Goldratt [4], proposes for each path to sum up its 50%-quantiles, then take that result as the path's base runtime and add to it an additional buffer of 50% of that. Alternatively, some authors [3] recommend the **Root Square Error Method (RSEM)**, which takes as buffer the square root of the sum of squared differences between the 50%- and the 90% quantiles.

Table 1 shows the results of these two variants of the CCPM method applied to the SGM from Fig. 1 (without traversal edge A-D). Looking only at the medians, again path (S-A-B-E) with value 19.34 dominates path (S-C-D-E) which has value 16.75 (these values differ from the medians of Fig. 2 because they are simple sums of the single activity medians and not of the convoluted distributions). But taking into account the buffers, the RSEM method identifies (S-C-D-E) as the critical chain. If we compare the relating exact quantiles of the calculated finishing times to the desired 90% quantile of the accurate model evaluation (at 33.65 time units), we can conclude:

- C&PM yields an "in time"-completion probability of only around 80.5%,

Table 1. CCPM – total runtime (+ buffer) & associated quantiles (Color figure online)

	C&PM	RSEM
path A-B [time units]	19.34 (+9.67)	19.34 (+6.42)
path C-D [time units]	16.75 (+8.38)	16.75 (+15.60)
total/maximum [time units]	29.01	32.35
relating exact quantile	80.46%	88.01%

– with RSEM the problematic sub-critical side path (C-D) gains more importance by its dominating feeding buffer. That indeed increases the completion probability to around 88% (which is still below the desired 90% level).

2.5 Non-series-parallel SGMs

We now return to the SGM in Fig. 1, but this time we include the traversal edge A-D into the calculations. This is remarkable, since exactly that edge destroys the series-parallel reducibility of the graph. Therefore, in order to obtain the precise runtime distribution, we can no longer rely on the convolution and maximum operators, but we need indeed a more powerful computation method. We chose to employ pMC, in particular we use the probabilistic model checker PRISM [12], which under the hood performs a state space analysis (transient analysis by means of uniformization). Since PRISM provides no explicit calculation feature for discrete densities, we use it as follows:

1. First we choose an appropriate discretization step width, e.g. 1% of the smallest occuring mean or standard deviation, as well as an estimation of the upper interval limit, which – for instance – can be gained by a pathwise consideration, thereby taking each single distribution's upper limit.
2. Then for each discrete time value t_i we perform a PRISM call of the form $P(T < t_i) =?$, which delivers the cumulative distribution value for time t_i.
3. PRISM provides a feature to chain such calculations for entire intervals by only one call (given start time, end time and step width); this obviously leads to a tremendous decrease of the calculation effort (granting PRISM the reuse of prior results).
4. The desired density values are eventually gained by a simple numerical differentiation, taking the difference of each two neighbouring distribution values.
5. The calculation time can be further reduced by splitting the PRISM call intervals and distributing them onto several CPU threads.

Figure 3 depicts the influence of the traversal edge by comparing the overall densities with/without it. Note that curves (1) (blue) and (3) (green) equal those of Fig. 2. The exact overall density for the SGM with traversal edge, shown as curve (4) (magenta), deviates slightly from curve (3) (without traversal): Now the 90% quantile is reached at 35.01 time units (previously: 33.65). For the sake of comparison, the PERT based density (1), gained by concentrating on the critical path only, is displayed once again (after insertion of the traverse edge, the critical path is still the same!). The 90% quantile of the PERT view lies (at 25.9 time units) more than a quarter below the actual value.

In summary, it can be stated that an accurate calculation of the density offers significant advantages over the established methods PERT and CCPM. In principle, state-space-based methods such as implemented by PRISM are able to produce such accurate distributions, but they are limited to small or medium-sized models because of the arising state space explosion problem. Admittedly, the SGM considered in this section was an extremely simple case – the reality

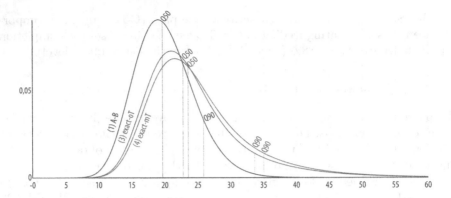

Fig. 3. Comparison of the densities: without (3) resp. with (4) the traverse A-D (Color figure online)

is usually much more complex. Therefore, in the following section, we develop a new method which combines series-parallel reduction and pMC, thereby making it applicable to larger SGMs of realistic size.

3   A New Reduction Method for Analysing Project Runtimes

Calculating the exact overall runtime density for arbitrarily structured graphs, equipped with general runtime distributions, is in general not feasible for the following reasons:

(a) Series-parallel reduction by use of the convolution and maximum operators quickly leads to very complicated mathematical expressions, if carried out symbolically. Those are difficult to handle, even with advanced tools such as Mathematica [14] or Maple [9].

(b) If the SGM at hand is not series-parallel reducible, purely analytic approaches fail if activity runtimes have general distributions, since state space analysis relies on the memoryless property of the exponential distribution.

(c) Even if all node execution times are exponentially or PH-type distributed, such that state space analysis would be possible in principle, one quickly reaches the limits of computability because of state space explosion.

Our proposed method, presented in this paper, overcomes these problems in the following way: Problem (a) is dealt with by representing general distributions not symbolically, but numerically. That means a general distribution is discretized and represented as a step function, and the convolution and maximum operators are performed on the basis of such numerical representations. A similar numerical approach had previously been described in [6]. Concerning problem (b), we enable state space analysis of SGMs (or subgraphs thereof) with non-exponential execution times by replacing those general distributions with fitted phase-type distributions (see, e.g. [2]). Finally, with regard to problem (c), our scheme avoids performing state space analysis on the overall model. Instead, it combines series-parallel reduction steps with state space analysis of small subgraphs in an iterative manner.

3.1   The Iterative Reduction Algorithm

Figure 4 illustrates our iterative reduction algorithm to calculate the overall runtime distribution. In each round, the algorithm searches for candidates for serial or parallel reduction and performs the respective reductions, as long as possible. If no further serial or parallel reduction is possible, the algorithm identifies a so-called "complex cluster", which is a generally structured subgraph whose runtime distribution will be analysed with the help of pMC. Since the vertices of the given SGM are associated with generally distributed execution times, which cannot be fed immediately into probabilistic model checkers such as PRISM, we need a suitable fitting method to approximate general distributions by exponential phases. This is explained in Sect. 3.3. There remains the problem of how to identify an appropriate cluster? We solve this problem by introducing so-called "syncpoints", as elaborated on in the following subsection.

3.2   An Efficient Algorithm Using Syncpoints

When performing the stepwise reduction, our algorithm needs to be able to identify appropriate starting points and end points of clusters to be reduced

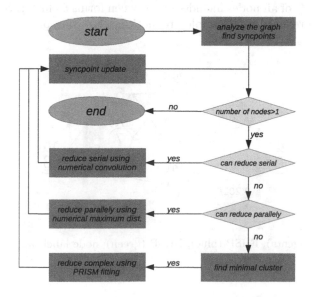

Fig. 4. Flow diagram of the iterative reduction algorithm

(either serially or parallelly or by complex reduction). This search can be directed by focusing on particular edge subsets which we call syncpoints:

Definition 5. *Given a SGM $G = (V, E, exec)$. A set $\mathcal{E} \subseteq E$ of edges is called a **full syncpoint** (FSP), if and only if for the node set $\mathcal{P}$ consisting of all the starting points of $\mathcal{E}$ and for the node set $\mathcal{S}$ consisting of all the end points of $\mathcal{E}$ the following three conditions hold:*

1. Each edge from $\mathcal{P}$ to $\mathcal{S}$ is in $\mathcal{E}$.
2. All nodes in $\mathcal{P}$ have the same set of successor nodes, namely $\mathcal{S}$.
3. All nodes in $\mathcal{S}$ have the same set of predecessor nodes, namely $\mathcal{P}$.

*If $|\mathcal{P}| = |\mathcal{S}| = 1$, we call $\mathcal{E}$ a 1-to-1-SP (11SP), a special subclass of FSPs. If only conditions (1) and (2) with $|\mathcal{P}| > 1$ resp. conditions (1) and (3) with $|\mathcal{S}| > 1$ hold, we call $\mathcal{E}$ a **backward** or **forward halfsyncpoint** (BHSP or FHSP). We denote $\mathcal{P}$ as the entrance side and $\mathcal{S}$ as the exit side of any syncpoint type.*

Figure 5 illustrates the concept of full syncpoints and halfsyncpoints. During SGM reduction, we make use of the syncpoint definition as follows:

– For each 11SP, we can combine its predecessor and its successor node to a single node by serial reduction, convolving the associated densities.
– If a FSP or a FHSP has got exit-sided node set M, which coincides with the entrance-sided node set of another FSP or BHSP, then there is a parallel reduction opportunity for all nodes of M, i.e. the associated densities can be combined by calculating their maximum.
– If all paths emerging from a FSP or FHSP lead to another FSP or BHSP and – vice versa – all paths reaching the latter come from the former one, then the set C of all nodes included in-between forms a cluster. Such a cluster can then be reduced by a complex reduction step using pMC.

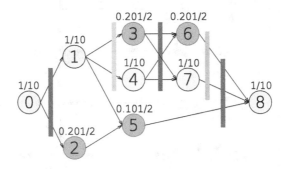

Fig. 5. FSPs (magenta), FHSP (blue), BHSP (green); node labels for later use (Color figure online)

Definition 6. *A FHSP resp. BHSP $\mathcal{E}$ is called a **subsyncpoint** (FSSP resp. BSSP), if and only if there is another set $\mathcal{E}'$ with $\mathcal{E} \subsetneq \mathcal{E}'$ and $\mathcal{E}'$ is either a FSP or a FHSP resp. BHSP.*

It is clear by definition, that the class of FHSPs decomposes in two subclasses, namely the FSSPs and the so-called **real FHSPs**, not beeing a subset of any superordinated syncpoint $\mathcal{E}'$; the same applies to BHSPs/BSSPs.

Each FSP with n entrance- resp. m exit-sided nodes implicitly defines $\binom{n}{k}$ BSSPs of cardinality k at its entrance side, resp. $\binom{m}{l}$ FSSPs of cardinality l at its exit side, with k, l the number of nodes attached at the respective side. Thus a FSP covers in sum $2^n - n - 2$ BSSPs and $2^m - m - 2$ FSSPs $(n, m > 1)$. From a hierarchical point of view, each FSP or HSP with cardinality k $(k > 2)$ immediately covers k SSPs of cardinality $k - 1$ (at the entrance resp. exit side).

The idea now is: The set of all syncpoints/halfsyncpoints in the given SGM induces a new, in general less complex DAG (regarding a kind of precedence to be defined), such that all described reduction steps can be performed on that structure, with a potentially tremendous reduction of calculation effort. In order to be able to identify and manage this, we need some more definitions:

Definition 7. *A set of edges $\mathcal{E} \subseteq E$ is called **Z-connected**, if and only if for each two edges $a, b \in \mathcal{E}$ there exist $k \geq 0$ and edges $e_1, e_2, \ldots, e_k \in \mathcal{E}$ such that for $(a, e_1, e_2, \ldots, e_k, b)$ each two neighboring edges have either the same start node or the same end node. $\mathcal{E}$ is called **max-Z-connected** if there is no other Z-connected set $\mathcal{E}' \subseteq E$ with $\mathcal{E} \subsetneq \mathcal{E}'$.*

Definition 8. *Reusing the notation of Definition 5, a set of edges $\mathcal{E} \subseteq E$ is called a **top-level manager(TLM)**, if and only if*

1. *$\mathcal{E}$ is max-Z-connected.*
2. *$\mathcal{E}$ contains at least one (full or half) syncpoint.*

Examples of TLMs are shown in Fig. 6. Remarks:

- Condition 2 of Definition 8 is for operational reasons, since TLMs without any contained syncpoint would not be useful.
- Since the SGM is acyclic, each FSP is also a TLM. In Fig. 5, for instance: $FSP(567 <> 8)$ or $FSP(0 <> 12)$.
- In Fig. 5, $FHSP(1 < 34)$ is neither a FSP nor a SSP, hence a real FHSP.

The next definition is needed, since it is possible to construct an acylic DAG without redundant edges where two max-Z-connected sets can be traversed in any order.

Definition 9. *Using the notation of Definition 5, we call a SGM **max-Z-acyclic** if*

1. *$\forall \mathcal{E} \subseteq E$ max-Z-connected and $\forall p = (e_1, \ldots, e_k) \in paths(E)$ holds: $|\mathcal{E} \cap p| \leq 1$.*
2. *$\forall \mathcal{E}, \mathcal{F} \subseteq E$ max-Z-connected, $\forall p_1, p_2 \in paths(E)$: If both p_1 and p_2 intersect with both $\mathcal{E}$ and $\mathcal{F}$, then they pass through $\mathcal{E}$ and $\mathcal{F}$ in the same order.*

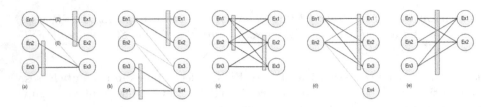

Fig. 6. TLM examples: only (e) is a FSP; (a)...(d) contain HSPs (green). (Color figure online)

Definition 10. *A TLM $\mathcal{E} \subseteq E$ is called a **predecessor** of another TLM $\mathcal{F} \subseteq E$ with $\mathcal{E} \neq \mathcal{F}$ if and only if there are edges $a \in \mathcal{E}, b \in \mathcal{F}$ and a directed path $(a, e_1, \ldots, e_k, b) \in paths(E)$. In this case $\mathcal{F}$ is called a **successor** of $\mathcal{E}$.*

Theorem 2. *For each max-Z-acyclic SGM there is a set of TLMs, fulfilling:*

1. *The TLMs are mutually disjoint and the set of TLMs is unique.*
2. *Definition 10 induces a well defined DAG on the TLMs, the **TLM-DAG**.*
3. *Each FSP resp. real HSP of the given SGM is covered by exactly one TLM.*

Proof (sketch): In a first step we use the max-Z-connected property of the TLMs to show that they are mutually disjoint and there is only one unique representation of them (independent of the discovering algorithm). Using the properties of Definition 9, we can show the precedence relationship on TLMs to be well defined. As remarked, each FSP is automatically a TLM; on the other hand each real HSP $\mathcal{E}$ is Z-connected and thus can be expanded to a max-Z-connected $\mathcal{E}'$, a TLM.

The following pseudocode sketches a recursive algorithm for identifying the TLMs and building the TLM-DAG:

```
# top-level call:
recFindTLMDAG(<SGM source node>, <emptyMap>, null)

# recursion method definition:
# @param curRecNode the current recursion subject node
# @param visitedNodeMap map of visited nodes -> already found following TLMs
# @param latestTLMOnStack the latest discovered TLM on recursion stack
# @returns the TLMGraph
TLMGraph recFindTLMDAG(curRecNode,visitedNodeMap,latestTLMOnStack)
  if (visitedNodeMap.contains(curRecNode))  # curRecNode already visited?
    if (visitedNodeMap.get(curRecNode).size()>0) # already found follow. TLMs?
      # update the result by adding edge(s):
      #   connect latest TLM on stack to already found TLMs beneath curRecNode
    else:
      # run recursion call for all successors of curRecNode
  else:
    # determine all Z-connected"children" (S) and "brothers" (P)
    #   by alternating fw/bw expansion, until there are no more new nodes.
    # group S and P by mutual dependencies in both directions:
    #   a) group P by common successors
    #   b) group S by common predecessors
    if (zBrotherGroupMap.size()==1 & zChildrenGroupMap.size()==1) # it's a FSP?
      # create new FSP and add it and the new connecting edge
      # (from latestTLMOnStack) to the result
      # remember all brothers (P) as visited, mapping to the found FSP
      # run recursion call for all children (S)
```

```
      else:  # it is not a FSP!
         # if there are HSP(s) between P and S, create a TLM to cover them:
         # each a)-group with at least 2 p, sharing same common successors
         # each b)-group with at least 2 s, sharing same common predecessors
         # induces such a HSP, to be kept by the TLM.
         # remember brothers (P) as visited: wrt. without a found following TLM
         # run recursion call for all children (S)
endMethod
```

Having constructed the TLM-DAG (including all FSPs), the next question is the integration of the halfsyncpoints, especially the subsyncpoints. At this so-called **micro level**, each TLM has to manage three issues:

(2) its subordinated HSPs, ideally in a hierarchical manner,
(2a) the relation between the entrance-/exit-sided nodes and the affecting HSPs,
(2b) the reachability (over paths) between its exit sided nodes and the entrance-sided nodes of the succeeding TLMs.

A solution for this is illustrated in Fig. 7, where the development of the micro linkage is shown based on the two rightmost examples of Fig. 6, assuming that TLM (d) is an immediate predecessor of TLM (e). Figure 8 illustrates the step-wise reduction of the example from Fig. 5 by relying on the corresponding sync-point TLM-DAG and its underlying micro structure. The transitions within the figure are reached by (P)arallel, (S)erial and (C)omplex reduction steps.

3.3 Numerical Reduction and Fitting of General Distributions

Whenever a serial or parallel reduction step is performed, our algorithm works on numerical representations of the operand densities. This avoids having to deal with complicated mathematical expressions (they arise very quickly after a couple of such operations have been performed) which could no longer be handled by formula manipulation packages. The precision of the numerical representation increases with increasing number of interpolation points, whereas the calculation effort for these numerical operations grows quadratically with the number of interpolation points. A similar numerical implementation of series-parallel reductions had been described in [6].

For each complex reduction, i.e. before a complex cluster can be analysed by means of pMC, an appropriate PRISM-Model needs to be generated. For

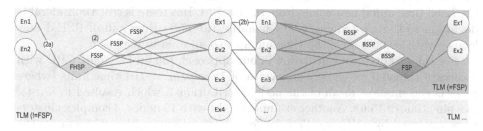

Fig. 7. Micro linkage of Fig. 6(d, e): each TLM manages its HSPs (2), their usage of entrance-/exit-sided nodes (2a) and the mutual reachability by these nodes (2b).

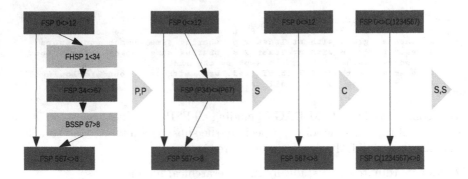

Fig. 8. Reduction steps along the syncpoint DAG: full SPs (orange) and half SPs (blue) (Color figure online)

this purpose, each single activity's runtime distribution is fitted by a phase-type distribution. At the moment, we advocate a fitting by two convolved Erlang distributions $Erl(\lambda_1, n)$ and $Erl(\lambda_2, m)$. Their parameters can be chosen in such a way that the first two moments match exactly to the original distribution and the error at the third moment is minimized. This fitting provides good results, as long as the coefficient of variation of the original distribution is at most 1, i.e. as long as $\sigma^2/\mu^2 \leq 1$. As a concluding remark on the issue of fitting, note that our algorithm presented in Sect. 3.1 is independent of the particular fitting method used, i.e. other fitting approaches (e.g. [1]) can easily be incorporated.

4   Complex Examples

Let us look once again at the example in Fig. 5 and assume that an Erlang distribution is chosen for all activity runtimes (find λ and n values printed directly on the nodes). Figure 9 displays the densities of the critical path (0–1–4–7–8), of the exact distribution (obtained by PRISM) and of the result by our algorithm. While our algorithm works quite accurately (submitting a real 89.57% confidence level as "90%"), PERT presents an actual level of about 50% as "90%"!

But what about CCPM? Table 2 considers the possible paths, assigned with their relating buffer. It is interesting that the more reliable quantile with 84.68% (at 72.53 time units) now comes from the C&PM. But there is still a considerable deviation to the wanted exact 90%-quantile (at 77.04 time units). RSEM now leads – with 79.96% (at 69.65 time units) – to an even worse result.

For the prototypical calculation in our approach (on an I7-2600K CPU with 4 cores/8 threads) we chose a stepwidth of $1\%_0 \cdot \mu_{min} = 0.01$ time units (where μ_{min} is the smallest mean of the node distributions), which resulted in a total machine time of 15.6 s. Another example (SGM with 19 nodes, 3 complex clusters to be analyzed by pMC, $\mu \in [5 \ldots 10]$, $\sigma = \lfloor \frac{\mu}{2} \rfloor$) finished within 35.5 s under the same conditions. Increasing the number of nodes (each one now having a normal N(10,4) distribution, with a $1\% \cdot \mu_{min}$ stepwidth) showed, as expected, that

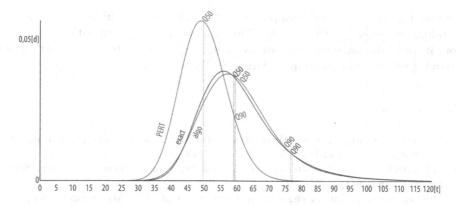

Fig. 9. Densities for the SGM from Fig. 5: PERT (red) vs. exact (blue) and our algorithm (green) (Color figure online)

Table 2. Pathwise CCPM end times (+buffer) and corresponding quantiles

path	0258	0158	01478	01468	01378	01368	overall	quantile
C&PM	44.31 + 22.16	45.63 + 22.81	**48.35+ 24.18**	47.03 + 23.52	47.03 + 23.52	45.71 + 22.86	**72.53**	**84.68%**
RSEM	**44.31+ 25.34**	45.63 + 23.26	48.35 + 10.15	47.03+ 14.27	47.03 + 14.27	45.71 + 17.44	**69.65**	**79.96%**

the runtime grows strongly with the complexity of the contained clusters: For instance, a randomly generated SGM with 137 nodes (73 caught in clusters, max. cluster size: 5) was finished within 182 s. Another example with 133 nodes (77 caught in clusters, max. cluster size: 16) required 2,269 s to complete.

5   Summary and Future Work

The presented method offers a remarkable chance to improve the accuracy of established project planning methods by the combined use of exact calculations and heuristic approximations, using pMC. Estimates of the overall runtime distribution – even for complex graph structures – can be calculated more accurately and also – compared to the customary simulation-based approaches – with feasable computation effort. For instance, one of the leading management methods, CCPM, can be enhanced in several ways: For a given project schedule, one obtains a handy calculus of the time-to-finish distribution. Furthermore – reversely – for a given project finalization confidence level (e.g. a 90% quantile) we can determine all buffer sizes (on the critical path and all side paths) in an analytical manner.

This holds even if the scheduling complexity of the CCPM is increased by an additional calculus regarding the resource- or skill-dependencies (an extension already envisaged in our work plan). In that context, we will also use the presented method to solve resource conflicts of the "bad multitasking" type [4] by taking or hedging a founded decision for a particular prioritization. In addition, it might become necessary to use bounding methods (e.g. by inserting or removing edges) in order to make larger SGMs tractable.

References

1. Bobbio, A., et al.: Matching three moments with minimal acyclic phase type distrbutions. Stoch. Models **21**(2–3), 303–326 (2005)
2. O'Cinneide, C.A.: Characterization of phase-type distributions. Commun. Stat. Stoch. Models **6**(1), 1–57 (1990)
3. Dilmaghani, F.: Critical chain project management (CCPM) at Bosch Security Systems (CCTV) Eindhoven. Masters Thesis, University of Twente (2008)
4. Goldratt, E.M.: Critical Chain. The North River Press Publishing Corporation, Great Barrington (1997)
5. Kapici, S.: A stochastic risk model for complex projects (Dissertation) Otto-von-Guericke-Universität, Magdeburg (2005)
6. Klar, R., et al.: Messung und Modellierung Paralleler und Verteilter Rechensysteme. B.G. Teubner, Stuttgart (1995)
7. Kulkarni, V.G., Adlakha, V.G.: Markov and Markov-regenerative PERT networks. Oper. Res. **34**(5), 769–781 (1986)
8. Kwiatkowska, M., Norman, G., Parker, D.: Stochastic model checking. In: Bernardo, M., Hillston, J. (eds.) SFM 2007. LNCS, vol. 4486, pp. 220–270. Springer, Heidelberg (2007). doi:10.1007/978-3-540-72522-0_6
9. Maplesoft: Maple, a symbolic and numeric computing environment (2017). https://www.maplesoft.com/products/maple/
10. Melchiors, P., Kolisch, R.: Scheduling of multiple R&D projects in a dynamic and stochastic environment. In: Fleischmann, B., et al. (eds.) Operations Research Proceedings 2008, pp. 135–140. Springer, Heidelberg (2007). doi:10.1007/978-3-642-00142-0_22
11. Melchiors, P.: Dynamic and Stochastic Multi-Project Planning. LNEMS, vol. 673. Springer, Cham (2015). doi:10.1007/978-3-319-04540-5
12. The PRISM Model-Checker website. http://www.prismmodelchecker.org
13. Shtub, A., et al.: Project Management: Processes, Methodologies and Economics, 2nd edn. Pearson Education Limited, Essex (2014)
14. Wolfram Mathematica. https://www.wolfram.com/mathematica/

Cyber-Physical Systems

Learning-Based Testing of Cyber-Physical Systems-of-Systems: A Platooning Study

Karl Meinke[(✉)]

School of Computer Science and Communication,
KTH Royal Institute of Technology, 100 44 Stockholm, Sweden
karlm@kth.se

Abstract. Learning-based testing (LBT) is a paradigm for fully automated requirements testing that combines machine learning with model-checking techniques. LBT has been shown to be effective for unit and integration testing of safety critical components in cyber-physical systems, e.g. automotive ECU software.

We consider the challenges faced, and some initial results obtained in an effort to scale up LBT to testing co-operative open cyber-physical systems-of-systems (CO-CPS). For this we focus on a case study of testing safety and performance properties of multi-vehicle platoons.

Keywords: Cyber-physical system · System-of-systems · Platooning · Model-based testing · Learning-based testing · Machine learning · Requirements testing

1 Introduction

A *cooperating cyber-physical system-of-systems* can be characterised by the use of wireless communication, multiple stakeholders, dynamic system definitions, and unpredictable operating environments. Such systems-of-systems have been termed *Cooperative Open Cyber-Physical Systems* (CO-CPS) [33]. It is assumed that no single stakeholder has overall system responsibility, and that cooperation relies on wireless communication to perform safety-relevant functions.

CO-CPS are emerging around the world, due to rapid progress in telecommunications, robotics and AI. Many examples can be found in Cooperative Intelligent Transport Systems (C-ITS) and intelligent manufacturing. However, they represent a great challenge to the software quality assurance (SQA) community. Not least, the cyber-physical character of CO-CPS means that the impact of safety and security incidents (malicious or unintended) is potentially very high. However, if we survey the range of current technologies available for SQA, we can find significant limitations in many current approaches to quality assurance of CO-CPS.

On the one hand, the dynamic and heterogeneous nature of CO-CPS makes a full static analysis technically difficult. The sheer scale of many proposed CO-CPS suggests that a full system-of-systems analysis would even be technically

© Springer International Publishing AG 2017
P. Reinecke and A. Di Marco (Eds.): EPEW 2017, LNCS 10497, pp. 135–151, 2017.
DOI: 10.1007/978-3-319-66583-2_9

infeasible. Furthermore, it is unclear (for commercial reasons) whether all source code in a CO-CPS would ever be made available for this. Static analysis of the individual components by their vendors might be technically feasible. However, it is difficult to see how such low-level component analysis could take into consideration unpredictable environment factors and high-level emergent phenomena (such as physical collisions). For this reason, software testing, laboratory simulations and field tests are the de-facto SQA standard used in industry today. Here the problem is that software testing traditionally focuses on unit, integration and system level testing. Simulation and field testing can be reliable and decisive at the level of systems-of-systems, but tend to be slow and unsystematic in their coverage. There is thus a great need to perform systematic and fully automated requirements testing on CO-CPS.

The scalability problem for quality assurance of CO-CPS might be made more tractable by taking a *model-based approach*, using judicious abstraction to suppress irrelevant technical detail. However, one is still faced with the fact that not all software vendors will take a model-driven approach, let alone exchange their models, to protect intellectual property (IP). Therefore, in the worst case one would be left to perform a model based analysis where some component models are known, but others are missing, inconsistent with code, or out of date.

Against this background situation for CO-CPS, within the EU ECSEL project SafeCOP[1], we are evaluating the potential of a technology known as *learning-based testing* (LBT) [23,24]. LBT is a paradigm which combines techniques from *model-driven development* (e.g. model-based testing, model checking of safety requirements etc.) with *machine learning*. The basic idea is to use machine learning to *reverse engineer* a behavioral system model from runtime observations of a system under test (SUT). Since LBT is a black-box technique, it is code and platform independent, potentially scalable, and need not infringe upon component IP rights. The runtime SUT observations can be made either by laboratory simulation (e.g. software-in-the-loop SIL, hardware-in-the-loop HIL) or field testing. The learned model can then be used to analyse safety properties [11], and even security properties [14], by using appropriate tools such as model checkers. Potential system anomalies discovered during model analysis are confirmed by executing the corresponding test cases on the SUT.

We present here some initial results of applying LBT to a case study of testing co-operative vehicle platoons [4]. One reason for choosing this case study is because the problem size can be scaled up uniformly by adding more vehicles. This allows us to measure the influence of different factors on the scalability of LBT technology.

The case study of platooning presented here is a first attempt to address two important questions about state-of-the-art LBT technology:

(1) how well does recently developed multi-core based LBT technology scale up to testing complex CO-CPS scenarios;
(2) how do problem size and other factors affect scalability?

---

[1] See www.safecop.eu.

The organisation of this paper is as follows. In Sect. 2 we review fundamental concepts and the state-of-the-art in learning-based testing. In Sect. 3, we consider the architecture and functionality of platooning as a CO-CPS. In Sect. 4 we present our case study of LBT applied to a platoon model. In Sect. 5 we survey related work in the literature. Finally in Sect. 6, we draw conclusions from our initial results, and comment on future research directions.

2   Learning-Based Testing

In this section, we review some fundamental principles of learning-based testing as these have been implemented in our research tool LBTest. The earliest version of this tool (LBTest 1.x) has been described in [26]. Therefore we will focus on the latest tool architecture LBTest 3.x, presented in Fig. 1. In Sect. 2.1 we use this architecture to explain the basic principles of LBT. Then, in Sects. 2.2 and 2.3, we show how concurrent aspects of this architecture contribute towards solving tool scalability issues[2].

2.1   Principles of LBT

LBTest uses active automaton learning aka. *regular inference* (see e.g. [13]) to generate queries about a black-box system, which can be used to infer a behavioral model in polynomial time [2].

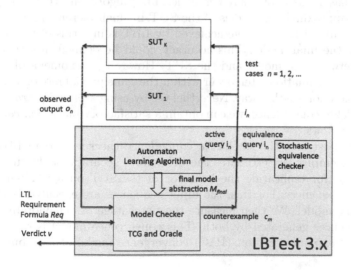

Fig. 1. LBTest 3.x concurrent learning architecture

---

[2] This architecture has been developed within the VINNOVA FFI project VIRTUES, http://www.csc.kth.se/~karlm/virtues/.

For requirements testing, partial models of the SUT can be subjected to model checking against a temporal logic requirement specification, even before the learning process is complete. In LBTest, propositional linear temporal logic[3] (PLTL) is used as a requirements modeling language. This particular logic has the advantage that test cases can easily be extracted from counterexamples generated by a model checker. LBTest makes use of a loosely integrated symbolic checker NuSMV [7]. We are also developing a more tightly integrated explicit state model checker for efficiency reasons. These two processes of learning and model checking may be interleaved, an idea first suggested in [27]. Then they incrementally build up a sequence $M_1, M_2, \ldots$ of models of the SUT, while generating and executing requirements test cases on each model M_i.

To separate true counter-examples (SUT errors) from false counter-examples (artifacts of an incomplete model) it is necessary to validate each counter-example derived from model checking. For this we can: (i) extract a test case representing the counter-example[4], (ii) execute it on the SUT, (iii) apply an equality test that compares the observed SUT behavior with the predicted bad behavior from the model, and (iv) automatically generate the test verdict (**pass**, **fail**) from step (iii). The soundness of this process relies on the soundness of the underlying model checker, and the soundness of equality testing.

The completeness of LBT relies on the completeness of the underlying model checker, as well as convergence results about the learning algorithms which are used (see [13]). However, within practical case studies of large complex systems it may not be possible for learning to be completed in any reasonable time frame (see e.g. [11]). This problem is significant for CO-CPS. Therefore, development of LBTest has focused on incremental learning algorithms that can generate incomplete approximating models of the SUT in small increments.

One measure of the coverage achieved by LBT is in terms of the behavioral accuracy of the final model. This accuracy could be defined in terms of trace inclusion between the model and the SUT. However, phenomena of both over and under approximation often occur within the same partial model, i.e. no strict trace inclusion holds either way. Nevertheless, by using a *probably exactly correct* (PEC) model of convergence, we can obtain a satisfactory black-box *convergence measure* as follows.

Figure 1 illustrates the stochastic equivalence checker used in LBTest 3.x. This checker empirically estimates the behavioral accuracy of the final learned model M_{final} for replicating the behavior of the SUT on a randomly chosen set of input sequences. For this, the input sequences are executed both on the SUT and the model. We then measure the percentage of behaviorally identical output sequences generated by both. This convergence model is related to the *probably approximately correct* (PAC) convergence model of [30], but for PEC

---

[3] Recall that propositional LTL extends basic propositional logic with the temporal modalities $G(\phi)$ (always ϕ), $F(\phi)$ (sometime ϕ) and $X(\phi)$ (next ϕ). Other derived operators and past operators may also be included. See e.g. [12] for details.

[4] Infinite counter-examples to LTL liveness formulas are truncated around the loop, and the weaker test verdict **warning** may be issued.

the probability of exact identity (not approximate equivalence) is estimated. PEC convergence aims at the needs of software safety analysis over the discrete data type partitions commonly employed in testing.

2.2 Towards Scalable LBT Architectures

From empirical studies such as [11, 20, 25] we have observed two important obstacles to scaling up LBT methods for large and complex SUTs. These are:

(i) *the tendency for learned model size to increase rapidly with SUT size*;

(ii) *the tendency for test latency (i.e. the time to execute a single test case) to increase with SUT size.*

   Even worse, these two problems compound one another, leading to long test session times and low final convergence measures. In benchmarking the architectural proposal of [27] we have also observed another significant problem:

(iii) *model checking each member M_i of a converging sequence of models $M_1, M_2, ...$ is highly inefficient, and does not seem to improve the rate of model convergence.*

We will consider each of these issues, and how it can be addressed, in turn.

(i) Model Size. The size of a learned model is a function of the code complexity of the underlying SUT, as well as the number of parameters of the SUT which the learning algorithm tries to stimulate and observe.

   One factor influencing model size is the number of SUT input variables and the number of test values chosen for each input variable. These parameters bound the number of *exit transitions* from each model state. The number of exit transitions is further influenced by the combinatorial strategy used to generate composite input test vectors from the individual input variable values. A judicious combinatorial choice is necessary to control the otherwise exponential explosion in the number of transitions. In LBTest 3.x, n-wise testing [17] is available as a combinatorial strategy.

   Another factor influencing model size is the number of observed SUT output variables, and the number of output value partition classes for each output variable. These factors influence the number of *states* in a learned model, since more output variables and finer output partitions lead to more easily distinguished SUT states.

   So, a judicious choice of model accuracy, combinatorial test strategy and model abstraction can all be applied to improve the efficiency of learning and testing.

   Besides these test configuration parameters, the problem of large model sizes has also been ameliorated by new research into machine learning algorithms. Since Anglium's seminal algorithm [2], many new learning algorithms, that can learn a model with fewer and/or shorter queries, have been derived, e.g. [16].

(ii) Test Latency. Improvements in learning and model checking algorithms are scarcely able to overcome a distinctive feature of large complex SUTs which is the tendency towards long test latency or execution times. For CO-CPS, communication network delays also become significant. Test latency times can become a significant component of an LBT test session duration.

Test latency can be ameliorated by executing test cases concurrently. With this aim we have conducted research into parallelized learning algorithms on multi-core platforms. Already in [15] certain improvements in learning performance by parallelization have been reported. An important challenge is to systematically characterize such improvements in terms of problem size parameters. Our work contributes to this area by studying a *parameterized and uniformly scalable learning problem* namely platooning. As the size (i.e. number of vehicles) of a platoon of identical vehicles scales up, the problem parameters:

(i) total number of lines of code under test, and
(ii) total number of program registers determining the global state space,

both increase linearly. Thus it becomes meaningful to compare testing results for different platoon sizes (c.f. the similar curves in Fig. 4). Without such uniform properties, benchmark results across an ad-hoc collection of SUTs can be very difficult to interpret.

(iii) Model Checking Overheads. Incremental learning generates a convergent sequence of models $M_1, M_2, \dots$. However, each model M_i will contain a good many structural features (states and transitions) that persist in model M_{i+1}. It is beyond the capability of any model checker we know of to identify these persistent features and avoid checking them twice in both M_i and M_{i+1}. Therefore, a long model generation sequence will contain significant redundant model checking effort. Our empirical observations with LBTest 2.x and NuSMV have shown that this redundant checking can consume more than 50% of the overall test session time. Furthermore, as reported in [21], model checker generated queries have not been observed to accelerate the convergence of learning in any case study so far.[5]

While it might be possible to introduce a sophisticated delta-oriented approach to model checking, the simplest solution seems to be to defer model checking until after machine learning.

2.3   Concurrent Multi-core LBT

Figure 1 illustrates a new architecture for LBT that significantly departs from the proposals of [23,27]. Two new features are prominent, and both are intended to counter the scalability bottlenecks described in Sect. 2.2.

---

[5] It seems possible to theoretically explain this observation for certain types of formulas by considering their semantics. However, this is outside the scope of our present discussion.

Firstly, the new architecture supports parallel execution of multiple instantiations of the SUT on a multi-core platform. The aim is to mitigate long SUT test latency. At the start of a test session, LBTest clones K versions of the executable SUT, each within its own external OS process. The value of K is chosen as a function of the number of SUT input values to be tested. Once started, each SUT process persists throughout the learning phase, and acts as a server to answer certain kinds of queries about SUT behavior. Different load balancing schemes on these query servers are used according to the learning strategy.

Of course, concurrent execution is a rather obvious solution to test latency. The real technical challenge here is to devise efficient parallel learning algorithms that can allow multiple threads to efficiently and safely perform concurrent updates on a single shared automaton model. At the same time we need to optimise multi-core usage on the hardware level. For this we have investigated concurrent implementations of Kearn's algorithm [19]. For reasons of space, these rather complex concurrent algorithms will be described elsewhere.

The second new feature of LBTest 3.x is its support for deferred model checking, as described in Sect. 2.2, using an iteration bound to terminate learning. Only when learning is terminated do model checking and counter-example validation of the final model M_{final} begin in a second phase. This minimises the redundant model checking identified in Sect. 2.2.

3   Platooning as a CO-CPS

In this Section we review some general features of platooning that characterise it as a CO-CPS. Then we discuss the particular platooning model that was tested in Sect. 4.

3.1   General Principles of Platooning

Platooning technology is sometimes called an "electronic towbar" between road vehicles, and this phrase gives much insight into the idea.

A *platoon* consists of a sequence of road vehicles $V_1, ..., V_n$ which (by means of sensors, wireless V2V communication and control algorithms for longitudinal or distance control) are able to maintain a fixed distance x_r between one another and a relative velocity $v_r = 0$ under normal cruising conditions. (See Fig. 2, adapted from [5].) The *lead vehicle*, V_1, is under manual control by a qualified platoon leader who needs to have the necessary technical skills to control the platoon. The vehicles $V_2, ..., V_n$ are its *followers*, and may be autonomous or semi-autonomous, depending on the extent to which lateral control (i.e. steering) is automated.

A platoon may be heterogeneous, consisting of different models from different vendors carrying different payloads. It should be possible to add and remove vehicles dynamically during a journey, and there are many safety critical use cases, such as lane change, emergency braking etc.

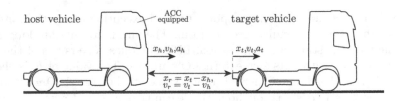

Fig. 2. Platoon vehicle pair: V_{i+1} (left) and V_i (right)

The interest in platooning technology, lies in the possibility to reduce fuel consumption and corresponding CO2 emissions, as well as to improve road usage and safety while reducing traffic congestion (see e.g. [29]). Platoons exploit the reduced aerodynamic drag that arises with short inter-vehicle distances. There is an important trade-off between fuel efficiency and safety in platoon design, since drag is reduced by shorter inter-vehicle spacing. System response times, component reliability, road hazards and the effects of safety critical uses cases such as emergency braking on the platoon and its environment all need to be evaluated during software design.

3.2 A Simple Platooning Model

For pragmatic reasons, our study of LBT scalability was restricted to software-in-the-loop (SIL) testing of a basic platoon simulator. The simulation is 1-dimensional, meaning that no steering model is used. The simulator is therefore only able to analyze certain use cases, such as straight-line cruising and emergency braking. Other use cases need a more complex simulation model, and this is the subject of ongoing research and industrial collaboration. However, our model includes many important physical characteristics such as maximum engine and brake torque, vehicle mass, aerodynamic drag etc. defined using a *point-mass* modeling approach. (See e.g. [34] for an introduction to vehicle modeling.)

The simulator consists of about 2000 Java LOC. However, to get a clearer impression of the underlying SUT complexity we provide here some details about its structure and function.

The block architecture of a single vehicle in the platoon simulator is illustrated in Fig. 3. This depicts a *brake-by-wire* BBW subsystem augmented with a *co-operative adaptive cruise controller* CACC. The latter is connected to an *odometry* unit ODOM (providing host vehicle position and velocity) and a *wireless communication* WCOM unit (relaying host and target positions and velocities). Odometry is based upon host velocity measurements[6]. The WCOM unit simulates a 2 ms inter-vehicle wireless message delay, without any transmission error model.

The CACC controller is a crucial component that provides longitudinal control of each follower vehicle. It dynamically issues accelerator and brake torque

---

[6] In practise, GPS localisation would be relied upon for greater accuracy.

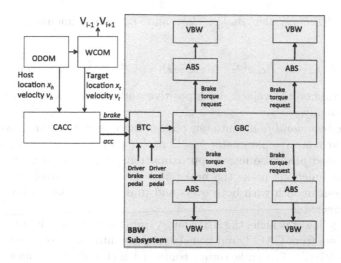

Fig. 3. Software architecture for platoon vehicle V_i

requests to maintain the position of the host vehicle within maximum and minimum distances from the target vehicle in front. A wide variety of CACC algorithms have been proposed in the literature. The controller tested here is a simple PD control algorithm with adaptive parameters, taken from [5]. For a general introduction to PID control theory one may consult e.g. [10]. The function of any PID controller in the context of an ACC problem is to maintain the relative position of the host vehicle V_{i+1} within the boundaries $x_{r,d,max}$ and $x_{r,d,min}$ (metres) from the target V_i, where

$$x_{r,d,max} = t_{hw} \cdot v_h + x_{r,0}, \qquad x_{r,d,min} = (t_{hw} - t_{hw,\delta}) \cdot v_h + x_{r,0}.$$

Here t_{hw} (seconds) is the *time headway* between V_{i+1} and V_i, and $t_{hw,\delta}$ causes a small difference in headway. The parameter $x_{r,0} > 0$ (m), maintains a safe relative inter-vehicle distance at $v_h = 0$ (m/s), to support so called *stop-and-go* functionality. The host position is maintained by two PD equations:

$$acc = K_{ACC}(k_{x_r} \cdot (x_r - x_{r,d,max}) + k_{v_r} \cdot v_r),$$

$$brake = K_{ACC}(k_{x_r} \cdot (x_r - x_{r,d,min}) + k_{v_r} \cdot v_r),$$

governing requested accelerator and brake torque. In the above formulas: (i) K_{ACC} (dimensionless) is a constant overall gain parameter. (ii) $x_r = x_t - x_h$ (metres) and $v_r = v_t - v_h$ (metres/second) are the relative distance and velocity to the target vehicle (c.f. Fig. 2). (iii) k_{x_r} is the P action: this gain is tuned to regulate the distance error to zero ($x_r - x_{r,d,max} = 0$ for *acc* and $x_r - x_{r,d,min} = 0$ for *brake*). (iv) k_{v_r} is the D action and the regulated error is v_r. (v) Since *acc* is smaller than *brake* (due to a different desired distance), it takes some time before the brakes are activated after the accelerator is released.

In this PD controller design, k_{x_r} and k_{v_r} are dimensionless *adaptive parameters*:

$$k_{x_r} = k_{x_r,1}(v_h) \cdot k_{x_r,2}(x_r - x_{r,d,max}), \qquad k_{v_r} = k_{v_r}(x_r - x_{r,d,max}).$$

All forces acting on the vehicle, both positive and negative, are resolved at each wheel individually.

To inject *behavioral faults* into our platooning model for testing, we replaced the non-linear adaptive parameter functions $k_{x_r,1}, k_{x_r,2}, k_{v_r} : \mathbb{R} \to \mathbb{R}$ of [5] with highly simplified piecewise linear approximations. These linear approximations to non-linear functions make the brake and accelerator control responses, *acc* and *brake*, less smooth with both over- and under-compensation for change, as we show in Sect. 4.2.

For each follower vehicle, the BBW subsystem takes the accelerator and brake torque requests from CACC, and translates these into forces on the four vehicle body wheels VBW[7]. The brake torque controller BTC calculates the global brake torque request (in Newton metres)

$$torqueRequest = (brake/100) \cdot maxBrakeTorque$$

and the global brake controller GBC distributes this brake request to each anti-locking brake system ABS_i, which controls wheel VBW_i.

The fundamental simulation cycle corresponds to 1 ms of real-world time, while the various architectural components have execution cycle times varying between 2 and 20 ms. Normally, vehicle software components would communicate periodically (but not necessarily deterministically) using the vehicle's CAN bus network, while the vehicles themselves communicate asynchronously. However, it is common industrial practise to perform SIL testing using a simplified *synchronous composition* of components to ensure reproducibility of test results. So our platoon simulator is also based on a synchronous composition of all architectural components, as well as the platoon vehicles themselves.

4   Test Experiment Design and Results

In this section, we first describe our testing experiment conducted on the platooning simulator described in Sect. 3, using the LBT tool architecture described in Sect. 2.3. We then describe the test results obtained, and interpret these from the perspective of LBT scalability.

4.1   Test Experiment Design

To test the primary use case of *high-speed cruising* for a platoon configuration of n vehicles, we focused on emulating the lead driver behavior, since in our

---

[7] For the lead vehicle, CACC is disabled and accelerator and brake pedal values are used by BBW instead. See Fig. 3.

simulator all follower vehicles autonomously adapt to this. Thus, each test case tc for an n-vehicle platoon consisted of a sequence $tc = (r_1, r_2, ..., r_\lambda)$ of lead driver accelerator and brake torque requests r_j. The continuous input spaces for each of these two input variables (accelerator and brake pedal angles) were sampled at 10% intervals, yielding $K = 21$ symbolic input values $0, a_1, ..., a_{10}, b_1, ..., b_{10}$ ranging from 0% to 100% pedal depression[8]. No assumptions were made about lead driver behavior, so both excessive and sporadic acceleration and braking could occur. The time headway t_{hw} between each successive pair of vehicles was nominally set to 2.0 s. A time headway of this size is normally quite safe for commercial CACC algorithms (see e.g. [5]).

For each test case $tc = (r_1, r_2, ..., r_\lambda)$, the length λ and torque requests r_j were chosen dynamically both by the learning algorithm and the equivalence checker. In the experiments of Sect. 4.2, λ typically took an average value around 12. The test case tc was then submitted to one of $K = 21$ SUT server processes S_p executing an n vehicle platoon simulator instance. The communication wrapper around S_p loaded and executed the request sequence $(r_1, r_2, ..., r_\lambda)$ sequentially. Each torque request value r_j was maintained constantly for a nominal 5 s (5000 simulation cycles). Thus the length of the simulation corresponding to tc was 5λ virtual seconds. The values chosen for λ were sufficient to reach high cruising speeds, in excess of 110 km/h.

Maintaining the torque request over a fixed number of seconds is a *temporal abstraction* technique necessary to achieve a balance between long simulation times and small final model size. This abstraction can be adjusted in the simulator. It also has the advantage that we can easily calculate the cumulative virtual simulation time for an entire test session.

The principle SUT output recorded for the test case tc was the time sequence of inter-vehicle gaps $x_{r,0}^i, ..., x_{r,\lambda}^i$, for each vehicle $i = 1, ... n-1$. Here, the time sequence term $x_{r,t}^i$, for $0 \leq t \leq \lambda$, represents the gap between the host-target pair, V_i and V_{i+1} measured at the end[9] of $5t$ virtual seconds (i.e. $5000t$ simulation cycles). The continuous values of each distance observation $x_{r,t}^i$ were partitioned within the communication wrapper into three discrete equivalence classes:

tooClose, tooFar, good,

based on the (host velocity dependent) distance boundaries $x_{r,d,min}^i$ and $x_{r,d,max}^i$. Thus the symbolic output good for $x_{r,t}^i$ represented the output partition class $x_{r,d,min}^i \leq x_{r,t}^i \leq x_{r,d,max}^i$.

To gain further insight into the physical state space covered by testing we also observed the lead vehicle velocity values $v_0^1, ..., v_\lambda^1$ and acceleration values $a_0^1, ..., a_\lambda^1$ at the same observation times. These continuous valued observations were partitioned into 1 km/h and 1 km/h^2 equivalence classes.

---

[8] Thus a_{10} represents 100% accelerator depression, a_9 represents 90% depression, etc. Simultaneous depression of both pedals is handled as a brake request by the BBW component.

[9] It is also possible to use SUT observations between the output cycles by thresholding. This can yield greater accuracy, but this approach was not taken here.

With regard to system-of-systems requirements, the most fundamental requirement is that all n platoon vehicles should always maintain a safe but fuel efficient distance between each other. This test requirement could be represented in PLTL for an $n + 1$-vehicle platoon (where $n \geq 1$) by the safety formula:

$$G(\, \texttt{Distance}_1 = good \,\&\, \texttt{Distance}_2 = good \,\&\, \ldots \,\&\, \texttt{Distance}_n = good \,). \quad (*)$$

Here $\texttt{Distance}_i$ represents the discretized gap between vehicles V_i and V_{i+1} corresponding to measurements $x^i_{r,t}$.

One experimental goal was to try to observe the injected errors in the CACC component, (described in Sect. 3.2) as violations of the test requirement (*). The other goal was to characterise the scalability of the tool.

4.2 Test Experiment Results

The test experiment described in Sect. 4.1 was conducted for platoon sizes $n = 2, \ldots, 6$ to investigate the scalability of the testing tool. To uniformise the results, each platoon vehicle in each configuration had identical physical parameters[10]. We measured the final model size for different platoon sizes and different test session durations. While test session duration is a platform dependent measurement[11], it was felt that this value gave good insight into tool usability and potential future improvements.

Figure 4 shows the growth of model size over time for platoon sizes $n = 2, \ldots, 6$ using concurrent learning. To analyse the benefit of concurrency, Fig. 4 also shows model growth for $n = 3$ under sequential learning. Note that the y-axis is in thousands of states (Kstates). The graph shows the effects of increasing test latency as the platoon size increases. The largest inferred model (for $n = 6$) had over 64,600 states and 1.35 million transitions achieved after 20 h and 25 min of learning. During this time, 1.5 million test cases tc were executed, with an average test case length of $\lambda = 10.6$. Since each step in tc corresponds to 5 virtual seconds, the total virtual testing time was over 22,000 h.

Notable in Fig. 4 is the gradual slowdown in rates of model growth over time. However, there is no sharp fall in tool performance. Furthermore, the vertical intervals between the curves are very similar, both for increasing n and t. These two characteristics seem to suggest good scalability properties for our approach as a function of the problem size n.

With regard to requirements errors, NuSMV developed a segmentation fault already with the smallest of our models for $n = 2$ (8826 states, 185 K transitions). However, using our explicit state model checker on the largest model for $n = 6$ (64,671 states, 1.35 million transitions), the error $\texttt{tooFar}$ was found to occur in 50,076 states (77% of all states), while the error $\texttt{tooClose}$ was found in just 101

---

[10] Non-homogeneous platoons could also be tested using our approach.

[11] The actual platform used was a 4-core MacBook Pro, Mid 2014, running Yosemite OS-X 10.10.5 with 2.8 GHz Intel Core i7, 16 GB 1600 MHz DDR3 and 1 TB static disk flash storage.

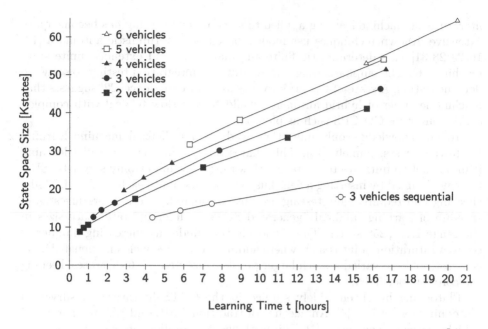

Fig. 4. Rates of model growth (state space size) over time for different platoon sizes.

states (0.0015% of all states) after 32.4 s of model checking. All errors proved to be valid SUT errors when corresponding test cases were executed on the SUT. The error `tooClose` was found only at low velocities, mainly at $v^1 = 0$, which seems to confirm the thesis of [5] that stop-and-go functionality is rather difficult to implement correctly. For the smallest model of $n = 2$ (8826 states), the error `tooFar` could also be found after 19 ms of model checking, but not error `tooClose`.

Through runtime monitoring, we estimated long term multi-core usage to range between 85%–95% over the problem size range $n = 2, \ldots, 6$, with approximately 10% fluctuations short term[12]. At peak core usage, CPU idle time was less than 1%, implying that further cores would have been of benefit.

For the experiments described in Fig. 4, the platoon models reached maximum convergence values of 9.4%, 9.4%, 8.8%, 7.1% and 6.0% for $n = 2, \ldots, 6$ respectively.

5   Related Work

The application of machine learning to testing has a somewhat long history, beginning with [32]. The architecture used in LBTest 2.x first appeared in [27] and was independently proposed in [24]. However, scalability and the effect of model checking on convergence, were not originally considered. Recently, the

---

[12] Based on 1 s sampling.

literature on machine learning applied to software engineering has become quite extensive. Known techniques use models based on deterministic automata [14, 16,23,28,31], non-deterministic finite automata [21], and extended finite state machines [6]. The emphasis ranges from unit and integration testing to software documentation. A state-of-the-art survey is [3]. Our experience [22] suggests that machine learning of hybrid automata would be too slow to deal with complex continuous state CO-CPS such as platoons.

To our knowledge, only one other study of parallelized machine learning for testing exists, namely [15]. This shares our premise that parallel learning is important to mitigate test latency. However, it evaluates only synthetic SUT latency obtained by inserting a 5 ms busy waiting loop into each SUT call. Model checking and requirements testing are not considered. The authors investigate speedup of learning randomly generated SUTs of different state space sizes in the range $1, \ldots, 256$ states. They conclude that under an increasing number of cores, a saturation point is met, where adding more cores yields no benefit[13]. By contrast, we have varied a much larger problem size $8K, \ldots, 64K$ states, keeping the core number fixed.

Platooning has been widely studied in the C-ITS literature. A survey of platooning research is [4]. An account of traditional SIL and HIL testing of a 3 vehicle platooning system is [1]. This work has very similar safety concerns to our own. Examples of static analysis applied to platooning are [8,9,18] where it is shown that verifying vehicle code does not scale to the whole system-of-systems, and a mixed top-down and bottom up strategy are applied.

6   Conclusions and Future Work

We have presented an initial assessment of the scalability of multi-core learning-based testing technology to cyber-physical systems-of-systems (CO-CPS). For this we have conducted testing experiments on a vehicle platooning simulator, where we have injected faults that violate safety and fuel efficiency requirements. Extensive testing experiments over different platoon sizes have demonstrated that learned model size scales well over the experimental time horizon and different platoon sizes. However, unsurprisingly perhaps, model convergence is low, at least according to the current PEC metric. Nevertheless effective testing, capable of finding valid SUT errors (both common and rare) was possible by learning large but incomplete models.

Future research needs to address several issues. Learning efficiency needs to be further improved to enhance coverage. Our study could be generalized by using more advanced simulators to test other use cases. We will also further consider how to scale up LBT to many-core platforms. Can the saturation effects cited in [15] be observed or avoided? The reliability questions surrounding incomplete model learning warrant further attention, e.g. the optimal choice of a learning convergence metric is an open question. Finally, equation (*) of

---

[13] Unfortunately our limited computing platform did not provide an opportunity to evaluate this result.

Sect. 4.1 represents a safety requirement that could be captured by a suitable *spatio-temporal logic*. Further study of spatio-temporal logics and model checking might be fruitful for CO-CPS use case testing.

This research has been funded by VINNOVA FFI project 2013-05608 VIRTUES and the Electronic Component Systems for European Leadership Joint Undertaking under grant agreement No. 692529 project SafeCOP.

References

1. Aki, M., Zheng, R., Yamabe, S., Nakano, K., Suda, Y., Suzuki, Y., Ishizaka, H., Kawashima, H., Sakuma, A.: Safety testing of an improved brake system for automatic platooning of trucks. Int. J. Intell. Transp. Syst. Res. **12**(3), 98–109 (2014)
2. Angluin, D.: Learning regular sets from queries and counterexamples. Inf. Comput. **75**(2), 87–106 (1987)
3. Bennaceur, A., Giannakopoulou, D., Hähnle, R., Meinke, K.: Machine learning for dynamic software analysis: potentials and limits (Dagstuhl seminar 16172). Dagstuhl Rep. **6**(4), 161–173 (2016)
4. Bergenhem, C., Shladover, S., Coelingh, E., Englund, C., Shladover, S., Tsugawa, S.: Overview of platooning systems. In: Proceedings of the 19th ITS World Congress, Vienna, October 2012
5. van den Bleek, R.: Design of a hybrid adaptive cruise control stop-&-go system. Master's thesis, Technische Universiteit Eindhoven, Department of Mechanical Engineering (2007)
6. Cassel, S., Howar, F., Jonsson, B., Steffen, B.: Active learning for extended finite state machines. Form. Asp. Comput. **28**(2), 233–263 (2016)
7. Cimatti, A., et al.: NuSMV 2: an opensource tool for symbolic model checking. In: Brinksma, E., Larsen, K.G. (eds.) CAV 2002. LNCS, vol. 2404, pp. 359–364. Springer, Heidelberg (2002). doi:10.1007/3-540-45657-0_29
8. Colin, S., Lanoix, A., Kouchnarenko, O., Souquières, J.: Using CSP‖B components: application to a platoon of vehicles. In: Cofer, D., Fantechi, A. (eds.) FMICS 2008. LNCS, vol. 5596, pp. 103–118. Springer, Heidelberg (2009). doi:10.1007/978-3-642-03240-0_11
9. El-Zaher, M., Contet, J., Gruer, P., Gechter, F., Koukam, A.: Compositional verification for reactive multi-agent systems applied to platoon non collision verification. Stud. Inform. Univ. **10**(3), 119–141 (2012)
10. Engelberg, S.: A Mathematical Introduction to Control Theory. Imperial College Press, London (2015)
11. Feng, L., Lundmark, S., Meinke, K., Niu, F., Sindhu, M.A., Wong, P.Y.H.: Case studies in learning-based testing. In: Yenigün, H., Yilmaz, C., Ulrich, A. (eds.) ICTSS 2013. LNCS, vol. 8254, pp. 164–179. Springer, Heidelberg (2013). doi:10.1007/978-3-642-41707-8_11
12. Fisher, M.: An Introduction to Practical Formal Methods Using Temporal Logic. Wiley, Hoboken (2011)
13. De la Higuera, C.: Grammatical Inference: Learning Automata and Grammars. Cambridge University Press, Cambridge (2010)
14. Hossen, K., Groz, R., Oriat, C., Richier, J.: Automatic model inference of web applications for security testing. In: Seventh IEEE International Conference on Software Testing, Verification and Validation, ICST 2014 Workshops Proceedings, pp. 22–23 (2014)

15. Howar, F., Bauer, O., Merten, M., Steffen, B., Margaria, T.: The teachers' crowd: the impact of distributed oracles on active automata learning. In: Hähnle, R., Knoop, J., Margaria, T., Schreiner, D., Steffen, B. (eds.) ISoLA 2011. CCIS, pp. 232–247. Springer, Heidelberg (2012). doi:10.1007/978-3-642-34781-8_18
16. Isberner, M., Howar, F., Steffen, B.: The TTT algorithm: a redundancy-free approach to active automata learning. In: Bonakdarpour, B., Smolka, S.A. (eds.) RV 2014. LNCS, vol. 8734, pp. 307–322. Springer, Cham (2014). doi:10.1007/978-3-319-11164-3_26
17. Jorgensen, P.C.: Software testing (2008)
18. Kamali, M., Dennis, L.A., McAree, O., Fisher, M., Veres, S.M.: Formal verification of autonomous vehicle platooning. CoRR abs/1602.01718 (2016)
19. Kearns, M., Vazirani, U.: An Introduction to Computational Learning Theory. MIT Press, Cambridge (1994)
20. Khosrowjerdi, H., Meinke, K., Rasmusson, A.: Automated behavioral requirements testing for automotive ECU applications. In: Proceedings of the 5th International Workshop on Model Based Safety Analysis. In: IMBSA 2017. LNCS. Springer (2017, to appear)
21. Meinke, K.: Recent progress in learning-based testing. In: Bennaceur, A., Hähnle, R., Meinke, K. (eds.) Machine Learning for Dynamic Software Analysis: Potentials and Limits: Proceedings of Dagstuhl Workshop, vol. 16172. Springer (2017, to appear)
22. Meinke, K., Niu, F.: An incremental learning algorithm for hybrid automata (2013)
23. Meinke, K., Sindhu, M.A.: Incremental learning-based testing for reactive systems. In: Gogolla, M., Wolff, B. (eds.) TAP 2011. LNCS, vol. 6706, pp. 134–151. Springer, Heidelberg (2011). doi:10.1007/978-3-642-21768-5_11
24. Meinke, K.: Automated black-box testing of functional correctness using function approximation. In: Proceedings of the 2004 ACM SIGSOFT International Symposium on Software Testing and Analysis (ISSTA 2004), pp. 143–153. ACM Press (2004)
25. Meinke, K., Nycander, P.: Learning-based testing of distributed microservice architectures: correctness and fault injection. In: Bianculli, D., Calinescu, R., Rumpe, B. (eds.) SEFM 2015. LNCS, vol. 9509, pp. 3–10. Springer, Heidelberg (2015). doi:10.1007/978-3-662-49224-6_1
26. Meinke, K., Sindhu, M.A.: Lbtest: a learning-based testing tool for reactive systems. In: Proceedings of the 2013 IEEE Sixth International Conference on Software Testing, Verification and Validation (ICST 2013), pp. 447–454. IEEE Computer Society (2013)
27. Peled, D., Vardi, M.Y., Yannakakis, M.: Black box checking. In: Wu, J., Chanson, S.T., Gao, Q. (eds.) Formal Methods for Protocol Engineering and Distributed Systems. IAICT, vol. 28, pp. 225–240. Springer, Boston (1999). doi:10.1007/978-0-387-35578-8_13
28. Raffelt, H., Steffen, B., Margaria, T.: Dynamic testing via automata learning. In: Yorav, K. (ed.) HVC 2007. LNCS, vol. 4899, pp. 136–152. Springer, Heidelberg (2008). doi:10.1007/978-3-540-77966-7_13
29. Sjöberg, K.: Platooning - challenges and opportunities (2016). https://docbox.etsi.org/Workshop/2016/201603_ITS_WORKSHOP/S04_TWDS_ACCIDENT_FREE_AUTOMATED_DRIVING
30. Valiant, L.G.: A theory of the learnable. Commun. ACM **27**(11), 1134–1142 (1984)

31. Walkinshaw, N., Bogdanov, K., Derrick, J., Paris, J.: Increasing functional coverage by inductive testing: a case study. In: Petrenko, A., Simão, A., Maldonado, J.C. (eds.) ICTSS 2010. LNCS, vol. 6435, pp. 126–141. Springer, Heidelberg (2010). doi:10.1007/978-3-642-16573-3_10

32. Weyuker, E.: Assessing test data adequacy through program inference. ACM Trans. Program. Lang. Syst **5**(4), 641–655 (1983)

33. Willke, T., Tientrakool, P., Maxemchuk, N.: A survey of inter-vehicle communication protocols and their applications. IEEE Commun. Surv. Tutor. **11**(2), 3–20 (2009)

34. Özguner, U., Acarman, T., Redmill, K.: Autonomous Ground Vehicles. Artech House Publishers, Boston (2011)

An Inspection-Based Compositional Approach to the Quantitative Evaluation of Assembly Lines

Marco Biagi[1], Laura Carnevali[1(✉)], Tommaso Papini[1], Kumiko Tadano[2], and Enrico Vicario[1]

[1] Department of Information Engineering, University of Florence, Florence, Italy
{marco.biagi,laura.carnevali,tommaso.papini,enrico.vicario}@unifi.it
[2] Data Science Research Labs, NEC Corporation, Kawasaki, Japan
k-tadano@bq.jp.nec.com

Abstract. We present a model-based approach to performance evaluation of a collection of similar systems based on runtime observations. As a concrete example, we consider an assembly line made of sequential workstations with transfer blocking and no buffering capacity, implementing complex workflows with random choices and sequential/cyclic phases with generally distributed durations and no internal parallelism. Starting from the steady state, an inspection mechanism is subject to some degree of uncertainty in the identification of the current phase of each workstation, and is in any case unable to estimate remaining times. By relying on the positive correlation between delays at different workstations, we provide stochastic upper and lower approximations of the performance measures of interest, including the time to completion of the local workflow of each workstation and the time until when a workstation starts a new job. Experimental results show that the approximated evaluation is accurate and feasible for lines of significant complexity.

Keywords: Workflow · Assembly lines · Inspection at steady state · Stochastic ordering · Partial observability · Semi markov process · Generalized semi markov process

1 Introduction

Quantitative approaches combining predictive models with runtime observations find a relevant application in monitoring and control of several manufacturing processes, including in particular assembly lines. Depending on the level of abstraction, these methods can support both high-level horizontal integration along a supply chain or scheduling of actions performed by humans [11] or end-effectors within a physical assembly line, notably fitting in the agenda of Industry 4.0.

Evaluation of performance measures for manufacturing processes has been widely addressed using models based on open queuing networks in sequential composition. In particular, blocking in the transfer of products across subsequent

© Springer International Publishing AG 2017
P. Reinecke and A. Di Marco (Eds.): EPEW 2017, LNCS 10497, pp. 152–166, 2017.
DOI: 10.1007/978-3-319-66583-2_10

stages of the line has been investigated and advocated as a way to account for buffering constraints that take relevance and may become restrictive in the handling of physical products [1, 20]. Numerical solution for the evaluation of the distribution of the lead time of a line subject to disturbances or fluctuations is addressed in [2], and impact of the lead time variability on the quality of deteriorating products is evaluated in [10]. The transfer blocking abstraction has been addressed also beyond the limits of manufacturing systems, notably in Software Architectures, considering different blocking schemes that may fit the expressive needs of the context or may allow product form solution [3]. As a common trait, in these works, stations are associated with exponentially distributed (EXP) service times, so that the underlying stochastic process falls in the class of Continuous Time Markov Chains (CTMCs), and the emphasis of complexity is instead focused on the structure and size of the composition of the line. Moreover, solution methods are applied off-line without actualizing models with respect to information that may become available during the runtime.

Methods for performability evaluation of stochastic systems have addressed the analysis of models conditional to the occurrence of unexpected changes such as an internal failure or a variation of external conditions [18, 23], usually assumed to occur when the model is in the steady state. This reduces complexity by decoupling the evaluation of performance under nominal conditions from a set of transient behaviors following a variety of rare unexpected events. Also in this case, most approaches develop on models with EXP durations, so that the analysis can be accomplished by restarting the system from the steady state distribution of the logical state, without taking into account remaining times of activities ongoing when the change occurs. In [6], the analysis is applied to models with Continuous Phase Type durations [15, 21] by relying on the EXP sojourn time occurring within individual phases.

When models are applied during the runtime [5, 9], actual observations can be used to restrain the set of plausible current states [8] and thus reduce the variability in the prediction of future behavior. In turn, this may become the basis for scheduling actions [4, 22] that adapt the system in response to changes in requirements or deviations from the nominal behavior of the environment.

In this paper, we address inspection-based evaluation of performance measures in collections of similar systems, and we illustrate the proposed approach with reference to the case of assembly lines. Specifically, we consider workstations implementing a complex workflow modeled as a state machine [19], with random choices and sequential phases with generally distributed (GEN) durations, with possible cycles but without internal parallelism. Workstations are composed in series with transfer blocking and no buffering capacity, so that, on completion of a job, a workstation is blocked until the subsequent workstation is ready to start a new job; though the approach is open to extensions to encompass buffering, this aspect is not considered here not to increase the complexity of solution. The structure of the overall model can be conveniently connected with the formalism of UML state charts [12] following the approach of [13].

We consider the case that some inspection mechanism is able to acquire a partial observation on the logical location of the state of all the workstations.

The observation is partial in a twofold sense: it does not capture remaining times of ongoing phases within active workstations, which are relevant due to the presence of GEN durations; and, it may be not able to distinguish some logical locations within the same workstation. Note that the assumption of partial observability is realistic in several applicative scenarios, given that keeping track of the starting time of activities requires event-driven monitoring or at least a very strict polling system, which may be not available in many manufacturing systems. Moreover, it may be convenient to perform operations on the system only sporadically, for instance to save costs, so that continuous monitoring is not adopted.

We assume that the time between consecutive inspections is significantly larger than average temporal parameters of the system, so that it can then be considered in a steady state at the time of observation. Starting from the assumption of steady state conditional to the acquired observation, we are interested in evaluating a suite of transient performance measures including: the time to completion of the local workflow of each workstation; the time when a ready product will be accepted by the subsequent workstation; and, the time until when a workstation starts a new job. Due to GEN durations, the underlying stochastic process of each workstation is in the class of Semi Markov Processes (SMPs); besides, due to the concurrent execution at multiple workstations, the overall composition falls in the class of Generalized Semi Markov Processes (GSMP), so that numerical solution is not practically feasible. To circumvent this complexity, we provide stochastic upper and lower approximations of the measures of interest by relying on the positive correlation between delays at different workstations.

The approach is implemented using the API of the ORIS Tool [7] for model construction and analysis, showing that the approximated evaluation is accurate and feasible for assembly lines of significant complexity. Overall, the obtained experimental results support the definition of high-level strategies for the optimization of system operation and maintenance, for instance with the goal of limiting the time during which workstations are blocked.

In the rest of the work, we describe the considered class of assembly lines (Sect. 2.1) and we define transient performance measures (Sect. 2.2); then, we provide a Petri net model of an assembly line (Sect. 3.1) and we present a solution method (Sects. 3.2 and Sect. 3.3); finally, we present the experimental results (Sect. 4) and we draw our conclusions (Sect. 5).

2   Problem Definition

2.1   Production Line Model

We consider an *assembly line* made of N sequential *workstations* $WS_1, \ldots, WS_N$, each performing a specific *job* that must be completed before the assembly is moved to the next workstation in the line. An additional initial block termed *generator* issues requests to start the manufacturing of a new product, and an additional final block termed *sink* collects the finished products. Figure 1 shows the state diagrams of the generator, the n-th workstation WS_n, and the sink. The generator is in the *producing* state while generating a request; it moves from the

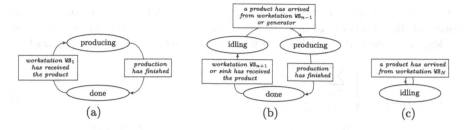

Fig. 1. State diagrams of: (a) the generator, (b) the n-th workstation, (c) the sink.

producing state to the *done* state whenever a request is issued; and, it moves back to the producing state as soon as workstation WS_1 has received the request and has started manufacturing the corresponding product.

A workstation is *producing* while performing its job on a product, implementing a specific *workflow* with known *phases* (sequential, alternative, cyclic) and durations. A workstation moves to the *done* state as soon as its task on the current product has been completed; next, it enters the *idling* state when the product has been received by the next workstation or by the sink (if this is the last workstation); finally, it moves back to the producing state when a product has arrived from the previous workstation or from the generator (if this is the first workstation). A workstation can process only one product at a time, and it is termed *busy* if it is in the producing or in the done state.

The sink is always in the *idling* state, collecting finished products received from workstation WS_N, which have traversed the whole assembly line.

An assembly line is monitored by external observers, e.g., human inspectors, polling sensors. If the time between consecutive inspections is significantly larger than average temporal parameters of the system, the line can be considered in a steady state at the observation time. An observation is a tuple $\omega = \langle \omega_0, \omega_1, \ldots, \omega_N \rangle$ where: (*i*) $\omega_0 = \langle \sigma_0 \rangle$ identifies the state $\sigma_0 \in \{producing, done\}$ of the generator; and, (*ii*) $\forall n \in [1, N]$, $\omega_n = \langle \sigma_n, \phi_n \rangle$ identifies the state $\sigma_n \in \{producing, done, idling\}$ of workstation WS_n and the set ϕ_n of its possible current phases. In particular, given that ambiguity affects only the observation of producing states, ϕ_n is equal to the power set $\mathbb{P}(\Gamma_n)$ of the set of producing phases $\Gamma_n = \{\gamma_n^1, \ldots, \gamma_n^{M_n}\}$ if $\sigma_n = producing$, and $\phi_n = \emptyset$ if $\sigma_n \in \{done, idling\}$. In so doing, the current producing phase is not identified unambiguously if $|\phi_n| > 1$. Given that observations are subject to ambiguity in the identification of the current producing phase, and that, in any case, they do not provide information on the time spent in the current state or phase, the assembly line turns out to be *partially observable*.

2.2   Transient Performance Measures

Observations can be used to predict the future behavior of the system and to improve the production process in individual workstations and in the overall line.

To this end, the **time to done** $\mathrm{TTD}(n,\omega)$ is defined as the time to complete the manufacturing of a product in workstation WS_n (making it available for workstation WS_{n+1} or the sink), conditioned to the observation $\omega = \langle \omega_0, \omega_1, \ldots, \omega_N \rangle$:

$$\mathrm{TTD}(n,\omega) := \begin{cases} \displaystyle\sum_{\gamma \in \phi_n} P_{n,\gamma,\omega} \cdot (R(n,\gamma) + Z(n,\gamma)) & \text{if } \sigma_n = producing \\ \mathrm{TTD}(n-1,\omega) + V(n) & \text{if } \sigma_n = idling \\ 0 & \text{if } \sigma_n = done \end{cases} \quad (1)$$

where: ϕ_n is the set of producing phases identified by ω_n; $P_{n,\gamma,\omega}$ is the steady-state probability that WS_n is in phase γ conditioned to observation ω; $R(n,\gamma)$ is the remaining time in phase γ of WS_n; $Z(n,\gamma)$ is the execution time of the producing phases of WS_n that follow γ; $V(n)$ is the producing time of WS_n; and, $\mathrm{TTD}(0,\omega)$ is equal to 0 if the generator state is $\sigma_0 = done$ and equal to the remaining time to the arrival of a new request if $\sigma_0 = producing$.

Note that, if WS_n is producing, $\mathrm{TTD}(n,\omega)$ is the time to complete the job on the product currently being processed; otherwise, if WS_n is idling, $\mathrm{TTD}(n,\omega)$ is the time until a product is moved to WS_n plus the time required to process it; finally, if WS_n is in the done state, $\mathrm{TTD}(n,\omega)$ is equal to zero, given that a product is waiting to be moved from WS_n to WS_{n+1}. Equation 1 can be solved by recursively computing the TTD of workstations that precede WS_n in the line, until a busy workstation or the generator is reached.

In the applicative perspective, $\mathrm{TTD}(n,\omega)$ could be used to optimize production in workstations, reducing the time during which they are blocked.

The **time to idle** $\mathrm{TTI}(n,\omega)$ is the time to complete the manufacturing of a product in workstation WS_n and move the product to workstation WS_{n+1}, conditioned to the observation $\omega = \langle \omega_0, \omega_1, \ldots, \omega_N \rangle$:

$$\mathrm{TTI}(n,\omega) := \begin{cases} \max\{\mathrm{TTD}(n,\omega), \mathrm{TTI}(n+1,\omega)\} & \text{if } \sigma_n \in \{producing, done\} \\ 0 & \text{if } \sigma_n = idling \end{cases}$$
$$(2)$$

Equation 2 can be solved by recursively computing the TTD of workstations that follow WS_n in the line, until an idling workstation or the sink is reached, i.e., $\mathrm{TTI}(n,\omega) = \max\{\mathrm{TTD}(n,\omega), \ldots, \mathrm{TTD}(n+k,\omega)\}$, where either WS_{n+k} is the last workstation (i.e., $n+k = N$) or it is idling. According to this, WS_n becomes idling when the slowest workstation WS_j among the following busy workstations $\mathrm{WS}_n, \ldots, \mathrm{WS}_{n+k}$ becomes idling (i.e., WS_j is the bottleneck).

In the applicative perspective, $\mathrm{TTI}(n,\omega)$ could be used to determine whether the considered workstation is early or late, and consequently switch the preceding workstations to faster or slower operation modes, respectively, reducing energy consumption while preserving the throughput of final products.

The **time to start next** $\mathrm{TTSN}(n,\omega)$ is the time to start the manufacturing of a new product in WS_n, conditioned to the observation $\omega = \langle \omega_0, \omega_1, \ldots, \omega_N \rangle$:

$$\mathrm{TTSN}(n,\omega) := \max\{\mathrm{TTI}(n,\omega), \mathrm{TTD}(n-1,\omega)\} \quad (3)$$

Solution of Eq. 3 requires the evaluation of both the preceding and the following workstations to compute $TTD(n-1,\omega)$ and $TTI(n,\omega)$, respectively. In the applicative perspective, $TTSN(n,\omega)$ can be compared with $TTD(n,\omega)$ to evaluate the time interval during which workstation WS_n is not working on a product. Moreover, $TTSN(n,\omega)$ could be exploited to temporarily switch off workstations that are not likely to process a new product in the near feature, permitting to reduce energy wasted by the assembly line.

For $TTD(k,\omega)$, $TTI(k,\omega)$, $TTSN(k,\omega)$, we evaluate the Cumulative Distribution Function (CDF) $F_{TTD(k,\omega)}(t)$, $F_{TTI(k,\omega)}(t)$, $F_{TTSN(k,\omega)}(t)$, respectively.

3   Modelling and Solution Technique

3.1   Partially Observable Stochastic Time Petri Nets

PO-STPN syntax. Petri nets are widely used to specify workflow models [26], given that they inherently support the representation of concurrency. To model partially observable assembly lines with stochastic execution times, we extend Stochastic Time Petri Nets (STPNs) [27] with observation symbols. Specifically, a *Partially Observable Stochastic Time Petri Net* (PO-STPN) is a tuple $\langle P, T, A^-, A^+, m_0, F, W, O, H \rangle$ where: P and T are the (disjoint) sets of places and transitions, respectively; $A^- \subseteq P \times T$ and $A^+ \subseteq T \times P$ are the sets of precondition and postcondition arcs, respectively; $m_0 : P \to \mathbb{N}$ is the initial marking assigning a number of tokens to each place; $F : T \to [0,1]^{[EFT_t, LFT_t]}$ associates each transition t with a CDF $F(t) : [EFT_t, LFT_t] \to [0,1]$, where $EFT_t \in \mathbb{Q}_{\geq 0}$ and $LFT_t \in \mathbb{Q}_{\geq 0} \cup \{\infty\}$ are termed *earliest* and *latest firing time*, respectively; $W : T \to \mathbb{R}_{>0}$ associates each transition with a weight; O is the set of observable symbols; and, $H : P \to O$ associates each place with an observation symbol.

A place p is an *input* or an *output* place for a transition t if $\langle p,t \rangle \in A^-$ or $\langle t,p \rangle \in A^+$, respectively. A transition t is *immediate* (IMM) if $EFT_t = LFT_t = 0$ and *timed* otherwise; a timed transition t is *exponential* (EXP) if $F_t(x) = 1 - e^{-\lambda x}$ over $[0, \infty]$ with $\lambda \in \mathbb{R}_{>0}$, and *general* (GEN) otherwise; a GEN transition t is *deterministic* (DET) if $EFT_t = LFT_t > 0$ and *distributed* otherwise; for each distributed transition t, F_t is the integral function of a Probability Density Function (PDF) f_t, i.e., $F_t(x) = \int_0^x f_t(y)dy$. IMM, EXP, GEN, and DET transitions are represented by thin black, thick white, thick black, and thick gray bars, respectively; weights are annotated next to transitions as weight $= value$; observation symbols are annotated next to places as obs $= value$.

PO-STPN semantics. The state of a PO-STPN is a pair $\langle m, \tau \rangle$, where m is a marking and $\tau : T \to \mathbb{R}_{\geq 0}$ associates each transition with a time-to-fire. A transition is *enabled* by a marking if each of its input places contains at least one token; an enabled transition t is *firable* in a state if its time-to-fire is equal to zero. When multiple transitions are firable, one of them is selected to fire with probability $\mu = W(t)/\sum_{t_i \in T_{f,s}} W(t_i)$, where $T_{f,s}$ is the set of firable transitions in s. When t fires, $s = \langle m, \tau \rangle$ is replaced by $s' = \langle m', \tau' \rangle$, where: m' is derived from m by removing a token from each input place of t, which yields

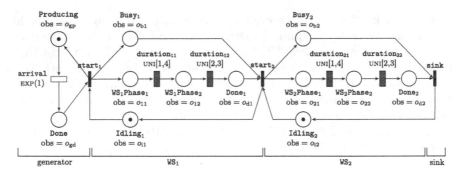

Fig. 2. PO-STPN of an assembly line with two workstations WS_1 and WS_2.

an intermediate marking m_{tmp}, and by adding a token to each output place of t; τ' is derived from τ by: (i) reducing the time-to-fire of each *persistent* transition (i.e., enabled by m, m_{tmp} and m') by the time elapsed in s; (ii) sampling the time-to-fire of each *newly-enabled* transition t_n (i.e., enabled by m' but not by m_{tmp}) according to F_{t_n}; and, (iii) removing the time-to-fire of each *disabled* transition (i.e., enabled by m but not by m').

Remark. For each symbol $o \in O$, we define $U(o) := \{p \in P \mid H(p) = o\}$ as the set of places associated with o. If $|U(o)| = 1 \ \forall o \in O$, each symbol identifies one and only one place; conversely, if $|U(o)| > 1$ for some $o \in O$, multiple places of the same workstation are associated with the same symbol, leading to ambiguity in the identification of the current producing phase (places belonging to different workstations are associated with different symbols).

An example. Figure 2 shows the PO-STPN of an assembly line made of two workstations WS_1 and WS_2, each implementing two sequential producing phases. The firing of the EXP transition `arrival` with rate 1 represents the generation of a request to start the manufacturing of a new product, moving a token from place `Producing` to place `Done`, which encode the namesake states of the generator shown in Fig. 1(a). For each workstation WS_n: places Idling_n and Done_n encode the namesake states shown in Fig. 1(b); place $WS_n\text{Phase}_i$ encodes the i-th phase of the producing state of Fig. 1(b); and, place Busy_n encodes either the producing or the done state. In both workstations, the first phase has a uniform duration within $[1, 4]$, while the second phase has a uniform duration over $[2, 3]$. Given that WS_2 is the last workstation, when its job is completed, which is represented by the firing of duration_{22}, the product is directly sent to the sink, which is modeled by the firing of the namesake IMM transition.

3.2   Evaluation of the Steady State Probability of a Producing Phase Conditioned to an Observation

The steady-state probability $P_{n,\gamma,\omega}$ that workstation WS_n is performing phase γ conditioned to observation $\omega = \langle \omega_0, \omega_1, \ldots, \omega_n \rangle$ is derived from *unconditional* steady state probabilities of the producing phases of $\mathtt{WS}_n$ through the Bayes theorem, i.e., $P_{n,\gamma,\omega} = P_\omega \mid P_{n,\gamma} \cdot P_{n,\gamma}/P_\omega$, where P_ω is the steady state probability of ω and $P_{n,\gamma}$ is the steady state probability that $\mathtt{WS}_n$ is performing phase γ (not conditioned on ω). Given that $P_\omega \mid P_{n,\gamma}$ is equal to 1, $P_{n,\gamma,\omega}$ can be rewritten as $P_{n,\gamma,\omega} = P_{n,\gamma}/\sum_{\eta \in \phi_n} P_{n,\eta}$, where ϕ_n is the set of phases identified by ω_n.

In principle, for each workstation $\mathtt{WS}_n$ and for each phase γ of its job, $P_{n,\gamma}$ could be evaluated from steady state analysis of the PO-STPN model of the assembly line, yielding $P_{n,\gamma,\omega} = \pi(m_p)/\sum_{q \in U(o)} \pi(m_q)$, where p is the place that encodes the execution of phase γ (e.g. in Fig. 2, place $\mathtt{WS}_1\mathtt{Phase}_1$ encodes the execution of the first of the two sequential phases of $\mathtt{WS}_1$), m_p is any marking where place p contains a token, o is the symbol observed when $\mathtt{WS}_n$ is performing phase γ, and $U(o)$ is the set of places associated with o (which includes p).

However, the complexity of steady state evaluation grows with the number of workstations and with the complexity of their manufacturing phases. Given that uncertainty affects the identification of the logical location within individual workstation (i.e., $\forall o \in O$, all places in $U(o)$ belong to the same workstation), $P_{n,\gamma}$ can be derived through steady state evaluation of an isolated model of $\mathtt{WS}_n$, obtained by chaining transition $\mathtt{start}_{n+1}$ (transition $\mathtt{sink}$ for the last workstation $\mathtt{WS}_N$) with transition $\mathtt{start}_n$ through a single place, which turns out to be the unique output place of $\mathtt{start}_{n+1}$ and the unique input place of $\mathtt{start}_n$:

Theorem 1. *The steady-state probability $P_{n,\gamma,\omega}$ that workstation $\mathtt{WS}_n$ is performing phase γ conditioned to observation $\omega = \langle \omega_0, \omega_1, \ldots, \omega_n \rangle$ can be evaluated from the steady state probabilities $\tilde{\pi}$ of the markings of the isolated model of $\mathtt{WS}_n$:*

$$P_{n,\gamma,\omega} = \frac{\tilde{\pi}(m_p)}{\displaystyle\sum_{q \in U(o)} \tilde{\pi}(m_q)} \tag{4}$$

where p is the place encoding the execution of phase γ, o is the symbol observed when $\mathtt{WS}_n$ is performing γ, and $U(o)$ is the set of places associated with symbol o.

3.3   Evaluation of Transient Performance Measures

Evaluation of the transient performance measures defined in Sect. 2.2 requires, for each workstation $\mathtt{WS}_n$ with $n \in [1, N]$ and for each of producing phase $\gamma \in \Gamma_n$, the evaluation of: the CDF $F_{R(n,\gamma)}$ of the remaining time $R(n, \gamma)$ in γ, the CDF $F_{Z(n,\gamma)}$ of the execution time $Z(n, \gamma)$ of the producing phases that follow γ, and the CDF $F_{V(n)}$ of the overall producing time $V(n)$ of $\mathtt{WS}_n$.

Evaluation of $F_{R(n,\gamma)}$. In the PO-STPN model of an assembly line, the remaining times of the enabled GEN transitions (i.e., durations of manufacturing phases) are dependent random variables. As an example, consider the case of two subsequent single-phase workstations $\mathtt{WS}_1$ and $\mathtt{WS}_2$ that are in the done and in the producing state, respectively, with $\mathtt{WS}_1$ having an incoming product waiting to be processed; as soon as $\mathtt{WS}_2$ completes the manufacturing of its current product, WS$_2$ starts working on the last product processed by WS$_1$ which, in turn, starts processing a new product; in so doing, when the assembly line is subsequently inspected, the remaining times of WS$_1$ and WS$_2$ are dependent random variables.

A lower bound on $F_{R(n,\gamma)}$ can be derived assuming that $R(n,\gamma)$ is zero, as if the GEN transition modeling γ was newly enabled (*immediate approximation*), i.e., $\tilde{F}_{R(n,\gamma)}(t) = 1 \, \forall t$. Conversely, an upper bound can be computed assuming that $R(n,\gamma)$ is sampled from the distribution of the duration of phase γ, as if the corresponding transition was newly enabled (*newly enabled approximation*), i.e., $\tilde{F}_{R(n,\gamma)}(t) = F_g(t)$ where g is the transition modeling the duration of γ.

Another lower bound on $F_{R(n,\gamma)}$ can be obtained by considering the remaining times of the ongoing manufacturing phases of the active workstations as independent random variables (*independent remaining times approximation*):

Theorem 2. *If $\hat{R}$ is an independent version of the vector R of the remaining times of the ongoing phases at inspection, then $\hat{R} \geq_{st} R$, where $\geq_{st}$ is the usual stochastic order among random variables.*

The steady state distribution of each remaining time $\hat{R}(n,\gamma)$ in $\hat{R}$ can be derived according to the Key Renewal Theorem [17,24]:

$$\tilde{F}_{R(n,\gamma)}(t) = \frac{1}{\mu} \int_0^t [1 - F_g(s)]ds \tag{5}$$

where $F_g(s)$ is the sojourn time CDF in phase γ and μ is its expected value. In particular, if the sojourn time PDF in phase γ has a bounded support, i.e., $f(t) : [a, b] \rightarrow [0, 1]$, then $\tilde{f}_{R(n,\gamma)}(t)$ can be derived as:

$$\tilde{f}_{R(n,\gamma)}(x) = \begin{cases} \frac{1}{\mu} & \text{if } x < a \\ \frac{1}{\mu} - \frac{1}{\mu}F(x) & \text{if } a \leq x < b \\ 0 & \text{otherwise} \end{cases} \tag{6}$$

Evaluation of $F_{Z(n,\gamma)}$ and $F_{V(n)}$. Evaluation is performed through transient analysis of the isolated workstation model discussed in Sect. 3.2, computing $F_{Z(n,\gamma)}$ and $F_{V(n)}$ as the transient probability of the marking where place Done$_n$ contains a token, assuming WS$_n$Phase$_1$ and WS$_n$Phase$_k$ as initial marking, respectively (where WS$_n$Phase$_1$ encodes the execution of the first phase of WS$_n$ and WS$_n$Phase$_k$ encodes the execution of the phase of WS$_n$ that follows phase γ).

Values of $\tilde{F}_{R(n,\gamma)}$, $F_{Z(n,\gamma)}$, and $Z(n,\gamma)$ are combined according to Eqs. 1, 2, and 3. Given that the convolution and the max operations preserve the usual stochastic order, we obtain bounds on TTD(n,ω), TTI(n,ω) and TTSN(n,ω).

4   Computational Experience

The approach is first experimented on a simple assembly line, permitting to validate results against stochastic simulation (Sect. 4.1), and then applied to a more complex case study, for which simulation would be too computationally expensive, showing the scalability of the solution method (Sect. 4.2). In both models, durations are associated with example CDFs selected so as to highlight the capability to manage GEN timers and bounded supports; in general, any kind of expolynomial approximant could be used depending on the available statistics. Evaluation is performed using the API of the ORIS Tool [7] to implement the PO-STPN model of the assembly line and to perform transient and steady-state analysis of isolated models of workstations. Experiments were performed on a single core of an Intel Core i5-6600K processor equipped with 16 GB RAM, computing transient performance measures with a time step equal to 0.01, which is by far lower than temporal parameters of the two models.

4.1   A Simple Assembly Line

Using simulation to evaluate the considered performance measures, which are conditional to an observation ω at an inspection time t_i, requires: (i) estimation of the time t_s beyond which marking probabilities reach a steady state; (ii) selecting of t_i from time t_s on; (iii) execution of a large number of runs, discarding those where the marking of the PO-STPN model at time t_i is not the one identified by ω. According to this, we consider the model of Fig. 2, which underlies a Markov Regenerative Process (MRP) and permits to evaluate time t_s through regenerative transient analysis based on the method of stochastic state classes [14]. Specifically, $t_s = 85$ with a tolerance of ± 0.001; thus, we select $t_i = 90$. Moreover, we assume that, at the inspection time, a new product is waiting to be served and both WS_1 and WS_2 are performing their first phase, which corresponds to marking Done WS_1Phase_1 WS_2Phase_1 having probability nearly equal to 0.208 at time t_i, causing about 79.2% of the 2 000 000 simulation runs to be discarded.

Figures 3(a), (b), and (c) plot $F_{TTD(1,\omega)}(t)$, $F_{TTI(1,\omega)}(t)$, and $F_{TTSN(2,\omega)}(t)$, respectively, computed through simulation in nearly 41 min, 45 min, and 42 min, respectively, and the corresponding bounds computed through the approximate methods of Sect. 3, requiring nearly 0.15 s, 0.18 s, and 0.10 s, respectively. As expected, the *immediate* and the *newly enabled* curves represent a lower and an upper bound on the *simulation* curve, respectively; moreover, the *independent remaining times* curve is also a lower bound on the simulation curve, significantly tighter than the immediate curve. In particular, for $F_{TTD(1,\omega)}(t)$ note that: the *immediate* curve is a uniform CDF with support [2, 3] like the distribution of the second phase of WS_1, given that this approximation assumes that the first phase has just been completed; the newly enabled curve represents the convolution between the CDF of the two phases of WS_1, due to the fact that this approximation assumes that the first phase has just begun.

4.2   A Complex Assembly Line

We consider three types of workstation: (i) two sequential phases (see Fig. 4(a)); (ii) an initial phase followed by two alternative phases with probability 0.7

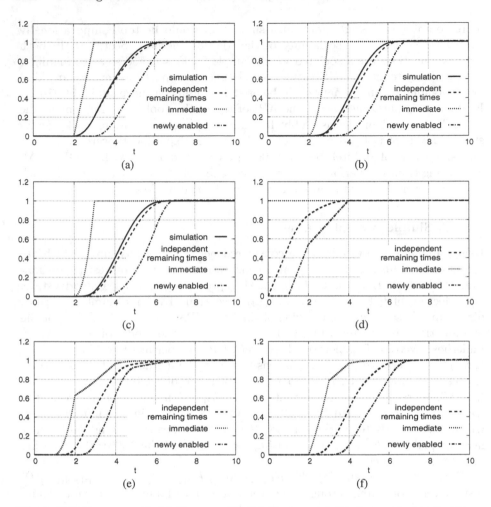

Fig. 3. Simple case study: (a) $F_{\mathrm{TTD}(1,\omega)}(t)$, (b) $F_{\mathrm{TTI}(1,\omega)}(t)$, (c) $F_{\mathrm{TTSN}(2,\omega)}(t)$. Complex case study: (d) $F_{\mathrm{TTD}(5,\omega)}(t)$, (e) $F_{\mathrm{TTI}(5,\omega)}(t)$, (f) $F_{\mathrm{TTSN}(5,\omega)}(t)$.

and 0.3, respectively (see Fig. 4(b)); (*iii*) an initial phase that can be cyclically repeated, each time with probability 0.1, followed by a second final phase (see Fig. 4(c)) Note that the alternative workstation is affected by ambiguity in the observation of the two alternative phases, which are in fact associated with the same symbol.

We consider an assembly line composing three times a serial, an alternative, and a cyclic workstation, for a total of 9 workstations, plus a generator with exponentially distributed arrivals with rate 1. To stress evaluation complexity, we assume that workstations are busy and the generator is blocked at the inspection time, and we compute performance measures for WS_5, an alternative workstation affected by observation ambiguity (simulation would be too computationally

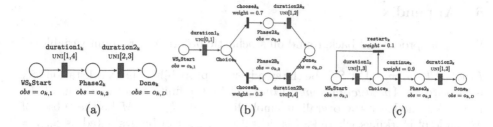

Fig. 4. PO-STPN model of 3 workstation types: (a) serial, (b) cyclic, (c) alternative

expensive due to the small probability of the considered marking). Figures 3(d), (e), and (f) plots bounds on $F_{\mathrm{TTD}(5,\omega)}(t)$, $F_{\mathrm{TTI}(5,\omega)}(t)$, and $F_{\mathrm{TTSN}(5,\omega)}(t)$ respectively, computed in nearly 0.126 s, 0.123 s, and 0.75 s, respectively, showing that the approach is feasible for models of real scale. As expected, the *independent remaining times* curves lie between the *immediate* and the *newly enabled* curves, comprising a tighter lower bound with respect to the *newly enabled* curves.

5   Conclusions

We have proposed a novel approach for the evaluation of transient performance measures in collections of similar systems on the basis of runtime observations. The approach is illustrated with reference to the case of assembly lines made of a sequence of workstations, with transfer blocking and no buffering capacity, each implementing a complex workflow with random choices, sequential/alternative/cyclic phases, GEN durations, without internal parallelism. Some inspection mechanism partially identifies the current logical location of each workstation, but is not able to determine the remaining times of the ongoing phases. Starting from the steady state, we consider transient performance measures conditional to an observation, including the time to complete the specific job of each workstation. Leveraging the positive correlation among the remaining times in different workstations, we compose the results of the analysis of individual workstations to derive stochastic bounds on the considered measures. The approach is experimented on a complex case study, showing that the obtained bounds are accurate and models of real scale can be effectively handled.

Future work includes encompassing more complex assembly lines, for instance with buffering, and evaluating other performance measures, including the total lead time until when the product that is presently at some intermediate workstation will reach the end of the line. Moreover, the approach could be exploited in the optimization of manufacturing processes, for instance to limit the wasted energy or to reduce the time during which workstations are blocked.

6 Appendix

We report proofs and background on stochastic ordering of random variables.

Proof of Theorem 1. Let $S_{n,\sigma}$ be the steady state probability that workstation WS_n is in state $\sigma \in \{producing, done, idling\}$, M be the finite set of markings of the PO-STPN model of the overall assembly line, and $M_{n,\sigma} \subset M$ be the subset of M made of markings where WS_n is in state σ. $S_{n,\sigma}$ can be evaluated as $S_{n,\sigma} = \sum_{m \in M_{k,\sigma}} \pi(m)$, where $\pi(m)$ is the steady state probability of marking m.

Let $M_{n,producing,\gamma} \subset M_{n,producing}$ be the set of markings where WS_n is performing phase γ. Given that the behavior of a workstation during the producing state does not depend on the other workstations (interactions occur only when it is in the idling state, waiting for a product from the preceding workstation, or in the done state, waiting for the following workstation to accept the product), the probability $P_{n,\gamma}$ that WS_n is performing phase γ can be computed as $P_{n,\gamma} = \sum_{m \in M_{n,producing,\gamma}} \pi(m) = \bar{\pi}_\gamma \cdot S_{n,processing}$ where $\bar{\pi}_\gamma$ is the steady state probability that WS_m is performing phase γ given that it is producing.

Hence, the steady-state probability $P_{n,\gamma,\omega}$ that WS_n is performing γ conditioned to $\omega = \langle \omega_0, \omega_1, \ldots, \omega_n \rangle$ can be derived as $P_{n,\gamma,\omega} = P_{n,\gamma} / \sum_{\eta \in \phi_n} P_{n,\eta} = \bar{\pi}_\gamma \cdot S_{n,processing} / \sum_{\eta \in \phi_n} \bar{\pi}_\eta \cdot S_{n,processing} = \bar{\pi}_\gamma / \sum_{\eta \in \phi_n} \bar{\pi}_\eta$. The conditional probability $\bar{\pi}_\gamma$ can be derived from the evaluation of the isolated model of WS_n, as the steady state probability $\tilde{\pi}(m_p)$ of any marking m_p where the place p that encodes the execution of phase γ contains a token. Similarly, $\sum_{\eta \in \phi_n} \bar{\pi}_\eta$ (where ϕ_n is the set of phases identified with ambiguity by ω_n through the observation of symbol o) can be derived as the sum of the steady state probabilities $\tilde{\pi}(m_q)$ of markings m_q where place $q \in U(o)$ encoding the execution of phase η contains a token. According to this, $P_{n,\gamma,\omega} = \tilde{\pi}(m_p) / \sum_{q \in U(o)} \tilde{\pi}(m_q)$. $\square$

To make the paper self-contained, we recall results on stochastic ordering of random variables [16, 25], which we use to prove Theorem 2.

Definition 1. *A real random variable X is stochastically smaller than another real random variable Y ($X \leq_{st} Y$) if $\Pr(X > z) \leq \Pr(Y > z)$ for all z.*

Definition 2. *Two random variables X and Y with covariance $cov(X, Y)$ are positively correlated if $cov(X, Y) \geq 0$.*

Definition 3. *Let $\hat{X} = \{\hat{X}_1, ..., \hat{X}_n\}$ be an independent version of random variables $X = \{X_1, ..., X_n\}$. The following properties hold: (i) $\{\hat{X}_1, ..., \hat{X}_n\}$ are mutually independent, and (ii) X_i and $\hat{X}_i$ have the same PDF $\forall i \in \{1, \ldots, n\}$.*

Theorem 3. *If $\hat{X}$ is an independent version of a vector X of positively correlated random variables, then $\hat{X} \geq_{st} X$.*

Corollary 1. $\max_i \hat{X}_i \geq_{st} \max_i X_i$

Leveraging the above results, we can prove the following statement:

Theorem 4. *At inspection, the remaining times of the ongoing phases of the producing workstations are positively correlated.*

Proof. Let a production line made of n workstations $\mathsf{WS}_1, \ldots, \mathsf{WS}_N$ be inspected at time i, and let $R = \{R_{a(1)}, R_{a(2)}, \ldots, R_{a(h)}\}$ be the vector of the remaining times of the ongoing phases of the $h \leq N$ processing workstations, where $a(l)$ is the physical number of the l-th processing workstation. Let x_n^j be the time at which WS_n starts processing the j-th product, y_n^j be the time at which WS_n completes its job on the j-th product, and Δ_n be the sojourn time of a product in WS_n. Based on the production line model of Sect. 2.1, $R_{a(l)}^j = \Delta_{a(l)} - i + x_{a(l)}^j$, $y_n^j = x_n^j + \Delta_n$, and $x_n^j = \max\{y_{n-1}^j, x_{n+1}^{j-1}\}$. According to this, random variables $\{x_1^j, \ldots, x_{a(h)}^j\}$ are positively correlated. Moreover, exploiting the fact that time i is a constant and random variables $\Delta_{a(1)}, \ldots, \Delta_{a(h)}$ are independent of each other and identically distributed for each manufactured product, it can be shown that the covariance between any two x_n^i x_n^j is equal to the covariance between $R_{a(l)}^i$ and $R_{a(l)}^j$. Therefore, the remaining times are positively correlated. $\square$

Finally, Theorem 2 is proved by leveraging Theorems 3 and 4.

Proof of Theorem 2. By Theorem 4, the remaining times stored in vector R are positively correlated. As a consequence, by Theorem 3, their independent version $\hat{R}$ is stochastically ordered with respect to R.

References

1. Akyildiz, I.F.: On the exact and approximate throughput analysis of closed queuing networks with blocking. IEEE Trans. Softw. Eng. **14**(1), 62–70 (1988)
2. Angius, A., Horváth, A., Colledani, M.: Moments of accumulated reward and completion time in markovian models with application to unreliable manufacturing systems. Perform. Eval. **75**, 69–88 (2014)
3. Balsamo, S., Personè, V.D.N., Inverardi, P.: A review on queueing network models with finite capacity queues for software architectures performance prediction. Perform. Eval. **51**(2), 269–288 (2003)
4. Biagi, M., Carnevali, L., Paolieri, M., Patara, F., Vicario, E.: A stochastic model-based approach to online event prediction and response scheduling. In: Fiems, D., Paolieri, M., Platis, A.N. (eds.) EPEW 2016. LNCS, vol. 9951, pp. 32–47. Springer, Cham (2016). doi:10.1007/978-3-319-46433-6_3
5. Blair, G., Bencomo, N., France, R.B.: Models@run.time. Computer **42**, 10 (2009)
6. Bruneo, D., Distefano, S., Longo, F., Puliafito, A., Scarpa, M.: Evaluating wireless sensor node longevity through Markovian techniques. Comput. Netw. **56**(2), 521–532 (2012)
7. Bucci, G., Carnevali, L., Ridi, L., Vicario, E.: Oris: a tool for modeling, verification and evaluation of real-time systems. Int. J. Softw. Tools Technol. Transf. **12**(5), 391–403 (2010)
8. Carnevali, L., Nugent, C., Patara, F., Vicario, E.: A continuous-time model-based approach to activity recognition for ambient assisted living. In: Campos, J., Haverkort, B.R. (eds.) QEST 2015. LNCS, vol. 9259, pp. 38–53. Springer, Cham (2015). doi:10.1007/978-3-319-22264-6_3

9. Cheng, B.H.C., et al.: Using models at runtime to address assurance for self-adaptive systems. In: Bencomo, N., France, R., Cheng, B.H.C., Aßmann, U. (eds.) Models@run.time. LNCS, vol. 8378, pp. 101–136. Springer, Cham (2014). doi:10.1007/978-3-319-08915-7_4

10. Colledani, M., Horvath, A., Angius, A.: Production quality performance in manufacturing systems processing deteriorating products. CIRP Ann. Manuf. Technol. **64**(1), 431–434 (2015)

11. Gorecky, D., Schmitt, M., Loskyll, M., Zühlke, D.: Human-machine-interaction in the industry 4.0 era. In: IEEE International Conference on Industrial Informatics (INDIN), pp. 289–294. IEEE (2014)

12. Harel, D., Naamad, A.: The statemate semantics of statecharts. ACM Trans. Softw. Eng. Methodol. (TOSEM) **5**(4), 293–333 (1996)

13. Homm, D., German, R.: Analysis of hierarchical semi-markov processes with parallel regions. In: Remke, A., Haverkort, B.R. (eds.) MMB&DFT 2016. LNCS, vol. 9629. Springer, Cham (2016)

14. Horváth, A., Paolieri, M., Ridi, L., Vicario, E.: Transient analysis of non-Markovian models using stochastic state classes. Perform. Eval. **69**(7–8), 315–335 (2012)

15. Horváth, A., Telek, M.: PhFit: a general phase-type fitting tool. In: Field, T., Harrison, P.G., Bradley, J., Harder, U. (eds.) TOOLS 2002. LNCS, vol. 2324, pp. 82–91. Springer, Heidelberg (2002). doi:10.1007/3-540-46029-2_5

16. Kulkarni, V.: Introduction to Modeling and Analysis of Stochastic Systems. Springer Texts in Statistics. Springer, Heidelberg (2011). doi:10.1007/978-1-4419-1772-0

17. Kulkarni, V.G.: Modeling and Analysis of Stochastic Systems. CRC Press, Boca Raton (2016)

18. Muppala, J.K., Woolet, S.P., Trivedi, K.S.: Real-time systems performance in the presence of failures. Computer **24**(5), 37–47 (1991)

19. Murata, T.: Petri nets: properties, analysis and applications. Proc. IEEE **77**(4), 541–580 (1989)

20. Perros, H.G., Altiok, T.: Approximate analysis of open networks of queues with blocking: tandem configurations. IEEE Trans. Softw. Eng. **3**, 450–461 (1986)

21. Reinecke, P., Krauß, T., Wolter, K.: Hyperstar: phase-type fitting made easy. In: International Conference on Quantitative Evaluation of Systems (QEST), pp. 201–202. IEEE (2012)

22. Salfner, F., Lenk, M., Malek, M.: A survey of online failure prediction methods. ACM Comput. Surv. (CSUR) **42**(3), 10 (2010)

23. Sanders, W.H., Meyer, I.F.: A unified approach for specifying measures of performance, dependability and performability. In: Avižienis, A., Laprie, J.C. (eds.) Dependable Computing for Critical Application. Dependable Computing and Fault-Tolerant Systems, vol. 4. Springer, Vienna (1991)

24. Serfozo, R.: Basics of Applied Stochastic Processes. Probability and Its Applications. Springer, Heidelberg (2009)

25. Stoyan, D., Daley, D.J.: Comparison Methods for Queues and Other Stochastic Models. Wiley, New York (1983)

26. Van der Aalst, W.M.: The application of petri nets to workflow management. J. Circ. Syst. Comput. **8**(01), 21–66 (1998)

27. Vicario, E., Sassoli, L., Carnevali, L.: Using stochastic state classes in quantitative evaluation of dense-time reactive systems. IEEE Trans. Softw. Eng. **35**(5), 703–719 (2009)

Performance, Energy and Security

Machine Learning Models for Predicting Timely Virtual Machine Live Migration

Osama Alrajeh$^{(\boxtimes)}$, Matthew Forshaw, and Nigel Thomas

School of Computing Science, Newcastle University, Newcastle upon Tyne, UK
{o.alrajeh1,matthew.forshaw,nigel.thomas}@ncl.ac.uk

Abstract. Virtual machine (VM) consolidation is among the key strategic approaches that can be employed to reduce energy consumption in large computing infrastructure. However, live migration of VMs is not a trivial operation and consequently not all VMs can be easily consolidated in all circumstances. In this paper we present experiments attempting to live migrate the Kernel-based VM (KVM) executing workload form the SPECjvm2008 benchmark. In order to understand what factors influence live migration we investigate three machine learning models to predict successful live migration using different training and evaluation sets drawn from our experimental data.

Keywords: VM consolidation · Live migration · Prediction · Energy efficiency

1 Introduction

Virtual Machine (VM) Live Migration has become an established technology used to consolidate virtualised workload onto a smaller number of physical machines, as a mechanism to reduce overall energy consumption. Live migration is attractive as it attempts to provide a seamless transfer of service between physical machines without impacting client processes or applications. Given the capacities of load balancing, fault tolerance, and energy management, data centres routinely employ live migration [1]. However, such transfer is not trivial and clearly impacts on the resources of the physical machines involved and the network which connects them. As such, depending on the resources available and the load experienced by the VM and on the physical machines, live migration might not always be feasible or even possible.

The primary factor that contributed to the rise and development of cloud computing was the relatively recent advent of novel technologies. Applying these developments facilitated the worldwide provision of opportune, functional, and cost-effective consumer products. A central feature of cloud computing is the virtualisation component, and this drives the cloud on account of an extensive range of advantages, including isolation, straightforward manageability, cost-effectiveness, adaptability, and partitioning.

This paper addresses the above need by presenting predictive models for VM live migration which can be employed to select the VMs that can be readily

© Springer International Publishing AG 2017
P. Reinecke and A. Di Marco (Eds.): EPEW 2017, LNCS 10497, pp. 169–183, 2017.
DOI: 10.1007/978-3-319-66583-2_11

migrated based on the characteristics of their workload. In order to build the models, we perform a live experiment to measure the VM live migration duration time, CPU utilisation, memory usage, and I/O activities during the migration process. The experiment involves employing two physical hosts to facilitate the migration of VMs from one to the other by using Kernel-based Virtual Machine (KVM) [2]. The SPECjvm2008 benchmark [3] is selected for the purpose of producing workloads on the VMs, which allows us to consider a range of different workload features. Following this, the migration times associated with identical workloads are considered for VMs characterised by various hardware constraints.

The remainder of this paper is organized as follows. Section 2 discusses the related work. In Sect. 3, we introduce the experiment environment. Section 4 presents the results of our preliminary experimentation. We explain and build the predictive models in Sect. 5. In Sect. 6, we evaluate the model with various datasets, before concluding and motivating future work in Sect. 7.

2   Related Work

The concept of VM live migration approach originated with Clark *et al.* [1], whose core idea was to facilitate the relocation of the VM from source to target host with minimal downtime. To achieve the goal, the researchers formulated a pre-copy algorithm. This algorithm operates in the following way: prior to transitioning the VM's execution host, it copies and transfers pages from the source memory host to the target memory host; this is carried out until there are relatively few uncopied pages. Following the transferal of the remainder of the pages from the source host, the source host VM engages in suspension and resumption on the target host.

The modelling and prediction of virtual machine migration has formed the basis for a number of previous works. Prior experimental work has sought to quantify the impact of network bandwidth [4] and workload characteristics [5] on the performance of VM live migration. Akoush *et al.* [6] develop predictive models of live migration performance based on experimentation with a number of SPEC benchmarks, and observe that network link speed exists as the most dominant factor in migration performance. Meanwhile, Hu *et al.* [7] develop predictive models of the performance and energy impact of virtual machine live migration on the Xen platform, and demonstrate potential savings of 73%.

Machine learning approaches have been used operationally to inform various resource scheduling decisions within large scale computing systems [8]. Uriarte *et al.* [9] apply a Random Forest method to service clustering in autonomic cloud environments.

3   Experiment Environment

A live experiment constitutes the basis of this paper, and this is carried out to measure the durations of VM live migration and resource utilisation of the VMs during the migration in the context of different workloads. In order to produce the different workloads, the SPECjvm2008 benchmark [3] is employed, and KVM [2] is used as a hypervisor. Section 3.1 describes the experimental setup; following this, Sect. 3.2 introduces the benchmark for VM workload generation and Sect. 3.3 details the experimental scenario used for the rest of the paper.

3.1 Experiment Set Up

The principal obligation which has influenced the experimental setup is that it should reproduce the VM live migration procedure as it takes place in real-world contexts. For this reason, the Kernel-based Virtual Machine (KVM), a common hypervisor for data centres, has been employed. VM live migration necessitates the storage of VM images in a dedicated area that can be accessed by every physical host. Consequently, VM images are stored using network attached storage (NAS), and this has created a situation in which the procedure of VM live migration is restricted to copying the memory pages and the CPU state among physical hosts; this method is referred to as pre-copy live migration. For NAS, we used the free software licensed Openfiler [10]. In addition, the setup involves two servers and one client computer, linked together by a 1,000 Mb switch. The servers run CentOS 7 Linux and KVM is installed on each. The hardware features of both servers as follows: server 1 employs 4 CPUs Core 2 Quad @ 2.66 GHz and 4 GB DDR2 SDRAM. Server 2 has 8 CPUs Intel Core i7 @ 2.80 GHz and 4 GB DDR3 SDRAM. The client computer operates Ubuntu 16.04 LTS and employs 1 CPU Intel Core i7 @ 3.20 GHz and 4 GB DDR3 SDRAM. The difference between the two servers is a factor that assists us to obtain various results. Finally, it should be noted that the client computer is employed to trigger VM live migration between the servers.

3.2 Benchmarks

39 distinct workloads are incorporated into the SPECjvm2008 benchmark [3], and these are used to examine the way in which hardware systems and Java VMs (JVMs) perform. The default runtime for every workload is four minutes, and the benchmark involves a pair of running configurations: namely, base and peak. The fixed runtime of the base configuration involves a warm-up period of two minutes and a general operational period of four minutes. In our experiment, we use 20 workloads from the SPECjvm2008 benchmark as shown in Table 1.

3.3 Experiment Scenario

We first formulated a VM in Sever 1 by using KVM. In turn, the image was stored in NAS. The VM runs Ubuntu 16.04 LTS, and it incorporates the SPECjvm2008 to generate a range of workloads. Several distinct VM hardware capacities have been employed. Each VM is denoted as VM_{ij} where $i, j = \{1, 2, 3\}$, i is the number of CPUs, and j is the RAM capability in Gigabytes (GB).

Table 1. SPECjvm2008 Benchmark workloads

Group name	Workloads
Compiler	compiler.compiler, compiler.sunflow
Compress	compress
Crypto	crypto.aes, crypto.rsa, crypto.signverify
Mpegaudio	mpegaudio
Scimark Large	scimark.fft.large, scimark.lu.large, scimark.sor.large,scimark.sparse.large, scimark.monte_carlo
Scimark Small	scimark.fft.small, scimark.lu.small, scimark.sor.small, scimark.sparse.small, scimark.monte_carlo
Serial	serial
Sunflow	sunflow
Xml	xml.transform, xml.validation

The experiment employs 20 SPECjvm workloads, and each is operated independently. Once the first minute of workload operation has been completed, the VM live migration procedure is started to transfer the VM from Server 1 to Server 2. After the VM has been migrated to Server 2, the VM gets restarted and starts the previous steps again between Server 2 and Server 1.

The presence of the client computer is important in facilitating the operation of the experiment in an automated way. We developed a bash script [11] that enables the client computer to run the benchmark on the VM and, following this, the initiation of the live migration between the two severs. The client accesses the VM and commences the operation of a single workload; following one minute of running the workload, the client facilitates the migration of the VM from the first to the second server; in turn, the commencement and completion times of the VM live migration are recorded in a logs file. In addition to this, the script maintains a log of the memory usage of each workload over the course of its runtime from second to second, and this is measured by the Memusg script [12]. Furthermore, top and sar are used to maintain a log of the CPU utilisation, total system memory, free memory, memory used, buffer cache, I/O activities, queue size, and load average over the course of the migration.

For future researchers who are interested in reproducing our test using their own hardware, a copy of the automated bash script can be found at [11]. It is possible to modify both the number of the test and the runtime of each workload.

4   Experimental Results

In this section, we discuss the results of our preliminary experimentation. We demonstrate the impact of different workload characteristics, and their impact on VM live migration time, on various VMs capacities.

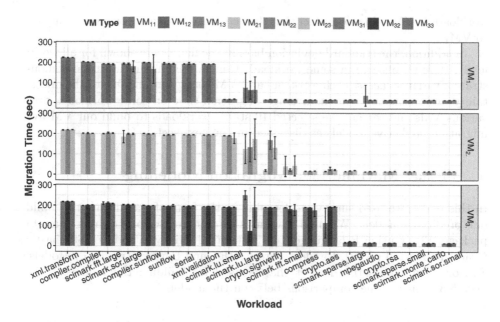

Fig. 1. Average migration time of VM_{ij} with various workloads

Figure 1 presents the results of VM migration from Server 1 to Server 2. The VMs migration starts after one minute of running the workload. We migrated each VM with each workload 10 times from Server 1 to Server 2 and from Server 2 to Server 1. The figure illustrates the average duration of the VM live migration of each workload from Server 1 to Server 2. The results of the migration time between Server 2 to Server 1 has drawn the same conclusion. For brevity, results of migration from Server 2 to Server 1 are not shown.

The results represent a clear discrepancy in average migration time between workloads due to the various workload characteristics. Some workloads such as `Compiler.compiler`, `xml.transform`, and `xml.validation` consume much memory than others which increases the time of VM live migration between servers. The reason is that VM live migration process requires copying the VM's memory pages from the source host to the destination host. At the point when there is a high rate of changing memory pages during the live migration with an insufficient network speed, the migration might never achieve or might take a long time. Also, the number of threads that produced by the workloads is another reason of delaying the VM live migration or making it impossible. For example, the threads that generated by `Serial` and `Sunflow` workloads put pressure in the memory which defined them as non-migratable workloads in our experiment.

The number of operations in the SPECjvm2008 benchmark is based on the available resources. When there are more resources available in the VM, the workloads generate more operations to test these resources. In our experiment, some workloads in VM_2, and VM_3 have longer migration time than the workloads in VM_1. due to the fact that they contain more memory pages compare to VM_1..

In addition, the standard deviation bars in the figures are shown for all workloads. Here we observe in some workloads that average migration time demonstrate a variance, across each workload and VM configuration. Due to the uncertainty and inconstancy of the average migration time on some workloads over various VMs and servers, we selected 10 stable workloads to build our VM live migration predictive models as discussed in the next section.

5   Virtual Machine Live Migration Modelling

In this paper, we used supervised learning in the form of classification to build our predictive models. We used classification and regression training (`caret`) package [13] by using R language [14]. Section 5.1 briefly introduces Stochastic Gradient Boosted (SGB), Random Forest (RF), and Bagged Tree (BT) Models. Section 5.2 discusses the dataset that is used for building the models, while Sect. 5.3 expresses the comparisons between the models.

5.1   Stochastic Gradient Boosted, Random Forest, and Bagged Tree

Stochastic Gradient Boosting (SGB) [15] is an advanced model of the Gradient Boosting model. The basic idea of the Gradient boosting method is to fit a classifier, typically a decision tree, at each iteration, so that the next classifier is trained to improve the existing trained ensemble. In the SGB, the subset of the training data is selected randomly in each iteration. Then, the random subset of the training set is used to fit the base learner and tune the model for the current iteration. The SGB model has many parameters [16] which can be defined by the user. The number of trees which determines what number of trees are to be built in the model. The interaction depth parameter can control the maximum size of each tree. The impact of each consecutive tree on the final predictions controlled by shrinkage also known as learning rate. The low learning rate with a few number of trees can give poor results, while a large number of trees, may improve results.

Random Forest (RF) [17] is a collection of decision trees that are different in the structure. RF selects the best splits of each node between a random subset of the features. Also, the bootstrap or subsampling methods used by the training set to grow each tree. The RF has two parameters which can be specified by the user; the total number of trees of the model, and the number of splits controls the number of features in each node split. Building RF models with a large number of trees does not lead to overfitting due to the "strong law of large numbers" [18].

Bagging [19] is a method that takes different samples datasets, creates a set of high-variance base learners (usually decision tree), and then averages the prediction results. In the Bagging, the subset of the training data is drawn with replacement from the training set which makes it different that SGB. Also, all features are considered for splitting a node, unlike RF where a subset of the features is randomly selected, and the best features of the subset are used to split the node.

5.2   Dataset

In order to create the predictive model for VM live migration decision, we need to select data from the experiment results which address the following question "what VMs are migratable or non-migratable?". The answer to the question assists choosing the VMs with a low migration time to be migrated. That will reduce the associated cost with the migration as well as the network traffic.

It is critical that we feed the model with the right data that solve the above question. Consequently, the dataset for training and testing the models is selected from 10 stable workloads' results with 10 features and marked with two class labels as illustrated in Table 2. When the workload can be migrated during its run time, we mark it as migratable workload. The migratable workloads take around 15 s to be fully migrated between hosts.

The dataset is determined to 20 s which is from 50 to 70 s where the migrations starting point occurred during this period. Each row of the dataset represents one second which is taken from the recorded logs of the experiment, and it has the values of the futures as well as the class labels. The dataset contains 18000 rows and is divided into two sets: a training set (75%) and testing set (25%). The training set is used to build the model, and the testing set is used to evaluate the performance of the model. We took advantage of the built-in function *createDataPartition* in `caret` package [13] to have stratified random splits within each class label of the data set.

5.3   Tuning the Models

We used `caret` package [13] to create our predictive models. The package depends on 27 packages, and the *train* is the main function that can be used to

Table 2. Dataset structure

Workload	Class label	Features
crypto.rsa	Migratable	Total system memory
scimark.monte_carlo		Memory used
scimark.sor.small		Free memory
scimark.sparse.small		Buffer cache
scimark.sparse.large		Number of CPUs
compiler.compiler	Non-migratable	CPU usage
serial		Load average
sunflow		Queue size
xml.transform		Blocked tasks
xml.validation		Transactions (I/O)

build and evaluate the models. The first step in creating a predictive model is to choose the model. In this paper, we selected the following models: Stochastic Gradient Boosted (SGB), Random Forest (RF), and Bagged Tree (BT). We used gbm package [16] for SGB, randomForest package for RF, and ipred package for BT.

Next, we need to set the tuning parameters of the models. There are no tuning parameters for BT model. However, for the SGB model, we tuned combinations of values of the number of trees, $n.tree = seq(100, 1000, by = 50)$, the depth of each tree, $interaction.depth = seq(1, 7, by = 2)$, and the learning rate (or shrinkage), $shrinkage = c(0.01, 0.1)$ which is in total generates 152 combinations. For RF model, the values of the number of trees to grow, $ntree = seq(100, 1000, by = 50)$, and the number of variables at each node splits, $mtry = seq(1, 7, by = 2)$.

Then, we need to specify the measures of performance for the models and pass it to the metric argument of the train function. We chose the area under the Receiver Operating Characteristic (ROC) curve, or simply AUC, to assess the performance of the models as advised in [20,21]. Finally, we need to specify the method of resampling such as cross-validation or bootstrap by using the *trainControl* function. We selected repeated K-fold cross-validation as resampling method where $K = 10$ which recommended in [22] and repeated 5 times. After resampling, the train function determines the best values of the tuning parameters and fit them to the final model.

5.4   Performance Evaluation of the Models

We evaluate and compare the performance of the models to determine which model performs best for our dataset. We explore the impact of model parameters on the performance, training time, and prediction time. We then identify the most important features of each model, before comparing the accuracy of the SGB, RF, BT models with seven alternative models.

Figures 2 and 3 exhibit the correlation between different tuning parameters of SGB and RF models and the resampled estimate of the AUC. For SGB model, the AUC value increases when the number of trees grows, and the value max tree depth increases. Also, the SGB performs better with 0.1 shrinkage value. The best parameters value of the SGB model for our dataset when the number of trees is 1,000, the depth is 7, and the shrinkage is 0.1. However, the best AUC value can be achieved with fewer trees and number of splits in RF model as illustrated in Fig. 3. Also, it should be noticed that the increasing of the number of the splits of each node could reduce the value of AUC which appeared in Fig. 3 when the number of splits is 5 or 7.

We compare the training time of each model with various tuning parameters values as exposed in Fig. 4. The training time of BT model is 62 s which is not in Fig. 4 because the model has no tuning parameters to compare between their values. Figure 4 shows a link between the training time and the number of trees and the number of splits or depth. When the number of trees, the number of splits, or the tree depth grows, the training time increases. Also, when the learning rate value of the SGB increases, the training time increases.

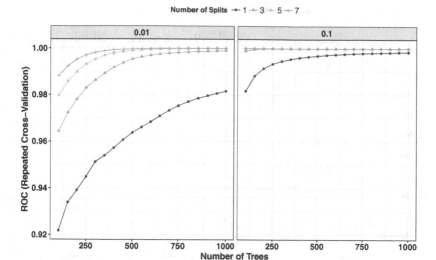

Fig. 2. The value of AUC with various SGB parameters' values

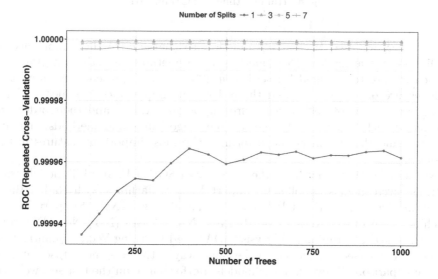

Fig. 3. The value of AUC with various RF parameters' values

Furthermore, the results show that the training time of SGB model can be faster than RF model when its tree depth is 1 or 3 and longer when its tree depth is 5 or 7. Moreover, the training time of different SGB models shows a wide variation while the training time of the RF models presents less variation. In addition, the prediction time of the models is very fast, between 1 to 15 ms.

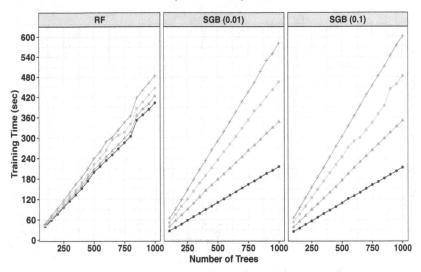

Fig. 4. Training time of SGB and RF

However, we used the *varImp* function to characterize the importance of predictors on the models. Each model has its method for estimating the link of each feature to the model described in [23]. Figure 5 presents the impact of each feature on the models and the order. The importance is scaled to have a maximum value of 100. The feature importance values and their order are distinct in each model due to the different algorithms and methods that are used to build the classifiers and determine the most important features in the models.

We went further to evaluate the accuracy of SGB, RF, and BT models with another seven models as follows: Support Vector Machines with Radial Basis Function Kernel (SVMRadial), K-Nearest Neighbors (KNN), Support Vector Machines with Linear Kernel (SVMLinear), Naive Bayes (NB), NeuralNetwork (NNet), Linear Discriminant Analysis (LDA), and Learning Vector Quantization (LVQ). We assessed the models in the same way with the same dataset to gain a fair comparison. To evaluate the models' performance on the test set, we used the confusionMatrix function to obtain a summary of the prediction results on our models. The function can provide information such as the accuracy rate, the confidence interval, the number of correct and incorrect predictions, the sensitivity, and the specificity. Figure 6 illustrates the estimated accuracy of each model as well as the 95% confidence interval. From Fig. 6, we can conclude that there is little difference between the SGB, RF, and BT in estimated accuracy and they have the highest accuracy compared to other models.

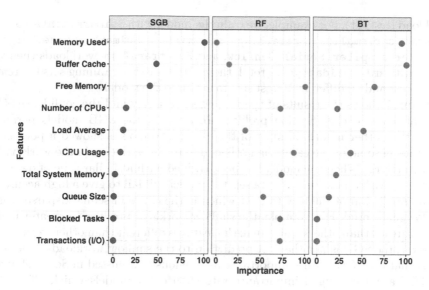

Fig. 5. Feature importance of SGB, RF, BT

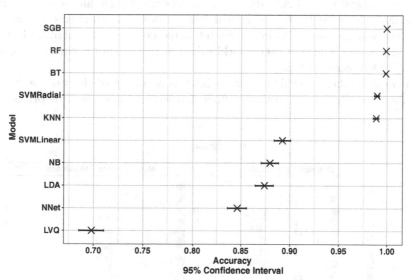

Fig. 6. Compare difference models

6 Predicting Migration Outcome

This section introduces the performance evaluation of SB, RF, BT models which created from various subsets of the original dataset. The aim is to understand the effect of one type of workloads or VMs on the estimated accuracy. First, we train the models with 9 workloads (dataset) and evaluate the classifiers with the 10th workload (test set). For example, we train the model with `Scimark.monte_carlo`, `Scimark.sor.small`, `Scimark.sparse.small`, `Scimark.sparse.largel`, `Compiler.compiler`, `Serial`, `Sunflow`, and `Xml.transform` workloads then we test it with `xml.validation`. In total, there are 10 distinct training sets to create the models and 10 different 10 test sets to evaluate the models.

Figure 7 shows the results of the estimated accuracy and the confidence 95% interval for each classification problem. In general, the SGB models perform better than other models in most of the cases. The models showed a poor performance in some of the cases than others due to the variation in the workloads' characteristics. When the model has been trained with a similar type of data set and tested with a new kind of data set, the model will fail to give a high accuracy on the prediction of the new test set which is the cause in scimark.sparse.large, sunflow, and serial. For example, the sunflow workload produces a greater number of threads than others which make its features different than other workloads. However, the SGB is not the best prediction to the sunflow workloads. The reason is that the SGB feature importance as previously discussed in Sect. 5.4 gives the load average a small importance rate than other models which affect the estimate accuracy in this case.

In the migratable workloads, `Scimark.sparse.large` has a poor performance over the models. The workload has similar characteristics of the non-migratable workloads which make the model predict it as a non-migratable workload. We can conclude that other workloads' features are not accurate to predict the `scimark.sparse.large`.

We went further to evaluate the estimated accuracy of the various type of VMs' datasets. Our aim is to know which characteristics of the VMs present the

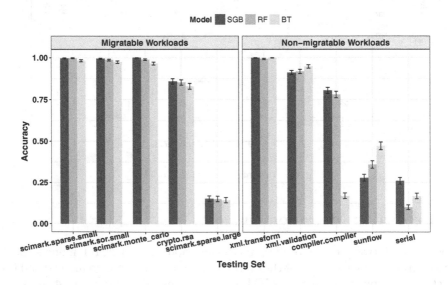

Fig. 7. Compare SGB, RF, and BT with various datasets

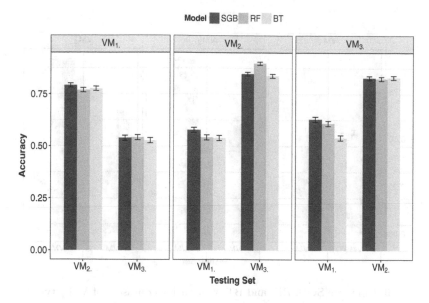

Fig. 8. Compare SGB, RF, and BT with various datasets of VM_i types

most impact on the estimated accuracy. Regardless the memory size of the VMs in our dataset, Fig. 8 shows the results of training the models with VMs that are different in the number of the CPUs while Fig. 9 present the estimated accuracy of VMs which are different in memory sizes.

In each case, the models created from 3 various training sets and validated with two testing sets. In Fig. 8, the experiment results of $VM_1.$, $VM_2.$, and $VM_3.$ are the data sets to build and evaluate the models. We used one of the datasets to create the models, and the other two datasets are used to assess the models. For example, the experiment results of the VMs with one CPU used to build the models, and the results of the VMs with two and three CPUs used to evaluate the model. The $VM_1.$ models are better in predicting the VMs with 2 CPUs, and the $VM_2.$ models perform better in predicting the VMs with 3 CPUs while $VM_3.$ models are much beneficial in predicting the VMs with 2 CPUs.

We used the experiment results of $VM_{.1}$, $VM_{.2}$, and $VM_{.3}$ to train and test the models as shown in Fig. 9. For example, the experiment results of the VMs with 1 GB RAM used to build the models, and the results of the VMs with 2 and 3 GB RAM used to evaluate the model. The $VM_{.1}$ models are not the best in predicting the VMs with 2 and 3 GB RAM compare to the other models. The $VM_{.2}$ models perform better in predicting the VMs with 3 GB RAM while $VM_{.3}$ models are much useful in predicting the VMs with 2 GB RAM.

The prediction results of the models are influenced by a number of factors. First, the differences in the sub-datasets that are used to train and test the models can affect the estimated accuracy of the prediction. In order to create a predictive model, the sub-datasets of the original dataset should reflect the

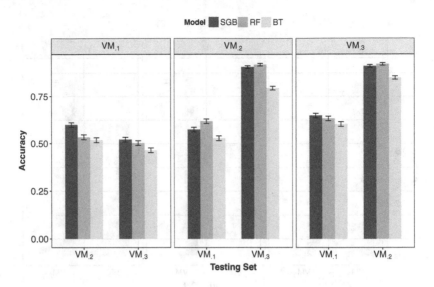

Fig. 9. Compare SGB, RF, and BT with various datasets of $VM_{.j}$ types

dataset. Second, the influence of predictors on the each model (importance of features) can affect the prediction results of the models. Finally, the process of building the models are different in each model. Each model has its algorithm and parameters to create the model. Any change on one of the parameters leads to amending the predictions results.

7   Conclusions

This paper has measured the live migration time for different workload character-istics on various VMs capacities. We used KVM as hypervisor and SPECjvm2008 benchmark to generate the workloads. The results show that some VMs can be migrated within a short time while others take a long time to migrate and some cannot be migrated during the workload execution. The paper presents the process of creating Stochastic Gradient Boosted (SGB), Random Forest (RF) and Bagged Tree (BT) models from the results of the experiment. We showed the effect of tuning the models with different values as well as training and eval-uating the models among the various sub-datasets from the original dataset. It is clear that there is no easy choice of the best model to employ and in practice a combination of the models presented could be used to gain a better prediction of which VMs to migrate. In our future work we aim to deploy these models into a trace driven simulation environment in order to experiment with different consolidation strategies under different workload assumptions.

References

1. Clark, C., Fraser, K., Hand, S., Hansen, J.G., Jul, E., Limpach, C., Pratt, I., Warfield, A: Live migration of virtual machines. In: NSDI (2005)
2. KVM. http://www.linux-kvm.org. Accessed 5 May 2017
3. SPECjvm2008. http://www.spec.org/jvm2008. Accessed 5 May 2017
4. Strunk, A., Dargie, W.: Does live migration of virtual machines cost energy? In: IEEE AINA (2013)
5. Rybina, K., Patni, A., Schill, A.: Analysing the migration time of live migration of multiple virtual machines. In: CLOSER, pp. 590–597 (2014)
6. Akoush, S., Sohan, R., Rice, A., Moore, A.W., Hopper, A.: Predicting the performance of virtual machine migration. In: IEEE MASCOTS, pp. 37–46 (2010)
7. Hu, W., Hicks, A., Zhang, L., Dow, E.M., Soni, V., Jiang, H., Bull, R., Matthews, J.N.: A quantitative study of virtual machine live migration. In: Proceedings of the 2013 ACM Cloud and Autonomic Computing Conference, p. 11. ACM (2013)
8. McGough, A.S., Moubayed, N.A., Forshaw, M.: Using machine learning in trace-driven energy-aware simulations of high-throughput computing systems. In: Proceedings of the 8th ACM/SPEC on International Conference on Performance Engineering Companion, pp. 55–60. ACM (2017)
9. Uriarte, R.B., Tiezzi, F., Tsaftaris, S.A.: Supporting autonomic management of clouds: service clustering with random forest. IEEE Trans. Netw. Serv. Manage. 13(3), 595–607 (2016)
10. Openfiler. http://www.openfiler.com. Accessed 5 May 2017
11. Alrajeh, O.: VM live migration script. http://www.github.com/oalrajeh/VM_Live_Migration. Accessed 5 May 2017
12. Memusg. https://gist.github.com/netj/526585. Accessed 5 May 2017
13. Kuhn, M.: Caret package. J. Stat. Softw. 28(5), 1–26 (2008)
14. R Core Team. R: A Language and Environment for Statistical Computing. R Foundation for Statistical Computing, Vienna, Austria (2016)
15. Friedman, J.H.: Stochastic gradient boosting. Comput. Stat. Data Anal. 38(4), 367–378 (2002)
16. Ridgeway, G.: Generalized boosted models: a guide to the GBM package. Update 1(1), 2007 (2007)
17. Breiman, L.: Random forests. Mach. Learn. 45(1), 5–32 (2001)
18. Feller, W.: An Introduction to Probability Theory and Its Applications. Vol. 1, vol. 3. Wiley, New York (1968)
19. Breiman, L.: Bagging predictors. Mach. Learn. 24(2), 123–140 (1996)
20. Hanley, J.A., McNeil, B.J.: The meaning and use of the area under a receiver operating characteristic (ROC) curve. Radiology 143(1), 29–36 (1982)
21. Ling, C.X., Huang, J., Zhang, H.: AUC: a better measure than accuracy in comparing learning algorithms. In: Xiang, Y., Chaib-draa, B. (eds.) AI 2003. LNCS, vol. 2671, pp. 329–341. Springer, Heidelberg (2003). doi:10.1007/3-540-44886-1_25
22. Kohavi, R., et al.: A study of cross-validation and bootstrap for accuracy estimation and model selection. In: International Joint Conference on Artificial Intelligence, vol. 14, pp. 1137–1145, Stanford, CA (1995)
23. Kuhn, M.: Building predictive models in R using the caret package. J. Stat. Softw. 28(1), 1–26 (2008)

Model-Based Simulation in Möbius: An Efficient Approach Targeting Loosely Interconnected Components

Giulio Masetti[1,2]([⊠]), Silvano Chiaradonna[1], and Felicita Di Giandomenico[1]

[1] ISTI-CNR, Pisa, Italy
giulio.masetti@isti.cnr.it
[2] University of Pisa, Pisa, Italy

Abstract. This paper addresses the generation of stochastic models for dependability and performability analysis of complex systems, through automatic replication of template models inside the Möbius modeling framework. The proposed solution is tailored to systems composed by large populations of similar non-anonymous components, loosely interconnected with each other (as typically encountered in the electrical or transportation sectors). The approach is based on models that define channels shared among replicas, used to exchange the values of each state variable of a replica with the other replicas that need to use them. The goal is to improve the performance of simulation based solvers with respect to the existing state-sharing approach, when employed in the modeling of the addressed class of systems. Simulation results for the time overheads induced by both channel-sharing and state-sharing approaches for different system scenarios are presented and discussed. They confirm the expected gain in efficiency of the proposed channel-sharing approach in the addressed system context.

1 Introduction and Related Work

Model-based approaches are well-suited to analyze complex systems in terms of a variety of metrics, such as dependability and security related indicators [12]. Modern systems exacerbate the challenges of complexity and scalability. In this prospect, we target systems made of large populations of similar and loosely interconnected non-anonymous components. This is a typical configuration of many critical infrastructures, e.g. in the transportation and energy sectors, and in general in cyber-physical systems where the cyber control manages a set of physical components arranged according to a topology. Our goal is to enhance efficiency of performance and dependability analyses for such kinds of systems.

Research on efficient solutions to modeling and analysis of large systems is active for a long time and numerous studies are available in the literature. In the case of analytical solvers, many papers focus on the anonymous replication of an atomic model, well suited to represent a population of identical components, each interacting with all the others or completely independent, following the

© Springer International Publishing AG 2017
P. Reinecke and A. Di Marco (Eds.): EPEW 2017, LNCS 10497, pp. 184–198, 2017.
DOI: 10.1007/978-3-319-66583-2_12

approach originally presented in [17]. This configuration favors the application of the strong lumping theorem [4,15], enhancing the state space generation based solvers [11].

The need to cope with systems where components, although belonging to the same typology, differ in their parameters setting and in their impact on the overall system behavior, depending on the position they hold in the system configuration, triggered studies offering solutions under specific conditions. Hierarchical modeling [20], where lower-level models capture the detailed behaviour of components and the topmost level model represents the topology of interactions among components, is commonly adopted when the directed graph of interactions is acyclic and measures capturing global system behaviour, e.g., system availability or reliability, are considered. Instead, in systems typically composed by many loosely interconnected components according to dependency topologies presenting cycles, fixed-point solution strategies [19] can be considered, but only if components are nearly independent and the measure of interest allows the application of Brouwer's theorem to guarantee the solution's existence.

From a wider perspective, our interest is in systems whose components are loosely interconnected but not nearly independent, such as in a monotone load sharing regime [2], and in performability measures tackling aspects of many individual components, not necessarily linked to a global Brouwer-compatible measure. In addition, targeted system components can have non-Markovian behaviour, e.g., because of deterministic time delays, thus efficient analytical approaches, such as the techniques presented in [3,8,13,16], are not applicable. Hence, we focus on simulation-based solvers. In particular, we developed our solution adopting the Möbius modeling and evaluation framework [10], a powerful and widely adopted environment, encompassing a variety of modeling formalisms, composition operators, analytical solvers and a simulation-based solver.

In [6,12], a solution, referred in the following as *state-sharing* approach, is proposed in the context of Möbius, where an indexing mechanism is exploited to build non-anonymous replicas of a given template model. It is a general solution, but its efficiency, also when a simulation-based solver is employed, remains limited by the fact that it assumes a complete graph of interactions among the replicated components. This is a pessimistic vision in the great majority of real-world systems, typically composed by many loosely interconnected components according to regular dependency topologies (tree, mesh, cycle, etc.).

To improve in efficiency when simulation based solvers are employed, we investigated solutions exploiting the dependency topology. In this paper, a new strategy, named channel-sharing approach is presented. It implements non-anonymous replication, but using a single channel shared among all the replicas to exchange values of the state variables following the actual system topology, thus enhancing simulation performance.

An alternative approach, presented in [7], is instead based on: (1) the automatic generation of the indexed replicas of a template model representing a set of system components, and (2) the automatic definition of the composed model joining the indexed replicas, where the state variables are shared among the replicas following the actual system topology, thus removing (or excluding) the unnecessary interactions among replicas.

The rest of the paper is organized as follows. Section 2 describes the formalism and tool used to define and evaluate our approach, while Sect. 3 presents the logical architecture of targeted systems. Section 4 is then devoted to present our *channel-sharing* solution. Adopting the case study introduced in Sect. 5, the results obtained from the new approach are discussed and compared with those of the existing state-sharing approach. Some final considerations are made in Sect. 7, while conclusions are drawn in Sect. 8.

2   Formalism and Tool

To describe, implement and evaluated our approach, the Möbius modeling framework [10] and its supporting tool Möbius [9] have been used. Among the formalisms available in Möbius, our models are defined using the Stochastic Activity Networks (SAN) formalism [18], a stochastic extension of Petri nets based on four primitives: places, activities (transitions), input gates, and output gates. Special places, called "extended places", allow the representation of the primitive data types of the programming language C++, like short, float, double, including structures and arrays of primitive data types or of extended place types. Input gates control when an activity is enabled. The marking changes occurring when an activity completes are defined by the input and output gates. The SAN primitives are defined by C++ statements. Each SAN place is interpreted as a State Variable (SV) and the Cartesian product of all the feasible SVs values is the *state space* of the SAN model.

In Möbius, submodels can be composed hierarchically by sharing one or more SVs among them. In particular, the *Join* and *Rep* compositional operators [17] can be used, respectively, to bring together two or more submodels or to automatically construct identical copies (replicas) of a submodel. In a *Join* composed model, a SV can be local to a submodel, if it cannot be directly accessed by other submodels, or shared among a subset of submodels, if each submodel of the subset can directly access it. In a *Rep* composed model, a SV can be either local to each replica or shared among *all* replicas. To simply illustrate the *Join* and *Rep* operators at work, consider two SAN models, M_1 and M_2, containing places A_1, B_1 the former and A_2, B_2 the latter. We can obtain 7 different composed models joining M_1 and M_2 together with different combinations of shared variables. In fact, if A_1 is shared with A_2 then the composed join model has 3 SVs, namely the shared variable $A_1 = A_2$ and the local variables B_1 and B_2; if A_1 and B_1 are shared with B_2 and A_2, respectively, then the composed join model has 2 SVs, namely $A_1 = B_2$ and $B_1 = A_2$; and similarly for the other combinations, counting also a join with only local SVs. We can obtain 4 composed models replicating n times the submodel M_1 with different combinations of SVs. In fact, sharing only one SV among n replicas the composed model has $n + 1$ SVs, namely: one shared variable $A_1^{(1)} = \cdots = A_1^{(n)}$ and n local variables

$B_1^{(1)}, \ldots, B_1^{(n)}$, or one shared variable $B_1^{(1)} = \cdots = B_1^{(n)}$ and n local variables $A_1^{(1)}, \ldots, A_1^{(n)}$, where the superscript (j) refers to the j-th replica; a composed *Rep* model sharing both A_1 and B_1 has 2 shared SVs and no local SVs, namely $A_1^{(1)} = \cdots = A_1^{(n)}$ and $B_1^{(1)} = \cdots = B_1^{(n)}$; and finally, a composed model sharing no SV has $2n$ local SVs, namely $A_1^{(1)}, \ldots, A_1^{(n)}, B_1^{(1)}, \ldots, B_1^{(n)}$.

A template model is an atomic or composed generic model identified as a building block. It represents a group of homogeneous components, as described in Sect. 3. All the formalisms and solvers supported by Möbius are based on and defined in terms of C++ code. Thus, the tool supports external C++ data structures, statically defined at compilation time, and can include and link external C++ libraries.

3   Logical Architecture of Targeted Systems

The focus is on systems composed of a large number of components, characterized by a low degree of connectivity (and therefore dependency) as resulting from the system topology. Let's take for example the case of the electrical grid: the grid portion under analysis includes a number of collection points called buses. All buses have the same aim, that is collecting and distributing electricity, so they belong to the same component category and their models are similar. However, each bus has individual peculiarities, namely the position occupied in the grid topology and the number and kind of attached electrical equipment.

In the regular semi-Markov context, such as in [8], the degree of similarity among system components, once a model for each component has been designed, can be defined in terms of the infinitesimal generator matrix structure, and has a direct impact on solver's performance. In our non-Markov context, when abstracting the system component for analysis purpose, a *generic* component can be assumed and a template model for it can be built, which is then replicated through an indexing function to model all the *specific* components included in the system, each with its individual peculiarities. Component models are considered similar because they are instances of an unique template. The solution proposed in this paper helps the modeler in modeling a generic component and then automatically building the set of its non-anonymous indexed replicas.

In general terms, the logical architecture of the given reference system can be seen as composed by:

- A large number n of connected components (called specific components).
- One or more generic components. Each generic component groups all the specific components that can be defined using a single template SAN model, where an index is used to refer each specific component. Doing so, each specific component maintains individual peculiarities.
- A topology that defines the connections among the generic or specific components. The directed graph representing interdependencies among components can have cycles.

The proposed approach can be adopted in modeling those systems for which a generic component can be defined to represent the set of its specific components, using all the features supported by the SAN models. A SAN model, in particular, supports the following features: initial state of each component, state changing, random choices and random exponential or not exponential times can be defined as a function of the index of the specific component. Thus, for example, the deterministic repair time of the i-th specific component can be defined in the SAN using the i-th entry $det[i]$ of the constant C++ array det, initialized with the repair times of all the components. In addition, structure, behavior and parameters of specific components defined by the same template can be different.

In order to allow automatic generation of the non-anonymous replicas, the structure, behavior and parameters of the generic component have to be defined as a function of the index i of the replica, with $i = 0, 1, \ldots, n - 1$. The state S_i of each component i is represented by the Cartesian product of v state variables

$$S_i = SV_0^{(i)} \times SV_1^{(i)} \times \cdots \times SV_{v-1}^{(i)}$$

with v being the same for each replica. The SVs values can be discrete or continuous. A specific component can depend on the SVs of other components. The SVs of each specific component that impact on the other components, called dependency-related SVs, are $SV_0^{(i)}$, $SV_1^{(i)}$, $\ldots$, $SV_{m-1}^{(i)}$, with $m \leq v$, being m the same for each replica. The other $v - m$ SVs are $local$ SVs. The $dependency$ $degree$ of the component i, called $\delta_i \in \{0, 1, \ldots, n - 1\}$, indicates that the structure, behavior and parameters of the component i depend on the SVs of δ_i other components. The list of the components from which the component i depends on is called $\Delta_i = \{j_0, j_1, \ldots, j_{\delta_i - 1}\}$. If $\delta_i = 0$ then the component i does not depend on any other components, although S_i can impact on S_j, if $i \in \Delta_j$. When $\delta_i = 0$ and $\nexists j \mid i \in \Delta_j$ the component i is said to be independent. From a topological point of view, Δ_i, for $i = 0, \ldots, n - 1$, defines an oriented graph that represents how the n components are connected and how they depend on each other to form the overall system. An independent component corresponds to a disconnected node of the graph, while a system where there is full connectivity among its components results in a complete graph.

4   Channel-Sharing Approach

The proposed channel-sharing approach models the interactions among similar components represented by non-anonymous replication of a given template model. Its formal definition relies on the Möbius framework, but it is a general mechanism applicable to any modeling and evaluation environment that supports:

- the SAN-like formalism,
- composition of submodels based on sharing of SVs,
- automatic replication operator.

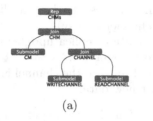
(a)

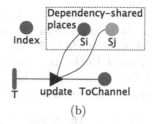
(b)

Fig. 1. Replicated model (a) of a generic component and SAN-based update (b), defined in *CM*, of a dependency-related SV (place) at completion of an activity.

Differently from the state-sharing approach, which relies on an n-sized array shared among all the replicas of the template model for each dependency-related SV (see [6,12]), the channel-sharing approach consists of: (1) defining as local to each replica all its dependency-related SVs, (2) using a small-sized channel, shared among all the replicas, to exchange the values of dependency-related SVs among the interested replicas, each time they are updated. The channel-sharing approach models similar components and the interactions among them, defining the composed model *CHMs* in Fig. 1a, where:

- The state-sharing Rep operator is used to automatically construct the overall model *CHMs*, composed by the indexed non anonymous replicas of the template *CHM*.
- The template *CHM* is the submodel defined by the operator Join composing the template model *CM*, that represents the generic component, with the submodel *CHANNEL*, that models the channel and the related read and write operations.
- All the dependency-related SVs, whose values are exchanged with other replicas, are shared with the submodel *CHANNEL* joined to the same replica, but they are local to the replica of the template *CHM*. All the other SVs can be local to the replica or shared among all the replicas, depending on the definition of the component.
- The model *CHANNEL* is used to send the values of the dependency-related SVs of a replica to other replicas, each time the dependency-related SVs are updated by the model *CM*.
- The channel used to transmit the values of the dependency-related SVs among replicas is defined in the model *CHANNEL* by a set of SVs (places) shared among the replicas.
- The model *CHANNEL* is defined by the operator Join composing the two atomic SAN models *WRITECHANNEL* and *READCHANNEL* that are used, respectively, by the sender replica to write the data into the channel and by all the destination replicas to read the data from the channel.
- The SAN models *WRITECHANNEL* and *READCHANNEL* include only immediate activities, thus they do not impact on the measures of interest.

Each time one or more dependency-related SVs of a replica are updated, the following steps are performed in the listed order:

(1) the model *WRITECHANNEL* writes the new values (the data) in the channel and locks the channel from writing (the channel is busy),
(2) the model *READCHANNEL* of each destination replica updates the SV(s) of the replica with the data received from the channel,
(3) when all destination replicas have received the data, the channel is unlocked from writing, and can be used to transmit new data.

Figure 1b depicts how the template model *CM* updates a dependency-related SV when an activity *T* completes, using the SAN formalism. Obviously, in the final complete model, all the SAN primitives (places, input and output gates) connected to *T*, are to be considered and defined in *CM*, depending on the actual component to model. The index of each replica is modeled by the place *Index*, that is defined at time 0, using the indexing mechanism proposed in [6,12]. The place *Index* is local to the replica, but is shared with the model *CHANNEL*. The place *Si* represents the dependency-related SV of a specific replica. The place *Sj* represents the dependency-related SVs of the other replicas that impact on *Si*. For each replica *i*, the places *Si* and *Sj* are local to the replica and the size of the array *Sj* is equal to δ, i.e., the number of replicas (components) from which the replica *i* depends. Thus, each dependency-related SV associated to a replica requires the definition of $\delta + 1$ SVs local to the replica. The places *Si* and *Sj* are shared with the model *CHANNEL*. The place *ToChannel* is shared with the model *CHANNEL* and it is used to trigger the activation of the submodel *WRITECHANNEL* each time the SV *Si* is updated. The SV *Si* is updated by the gate *update* at completion of the activity *T* as a function of the values of the array-type place *Sj*. The actual definition of the function, depending on the modeled component, is not shown being out of the scope of the paper. Figure 2a depicts the model used to write into the channel the data to be sent to the destination replicas, when one dependency-related SV *Si* is considered ($m = 1$). It can be easily extended when more than one dependency-related SV is updated at the same time, by adding fields to the record place *Channel* representing the channel data structure. The place *ReadyChannel*, initialized with 1 token, is shared among all replicas and is used to lock the channel (*ReadyChannel->Mark() == 0*) from writing until the data are read by all the destination replicas. The record-type extended place *Channel* represents the channel shared among

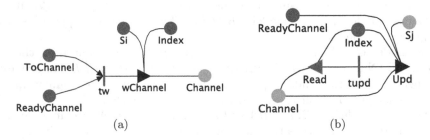

Fig. 2. SAN models used to write (a) and to read (a) the shared data into the channel.

all replicas used to transmit the values of the SVs. It is defined by 3 short-type fields: *data*, *index* and *ndest*, used to represent, respectively, the value of the dependency-related SV to send to the replicas, the index of the sender replica and the current number of destination replicas that have not yet read the data. In general, a field (a list or more efficiently an array) with the information about the destination replicas should be used, in order to trigger the destination replicas. But, depending on the topology of the dependencies, more simple data structures can be used to identify the destination replica. For example, considering a circular dependency topology, a simple function of the field *index* (the index of the sender) can be used to identify the destination replicas. The field *ndest* is used to get the current destination replica that have to read the data from the channel.

At completion of the immediate activity *tw*, that is enabled when there is 1 token in both places *ToChannel* and *ReadyChannel*, the gate *wChannel* updates: (1) the fields *data* and *index* with the values of the places respectively Si and *Index*, that are shared with the template model *CM*, and (2) the field *ndest* to δ. Figure 2b depicts the model used by a replica to receive data from the channel. At completion of the activity *tupd*, the output gate *Upd* performs the following steps in the listed order: (1) updates the value of Sj with the data of the channel, (2) decrements the field *ndest* of the channel, and (3) reset the shared place *ReadyChannel* to 1, if *Channel->ndest->Mark() == 0*, when the channel is ready to accept new data to send. The activity *tupd* is only enabled when the replica indexed by the place *Index* is the destination of the data in the channel, i.e., when the value of the place *Index* corresponds to the current value of the field *ndest* of the channel, as defined in the input gate *Read*. Thus the value of *ndest* is used to avoid multiple readings of the same data by a replica.

For the channel-sharing approach, each dependency-related SV of each replica is local to the replica. In addition, the channel definition requires only a small set of SVs to be shared among all the replicas (more precisely, there are $m+2$ shared SVs, i.e., m for data, 1 for the index and 1 for *ndest*). Therefore, a reduction in the time overhead during the initialization of the simulation solver is expected, with respect to approaches that share each dependency-related SVs among all replicas. However, the channel-sharing approach introduces a time overhead at simulation time, after the initialization of the data structures of the simulator, due to the new channel model, that is triggered each time the shared SVs are updated. In Sect. 6, comparisons between the channel-sharing approach and the state-sharing approach have been performed to understand the impact of these phenomena on the efficiency of the approaches, at varying both the number of considered replicas and the dependency degree.

5   Case Study

To illustrate the proposed approach, a simple but effective case study for our purpose is considered, having all the characteristics of the addressed category of systems as described in Sect. 3. This case study is based on systems where the failure of a component impacts on the failure of its neighbors, based on the interdependencies topology (failure correlation). Let's consider a population of n components working in parallel and properly functioning at time $t = 0$. At every time instant $t > 0$, each component can be either functioning or failed. No repair is considered. For the purpose of our analysis, the time to failure of each component is an exponentially distributed random time, although each probability density functions implemented in Möbius could be considered [10]. Whenever a component fails, the failure rate of its neighbors is updated. In particular, the failure of component i changes the failure rate of the components $(i + 1)\%n, \ldots, (i + \delta)\%n$, i.e., the first δ components that cyclically follow the failed component, with $\delta \ll n$.

The SAN model implementing the logical structure of the generic component, consists of two places, A and B, and a transition with a state-dependent rate. When a token is in A the component is properly working, when a token is in B the component is failed, and the two alternatives are mutually exclusive. In order to represent the interaction among components, either one or four shared dependency-related SVs are employed, i.e., $m = 1, 4$.

Being the focus of this paper on the efficacy of the channel-sharing modeling approach, further details are not relevant and therefore omitted.

6   Evaluation Results

In order to demonstrate the feasibility and utility of the proposed approach, a comparison of the performance results of the Möbius simulator induced by both channel-sharing and state-sharing approaches has been conducted. To this purpose, the terminating Möbius simulator [10] has been used to evaluate at each execution different measures of dependability (reward variables) for the proposed case study, like the cumulated time component i stays in a specific state and the probability that component i is failed at time t. As a form of validation, for some sample models and using state space generation of Möbius, it has been verified that the number of stable states obtained with both approaches is the same. In addition, also the results obtained for the defined measures have been the same for both channel-sharing and state-sharing approaches.

Each execution of the terminating Möbius simulator is defined for a specific setting of all the parameters of the considered models (corresponding to an experiment in the Möbius terminology). Each execution of the terminating simulator starts initializing the data structures, then runs k batches (replications in Möbius notation) with $k \geq 1$, each batch of $t = 100$ time units.

The following performance measures have been considered:

- $\tau(k)$: The total amount of CPU, in seconds, used by one execution of the Möbius simulator that runs k batches, with $k \geq 1$.
- τ_{init} or $\tau(0)$: The amount of CPU, in seconds, used by each execution of the Möbius simulator to initialize the data structures of the simulator. This is the CPU time used by the simulator to output the string "SIMULA-TOR::Preparing to run()". The definition of τ_{init} as a function of $\tau(k)$ is:

$\tau_{init} = \tau(1) - (\tau(2) - \tau(1)) = 2\tau(1) - \tau(2)$, where $\tau(2) - \tau(1)$ is the total amount of *CPU*, in seconds, used by one execution of the Möbius simulator to run a batch.

– $\Delta\hat{\tau}(k)$ and $\Delta\tau(k)$ to quantify the gain (if positive) or loss (if negative) of the channel-sharing over the state-sharing in percentage and in seconds, respectively:

$$\Delta\hat{\tau}(k) = \frac{\tau^{ss}(k) - \tau^{cs}(k)}{\tau^{ss}(k)},$$
$$\Delta\tau(k) = \tau^{ss}(k) - \tau^{cs}(k)$$

The considered *CPU* time includes both user and system *CPU* times. To exercise the approaches in relevant contexts, the following scenarios have been considered:

– for each specific component (replica) both one and four dependency-related SVs, i.e., $m = 1$ and $m = 4$,
– number n of replicas ranging from 100 to 1000,
– dependency degree δ varying from 1 (minimum connectivity) to 10,
– number of batches $k = 1000$, $k = 5000$ and $k = 10000$ (impacting on the precision required for the results).

Simulations were sequentially performed on Intel(R) Core(TM) i7-5960X with 3.00–3.50 GHz CPU, 20 M cache and 32 GB RAM.

Figure 3 depicts the execution times $\tau(1000)$, $\tau(5000)$ and τ_{init} for both approaches, as a function of the number of replicas n, for $\delta = 1$, when only one dependency-related state variable is used, $m = 1$. The values of τ_{init} for the state-sharing approach are always higher than the corresponding channel-sharing values. The difference between the two approaches increases greatly for

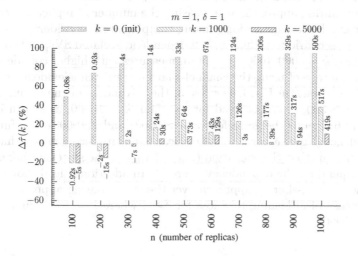

Fig. 3. Performance gain or loss of the channel-sharing over the state-sharing in percentage and in seconds, as a function of replicas number n, with $\delta = 1$ and $m = 1$.

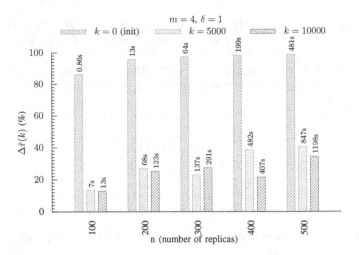

Fig. 4. Performance gain or loss of the channel-sharing over the state-sharing in percentage and seconds, as a function of replicas number n, with $\delta = 1$ and $m = 4$.

higher values of n, varying from about 0.08 s for $n = 100$ to about 500 s for $n = 1000$. This increase is caused by the n-sized array used by state-sharing to represent the dependency-related SV of each replica. Conversely, in the channel-sharing approach, the initialization time is negligible at varing of n. Values of $\tau(1000)$ and $\tau(5000)$ in Fig. 3 show how the execution time of both approaches is impacted differently by the number of batches and by the number of replicas. For $n \geq 500$ and for 1000 batches, the channel-sharing approach provides a significant performance increase over the state-sharing approach, whereas for 5000 batches the gain remains within 10%. As for τ_{init}, the worse overall performance of the state-sharing approach, at increasing the number of replicas, is caused by the higher number of connections among components it considers.

Changing scenario to include $m = 4$ dependency-related SVs, and considering more batches, $k = 5000, 10000$, to obtain measures with higher confidence intervals, the advantage of using the channel-sharing approach is evident for $n \geq 100$ and $\delta = 1$, as shown in Fig. 4. For $\delta = 3$, the advantage of the channel-sharing starts for $n \geq 300$, $k = 5000$, and for $n \geq 500$, $k = 10000$, as shown in Fig. 5. Finally, Fig. 6 depicts the gains for τ_{init}, $\tau(5000)$ and $\tau(10000)$ as a function of the dependency degree δ, for $n = 500$ replicas and considering $m = 4$ shared SVs. Figure 6 confirms that the execution times of τ_{init} of the state-sharing approach does not depend on the dependency degree δ. In addition, Fig. 6 shows how the gain of the channel-sharing approach over the state-sharing approach decreases at the increase of δ, leading to a loss for $\delta \geq 4$ where $k = 10000$ and for $\delta \geq 6$ where $k = 5000$.

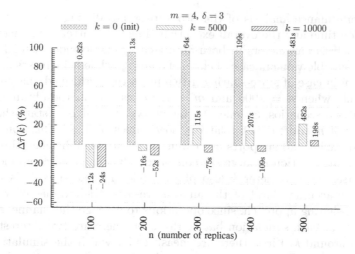

Fig. 5. Performance gain or loss of the channel-sharing over the state-sharing in percentage and seconds, as a function of replicas number n, with $\delta = 3$ and $m = 4$.

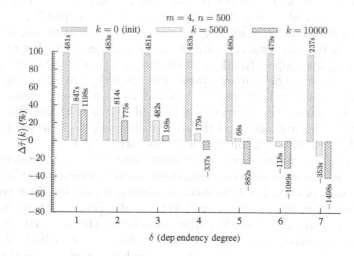

Fig. 6. Performance gain or loss of the channel-sharing over the state-sharing in percentage and seconds, as a function of replicas number δ, with $n = 500$ and $m = 4$.

7 Final Considerations

The performed analyses pointed out how the performance of the two models replication approaches is differently impacted by the four dimensions considered (n, m, δ, k). They provide useful support to understand which approach is better, given a system configuration in terms of number of replicas for each template model, number of dependency-related SVs, dependency degree among replicas, and the precision requested to the analysis results.

The performance analysis of the two approaches distinguishes between the initialization time and the overall execution time. The latter is impacted by all the four analysis parameters in both approaches. The former, instead assumes always a negligible value in case of channel-sharing, while it increases at increasing n and m in case of state-sharing, reaching in the performed analyses values around 8 min when $n = 1000$ and $m = 1$. Overall, from the obtained results in the analysed scenarios, channel-sharing shows preferable to state-sharing for high values of n and m, but relatively low values of δ, k. This confirms the expected behavior and supports our claim on promoting efficiency through the channel-sharing solution in large systems of loosely connected components.

Of course, the parameter k also plays a relevant role since it is representative of the accuracy level of the analysis results. High values of k penalize the channel-sharing approach, since operations to manage the channel structure are executed at each simulation batch. However, there are two interesting considerations around k. First, there are measures for which the simulation study converges in a relatively low number of batches (e.g., in the performed analyses, 1000 batches are always sufficient for the scenario with $m = 1$ to measure Mean Time To Failure of a specific component with a confidence interval narrower than $3 \cdot 10^{-5}$). Second, having the replication mechanism depending on a higher number of parameters gives more opportunities to the fine tuning of parameters values, in relation with the purpose of the analysis itself. For example, if employed as design support to take decision on system configuration options, less accurate results would be well acceptable since the objective is to comparatively evaluate alternatives and not provide definitive quantification of individual performances. In such a case, a low value of k could be a right choice, thus favoring the channel-sharing approach over the state-sharing approach, which always has to afford the same initialization time independently from k.

Concerning the values of the dependency degree δ and of the dependency-related SVs m, those adopted in the analyses are reasonable values in real contexts, such as in electrical systems configurations (e.g., the IEEE300 testbed [1] and the Illinois Center for a Smarter Electric Grid's Texas synthetic grid [14]).

Another feature of the discussed approaches is that they can actually co-exist in the same modeling framework, to gain in efficiency from both approaches by employing the one that fits better the peculiarities of the specific system components under analysis.

Finally, the channel-sharing approach, as well as the state-sharing approach and the one in [7], are solutions developed upon the primitives offered by the Möbius framework. A more radical approach is to implement the new operator directly inside Möbius, as preliminary proposed in [5]. However, it is still at embryonic level; more elaboration and, above all, implementation are required to be practically employed.

8   Conclusions

Moving from considerations on the opportunity to improve dependability and performability modeling framework by exploiting the generally limited connections among large population of system components, this paper proposed a novel modeling approach to automatically build (a possibly high number of) replicas of a template model. Differently from the already adopted state-sharing approach, the new channel-sharing approach is defined to account only for the actually existing dependencies among components of the system under analysis. To quantify the extent of the expected gain in performance and to better understand the interplay of the peculiarities of both solutions, an evaluation study involving both the state-shared and channel-shared approaches has been conducted.

Several scenarios, characterized by different values of the dependency degree, the number of replicas, the number of dependency-related SVs for each replica and the number of simulation batches, have been considered. Not surprisingly, the results show cases where the channel-shared solution is better, and others where the state-shared approach wins. Overall, the initial reasoning on the expected benefit brought by exploiting topology awareness has been confirmed. Although limited to the considered scenarios and parameters setting, the simulation outcomes demonstrate the superiority of the newly introduced method with respect to the state-sharing competitor, when the dependency degree is low, the number of replicas is high and few SVs are needed to represent the dependencies among components.

Among the most immediate advancements there is the adoption of the new approach in more complex system scenarios, as offered by the power grid sector. It would require relaxing the simplistic assumption on having just one template model with equal dependency degree among all its replicas, which has been made in this paper to easy the presentation of the channel-shared solution. It would be also the context to potentially exploit the co-existence of both channel-shared and state-shared approaches, to take advantage of the best performance shown by them, as discussed in Sect. 7.

References

1. Adibi, M.: Test systems task force: IEEE 300-bus test case (1993). http://icseg.iti. illinois.edu/ieee-300-bus-system
2. Amari, B.M.: Handbook of Performability Engineering. Springer, London (2008). doi:10.1007/978-1-84800-131-2
3. Brenner, L., Fernandes, P., Sales, A., Webber, T.: A framework to decompose GSPN models. In: Ciardo, G., Darondeau, P. (eds.) ICATPN 2005. LNCS, vol. 3536, pp. 128–147. Springer, Heidelberg (2005). doi:10.1007/11494744_9
4. Buchholz, P.: Exact and ordinary lumpability in finite Markov chains. J. Appl. Probab. **31**(1), 59–75 (1994)
5. Chiaradonna, S., Di Giandomenico, F., Masetti, G.: Efficient non-anonymus composition operator for modeling complex dependable systems. In: Tredan, G. (ed.) 12th European Dependable Computing Conference (EDCC 2016), Fast Abstracts Proceedings, Gothenburg, vol. abs/1608.05874, September 2016
6. Chiaradonna, S., Lollini, P., Di Giandomenico, F.: On a modeling framework for the analysis of interdependencies in electric power systems. In: 37th Annual IEEE/IFIP International Conference on Dependable Systems and Networks (DSN 2007), Edinburgh, pp. 185–195, June 2007

7. Chiaradonna, S., Masetti, G., Di Giandomenico, F.: A stochastic modeling approach for an efficient dependability evaluation of large systems with non-anonymous interconnected components. Conference submission - Under review (2017)
8. Ciardo, G., Trivedi, K.S.: A decomposition approach for stochastic reward net models. Perform. Eval. 18(1), 37–59 (1993)
9. Courtney, T., Gaonkar, S., Keefe, K., Rozier, E.W.D., Sanders, W.H.: Möbius 2.3: an extensible tool for dependability, security, and performance evaluation of large and complex system models. In: 39th Annual IEEE/IFIP International Conference on Dependable Systems and Networks (DSN 2009), Estoril, pp. 353–358 (2009)
10. Deavours, D.D., Clark, G., Courtney, T., Daly, D., Derisavi, S., Doyle, J.M., Sanders, W.H., Webster, P.G.: The Möbius framework and its implementation. IEEE Trans. Softw. Eng. 28(10), 956–969 (2002)
11. Derisavi, S., Kemper, P., Sanders, W.H.: Symbolic state-space exploration and numerical analysis of state-sharing composed models. Linear Algebra Appl. 386, 137–166 (2004). Special Issue on the Conference on the Numerical Solution of Markov Chains
12. Flammini, F.: Critical Infrastructure Security: Assessment, Prevention, Detection, Response. Information & Communication Technologies. WIT Press, Southampton (2012)
13. Hillston, J.: Compositional Markovian modelling using a process algebra. In: Stewart, W.J. (ed.) Computations with Markov Chains, pp. 177–196. Springer, Boston (1995). doi:10.1007/978-1-4615-2241-6_12
14. Illinois Center for a Smarter Electric Grid: June 2016 texas 2000 synthetic test case. http://icseg.iti.illinois.edu/synthetic-power-cases/texas2000-june2016
15. Kemeny, J.G., Snell, J.L.: Finite Markov Chains, 3rd edn. Springer, New York (1983)
16. Lam, V.V., Buchholz, P., Sanders, W.H.: A component-level path-based simulation approach for efficient analysis of large Markov models. In: Kuhl, M.E., Steiger, N.M., Armstrong, F.B., Joines, J.A. (eds.) 37th Conference on Winter Simulation, Orlando, pp. 584–590 (2005)
17. Sanders, W.H., Meyer, J.F.: A unified approach for specifying measures of performance, dependability and performability. In: Avizienis, A., Laprie, J. (eds.) Dependable Computing for Critical Applications. Dependable Computing and Fault-Tolerant Systems, vol. 4, pp. 215–237. Springer, Vienna (1991). doi:10.1007/978-3-7091-9123-1_10
18. Sanders, W.H., Meyer, J.F.: Stochastic activity networks: formal definitions and concepts*. In: Brinksma, E., Hermanns, H., Katoen, J.-P. (eds.) EEF School 2000. LNCS, vol. 2090, pp. 315–343. Springer, Heidelberg (2001). doi:10.1007/3-540-44667-2_9
19. Tomek, L.A., Trivedi, K.S.: Fixed point iteration in availability modeling. In: Cin, M.D., Hohl, W. (eds.) Fault-Tolerant Computing Systems. Informatik-Fachberichte, vol. 283. Springer, Heidelberg (1991). doi:10.1007/978-3-642-76930-6_20
20. Trivedi, K.S.: Probability and Statistics with Reliability, Queuing and Computer Science Applications, 2nd edn. Wiley, Chichester (2002)

Analysis of Performance and Energy Consumption in the Cloud

Mehdi Kandi[1], Farah Aït-Salaht[2,3], Hind Castel-Taleb[1(✉)], and Emmanuel Hyon[3(✉)]

[1] SAMOVAR, CNRS, Telecom SudParis, Université Paris-Saclay,
9, Rue Charles Fourier, 91011 Evry Cedex, France
medmehdikandi@gmail.com, hind.castel@telecom-sudparis.eu
[2] Crest-Ensai, Rennes, France
farah.ait-salaht@ensai.fr
[3] Sorbonne Universités, UPMC Univ Paris 06, CNRS,
LIP6 Paris UMR 7606, 4 place Jussieu 75005 Paris France,
Université Paris Nanterre, Nanterre, France
emmanuel.hyon@lip6.fr

Abstract. We analyze here a cloud system represented by hysteresis multi server queueing system. It is characterized by forward and backward thresholds for activation and deactivation of block of servers representing a set of VMs (Virtual Machines). The system is represented by a complex Markov Chain which is difficult to analyse when the size of the system is huge. We propose both analytical and numerical mathematical methods for deriving the steady-state probability distribution. We compute then performance and energy consumption measures and we define an overall cost taking into account both aspects. We compare the proposed methods with respect to the computation time and we analyse the impact of some parameters on the behaviour of the system.

1 Introduction

Improving the energy consumption of a cloud while guaranteeing a given quality of service is a problem encountered today by cloud providers. One way to achieve this is to adapt the capacities to demand which is made easier today with the virtualization of the servers. Hence, it is possible to modulate, in a transparent manner, the number of active Virtual Machines (VMs) over time. However, finding the policy that tailors resources to demand is a crucial point that requires accurate assessment of both the energy expended and the performance of the system. Multi server queuing models [2,3] or server farms models [6,15] have been proposed to represent dynamicity of a data center as well as to compute performance metrics. Multi-server threshold based-queueing system with hysteresis policy [4,9,13], in which activations and deactivations are governed by sequences of forward and reverse different thresholds, have been proposed, on the other hand, to efficiently manage the number of active VMs. For systems driven by hysteresis policies, the assessment of both performance and energy consumption requires the computation of the expected measures, but since cloud systems

© Springer International Publishing AG 2017
P. Reinecke and A. Di Marco (Eds.): EPEW 2017, LNCS 10497, pp. 199–213, 2017.
DOI: 10.1007/978-3-319-66583-2_13

are often defined on very large state spaces such a computation is difficult. When the system is represented by a complex Markov chain, we face up a computational complexity problem which makes exact analysis very cumbersome or even impossible.

Under some assumptions, evaluation of hysteresis multiserver systems has been already studied in the literature and different resolution methods have been presented to compute efficiently the performance measures of the system. Among the most significant works, we can mention the work of Lui and Golubchik [13] which is widely used in the literature. It solves the model using the concept of Stochastic Complement Analysis (SCA). It is based on partitioning the state space in disjoint sets in order to aggregate the Markov chain. In [12], Le Ny et al. propose an other way to compute the steady-state probabilities of a heterogeneous multi-server threshold queue with hysteresis by using a closed-form solution. Otherwise, in [1] an aggregated bounding approach is proposed to derive accurate bounds on performance measures. However, in these papers, it had been only considered the case where one VM is activated (resp. deactivated) according to the demand and the threshold sequences. On the other hand, Mitrani [14,15] defines server farm models in which several servers are activated at the same time. They are called activations by block. Such approaches allow to model more general practical models.

We propose in this paper to extend the current state of art and to couple the advantages of the activation by block and the advantages of hysteresis policy by considering a multi-server system with hysteresis in which activations/deactivations are made by block. This, up to our knowledge, has never been considered and studied previously in the literature.

This allows us to consider both performance and energy consumption in order to propose a trade-off between them. For the multi-server system with hysteresis and block activation, we establish and present three resolution methods. First method consists to adapt and extend the SCA aggregation method of [13]. The second investigated method is a numerical approach based on Level Dependent Quasi Birth and Death (LDQBD) method. At last, an analytical approach based on the balance equations method of [12] is presented in details. We adapt [12] and get closed form formulas for the steady-state probability distribution. Furthermore, by relaxing the former assumptions on the threshold sequences imposed by [13] or [12], we have generalized the set of threshold values. We then perform numerical results for Markov chains with large state space, as in cloud systems, and establish an overall cost taking into account both performance and energy consumption. Moreover, as we consider in this model more general assumptions for the thresholds, we can see in details the impact of their values on performance and energy consumption.

The paper is organized as follows: next (in Sect. 2), we describe the cloud system and present the considered queueing model. In Sect. 3, we detail the different methods to solve the model and compute the steady state probability vector. While part 4 presents the formulation used to express the expected costs in terms of performance and energetic consumption for the model, Sect. 5

presents numerical results of performance and energy consumption measures. Finally, achieved results are discussed in the conclusion and comments about further research issues are given.

2   Cloud System Description

We analyse a cloud system composed by a set of Virtual Machines (VMs). We model it using a multi-server queue, with C homogeneous servers representing the VMs. The service time of each VM is exponential with mean rate μ. In order to represent the dynamicity of resource provisioning, the VMs can be activated and deactivated over time. We assume that the job requests arrive at the system following a Poisson process with rate λ, and are enqueued in a finite queue. An arriving request can be rejected if it finds the system, which have a whole capacity of B, full. The servers management is governed by threshold vectors which control the operation of activating and deactivating the VMs. These thresholds depend on the number of customers waiting in the system.

We suppose the case where several VMs can be simultaneously activated or deactivated what is called activated or deactivated by block. We define K functioning levels, where each level corresponds to a given number of active servers. The number of active servers at level k is fixed and denoted by S_k, where $S_1 \leq S_2 \leq ... \leq S_K = C$. We suppose that $S_1 \geq 1$, so we have at least one active server by assumption.

The transition from functioning level k to level $k+1$ allows to allocate (turn on) one or more additional servers, going from S_k to S_{k+1} active servers, while the transition from level k to level $k-1$ allows to remove (turn off) one or more active servers, going from S_k to S_{k-1} active servers. Depending on the system occupancy, we transit from the level k to level $k+1$ when the occupancy in the system exceeds a threshold F_k, and from level k to level $k-1$ when the occupancy in the system falls below a threshold R_{k-1}. The model is then characterized by activation thresholds $F = (F_1, F_2, ..., F_{K-1})$ (called also forward thresholds), and deactivation thresholds $R = (R_1, R_2, ..., R_{K-1})$ (called also reverse thresholds). These thresholds are fixed and can not be modified during the system works. We furthermore assume that $F_1 < F_2 < ... < F_{K-1}$, that $R_1 < R_2 < ... < R_{K-1}$ and that $R_k < F_k, \forall k, 1 \leq k \leq K - 1$. We suppose that server deactivations occur at the end of the service, and when multiple servers are deactivated at the same time, all the customers who have not completed their service return to the queue.

The underlying model is described by the Continuous-Time Markov Chain (CTMC), denoted $\{X(t)\}_{t \geq 0}$. A state is represented by a couple (m, k) such that m is the number of customers in the system and k is the functioning level. The state space is denoted by A and is given by:

$$A = \{(m, k) \,|\, 0 \leq m \leq F_1, \text{ if } k = 1,$$
$$R_{k-1} + 1 \leq m \leq F_k, \text{ if } 1 < k < K,$$
$$R_{K-1} + 1 \leq m \leq B, \text{ if } k = K\}.$$

The transitions between states then follows:

$(m, k) \rightarrow (\min\{B, m + 1\}, k)$, with rate λ, if $m < F_k$;
    $\rightarrow (\min\{B, m + 1\}, \min\{K, k + 1\})$, with rate λ, if $m = F_k$;
    $\rightarrow (\max\{0, m-1\}, k)$, with rate $\mu \cdot \min\{S_k, m\}$, if $m > R_{k-1}+1$;
    $\rightarrow (\max\{0, m-1\}, \max\{1, k-1\})$ with rate $\mu \cdot \min\{S_k, m\}$, if $m = R_{k-1}+1$.

An example of the transitions is given Fig. 1.

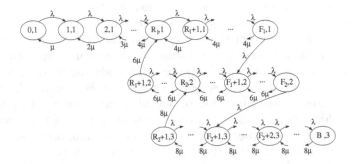

Fig. 1. Transition structure for $K = 3$, $S_1 = 4$, $S_2 = 6$, $S_3 = 8$, $R_1 \geq 5$ and $R_2 \geq 7$.

From a practical perspective, several other models fit with this block representation. For example, it can represent heterogeneous nodes of a cluster (possibly virtual), each node having a different number of cores. These nodes can be idle or activated. In this case, a node is represented by a level and S_k is the number of cores of the node. It can also represent a single physical component composed by many cores that can be activated or deactivated. On each core (represented by a level) a given number of S_k VMs are placed that share the CPU. These models follow the same markovian representation than the model studied here but their costs are different.

3   Resolution Approaches

We expose hereafter three techniques to solve the CTMC and compute the steady-state probability vector. These resolution methods are either numerical or analytical or both analytical and numerical. They have been developed for the model and their correctness is shown. Some comparisons are presented Sect. 5.

3.1   Stochastic Complement Analysis (SCA)

To solve the $\{X(t)\}_{t \geq 0}$ Markov chain, the first approach, proposed by Lui et al. [13], consists to aggregate the underlying Markov chain and uses a numerical method to compute the steady-state distribution. The different restrictions of [13] (i.e. $R_k < F_{k-1}$, $\forall k$ and activation deactivation of a single server) can be relaxed without substantially modifying the framework. Our approach considers block activations and deactivations as well as different orders of the thresholds: $R_k < F_{k-1}$ and $R_k \geq F_{k-1}$, for all k. It is presented below and some details can be found in [8].

First, we aggregate the state space of the underlying Markov chain by partitioning the set A into disjoint subsets. These subsets depend here on the functioning levels. Hence, the state space A is partitioned into K distinct sets denoted A_k, where, for any k in $1, \ldots, K$, we have $A_k = \{(i, j) \mid (i, j) \in A, j = k\}$. The set A_k contains the states belonging the level k.

From each subset, we define a corresponding Markov chain. Let $\{X_k(t)\}_{t \geq 0}$ be the Markov chain defined on state space A_k. These derived Markov chains are defined on reduced state spaces which makes their analysis less complex. The resolution of each of the derived Markov chain defines a conditional steady state probabilities. For the whole chain $\{X(t)\}$, by applying the state aggregation technique, each subset A_k is now represented by a single state, and an aggregated process is defined. A resolution of this aggregated process is performed, i.e., the probabilities of the system being in any given set are computed. At last, a disaggregation technique is applied to compute the individual steady state probabilities for the original Markov process. The method correctness is based on the following theorem stated by Lui et al. in [13].

Theorem 1. *Given an irreducible Markov process with state space A, let us partition this state space into two disjoint sets A_1 and A_2. Then, the transition rate matrix (denoted by Q) is given as follows:*

$$Q = \begin{pmatrix} Q_{A_1 A_1} & Q_{A_1 A_2} \\ Q_{A_2 A_1} & Q_{A_2 A_2} \end{pmatrix},$$

where $Q_{i,j}$ is the transition rate sub-matrix corresponding to transitions from partition i to partition j.

We point out that the computation of the steady state probabilities of the derived Markov chains $\{X_k(t)\}$ is performed using a numerical resolution method.

3.2   Level Dependant Quasi Birth and Death Process

The particular form of the generator of $\{X(t)\}_{t \geq 0}$ suggests us to use the Quasi Birth and Death (QBD) processes in order to benefit from the numerous numerical methods to solve them [16]. For short, a QBD process is a stochastic process in which the state space is two dimensional and can be decomposed in disjoint sets such that transition may only occur inside a set or occur towards only two other sets. This results in a generator with a tridiagonal form (as the birth and death process) in which the terms on the diagonals are matrices. When the matrices are identical for each level, it is said *level independent* but when the matrices are different the QBD is said *level dependant* (LDQBD).

Let us define $Q_{k,k'}(i, j)$ that denotes the i-th line and j-th column element of matrix $Q_{k,k'}$. We have:

Proposition 1. *The Markov Chain $\{X(t)\}_{t \geq 0}$ is a Level Dependant QBD with K levels, corresponding to the functioning levels. Its generator Q is decomposed in:*

$$Q = \begin{pmatrix} Q_{1,1} & Q_{12}, & & & & \\ Q_{2,1} & Q_{2,2} & Q_{2,3} & & & \\ & Q_{3,2} & Q_{3,3} & Q_{3,4} & & \\ & & \ddots & \ddots & \ddots & \\ & & & Q_{K-1,K-2} & Q_{K-1,K-1} & Q_{K-1,K} \\ & & & & Q_{K,K-1} & Q_{K,K} \end{pmatrix}.$$

For all k, the inner matrices $Q_{k,k-1}$, $Q_{k,k}$ and $Q_{k,k+1}$ are respectively of dimension $d_k \times d_{k-1}$, $d_k \times d_k$ and $d_k \times d_{k+1}$, letting $d_k = F_k - R_{k-1}$, $R_0 = -1$ and $F_K = B$.

For $k = 1$ we have:

$$Q_{1,1}(i,j) = \begin{cases} \lambda & \text{if } j = i+1 \\ \mu \min\{S_1, i\} & \text{if } j = i-1 \\ -\lambda & \text{if } i = 1 \text{ and } j = 1 \\ -(\lambda + \mu \min\{S_1, i\}) & \text{if } i = j \text{ and } i \neq 1 \\ 0 & \text{otherwise} \end{cases},$$

and

$$Q_{1,2}(i,j) = \begin{cases} \lambda & \text{if } i = d_1 \text{ and } j = F_1 - R_1 + 1 \\ 0 & \text{otherwise} \end{cases}.$$

For $k \in \{2, \ldots, K-1\}$, we get:

$$Q_{k,k-1}(i,j) = \begin{cases} \mu \min\{S_k, R_{k-1}+1\} & \text{if } i = 1 \text{ and } j = R_{k-1} - R_{k-2} \\ 0 & \text{otherwise} \end{cases},$$

also

$$Q_{k,k}(i,j) = \begin{cases} \lambda & \text{if } j = i+1 \\ \mu \min\{S_k, R_{k-1}+i\} & \text{if } j = i-1 \\ -(\lambda + \mu \min\{S_k, R_{k-1}+i\}) & \text{if } i = j \\ 0 & \text{otherwise} \end{cases},$$

and

$$Q_{k,k+1}(i,j) = \begin{cases} \lambda & \text{if } i = d_k \text{ and } j = F_k - R_k + 1 \\ 0 & \text{otherwise} \end{cases}.$$

Finally for $k = K$, it follows

$$Q_{K,K-1}(i,j) = \begin{cases} \mu \min\{S_K, R_{K-1}+1\} & \text{if } i = 1 \text{ and } j = R_{K-1} - R_{K-2} \\ 0 & \text{otherwise} \end{cases},$$

and

$$Q_{K,K}(i,j) = \begin{cases} \lambda & \text{if } j = i+1 \\ \mu \min\{S_K, R_{K-1}+j\} & \text{if } j = i-1 \\ -(\lambda + \mu \min\{S_K, R_{K-1}+j\}) & \text{if } i = j \text{ and } j \neq d_K \\ -\mu \min\{S_K, R_{K-1}+j\} & \text{if } i = j = d_K \\ 0 & \text{otherwise} \end{cases}.$$

The proof can be found in [8].

Numerically solving QBD is a hard computational task requiring to solve matrix equations and is often based on matrix geometric methods [11,16] or kernel methods [7]. This is even more the case for LDQBD. Here, among the numerical existing methods to solve them, this one proposed in [5] is used since it is shown that this method is efficient and numerically stable.

3.3   Closed Form Solution Using Balance Equations

We follow the approach of [12] and give a closed form for the steady state probability using balance equations and cuts on the state space. The relevance of our work is that we can take more general cases than [12] for the thresholds. We assume not only the case $R_k \leq F_{k-1}$, but also the case $R_k > F_{k-1}$ for each level $2 \leq k \leq K$. In this method, the probabilities are computed level by level, from level 1 to level K. For states of level 1, the steady-state probabilities are expressed in terms of $\pi(0,1)$. For a level $k \in \{2 \ldots K\}$, the process has two steps. First, the steady-state probability of the first state of the level: $\pi(R_{k-1}+1, k)$ is expressed in terms of the last state of the precedent level $\pi(F_{k-1}, k-1)$ which has been already computed and which can be expressed in terms of $\pi(0,1)$. After that, the other probabilities of the level k are computed in terms of $\pi(R_{k-1}+1, k)$. Henceforth, it results that all the probabilities are computed in terms of $\pi(0,1)$. At the end, from the normalizing condition, $\pi(0,1)$ can be derived. From now on, for any $k \in \{1 \ldots K\}$, we define $\mu_k = \mu S_k$, $\rho = \frac{\lambda}{\mu}$ and $\rho_k = \frac{\lambda}{\mu_k}$. Next, we give the formulas for the level 1 probabilities.

Level 1. The following lemma gives the steady-state probabilities for level 1.

Lemma 1 (Level 1 probabilities). *In level one, the service rate depends on the number of customers in the system. So, for a state $(m,1)$, if $1 \leq m < S_1$, then the service rate is $m\mu$ and if $m \geq S_1$ it is $S_1\mu$. We can deduce $\pi(m,1)$ by :*

$$\pi(m,1) = \begin{cases} \dfrac{\rho^m}{m!}\pi(0,1) & \text{if } 0 \leq m \leq S_1, \quad (1) \\[3mm] \rho_1^{m-S_1}\dfrac{\rho^{S_1}}{S_1!}\pi(0,1) & \text{if } S_1 < m \leq R_1, (2) \\[3mm] \dfrac{\rho^{S_1}}{S_1!}\left(\rho_1^{m-S_1} - \dfrac{\rho_1^{F_1-S_1+1}(1-\rho_1^{m-R_1})}{1-\rho_1^{F_1-R_1+1}}\right)\pi(0,1) & \text{if } R_1+1 \leq m \leq F_1 (3) \end{cases}$$

The proof uses special cuts on state space from which one derives local balance equations. It is given in [8].

Level k. Let us consider now k such that $2 \leq k \leq K - 1$. We assume that $R_{k-1}+1 \geq S_k$, and thus the service rate for each level is $\min(R_{k-1}+1, S_k) = S_k\mu$. In order to express the relationship between level $k - 1$ and level k, we should consider the cut of the state space between states of level $k - 1$ and states of level k. This gives us the following evolution equation: $\pi(F_{k-1}, k - 1)\lambda = \pi(R_{k-1} + 1, k)\mu_k$, which is equivalent to:

$$\pi(R_{k-1} + 1, k) = \rho_k \pi(F_{k-1}, k - 1). \tag{4}$$

All probabilities of level k can be expressed with respect to $\pi(R_{k-1} + 1, k)$. However, these probabilities depend also of the level $k+1$ by the threshold value R_k. Therefore two cases should be considered: either $R_k \leq F_{k-1}$ or $R_k > F_{k-1}$. We present now the case where $R_k > F_{k-1}$.

Lemma 2. *When $R_k > F_{k-1}$, for any $k \in \{2 \ldots K - 1\}$, we have:*

$$\pi(m, k) = \frac{1 - \rho^{m-R_{k-1}}}{1 - \rho_k}\pi(R_{k-1} + 1, k) \text{ if } R_{k-1} + 2 \leq m \leq F_{k-1} + 1, \tag{5}$$

$$\pi(m, k) = \frac{\rho_k^{m-F_{k-1}-1} - \rho_k^{m-R_{k-1}}}{1 - \rho_k}\pi(R_{k-1} + 1, k) \text{ if } F_{k-1} + 2 \leq m \leq R_k, \tag{6}$$

$$\pi(m, k) = \frac{\rho_k^{m-F_{k-1}-1} - \rho_k^{m-R_{k-1}}}{1 - \rho_k}\pi(R_{k-1} + 1, k) \tag{7}$$

$$- \frac{\rho_k}{\rho_{k+1}}\frac{1 - \rho_k^{m-R_k}}{1 - \rho_k}\pi(R_k + 1, k + 1) \text{ if } R_k + 1 \leq m \leq F_k.$$

with

$$\pi(R_{k+1}, k + 1) = \rho_{k+1}\frac{\rho_k^{F_k-F_{k-1}-1} - \rho_k^{F_k-R_{k-1}}}{1 - \rho_k^{F_k-R_k+1}}\pi(R_{k-1} + 1, k). \tag{8}$$

The proof of Lemma 2 is in [8].

We deduce from Eq. (8), that $\pi(R_k + 1, k + 1)$ is also expressed in terms of $\pi(R_{k-1}+1, k)$. Thus all the probabilities in Lemma 2 can be expressed in terms of the steady-state $\pi(R_{k-1} + 1, k)$ which is the first state of the level. Since, furthermore, $\pi(R_{k-1} + 1, k)$ is computed from $\pi(F_{k-1}, k - 1)$, then it can be expressed in terms of $\pi(0, 1)$. So from the normalizing condition we derive all the probabilities.

Since the case $R_k \leq F_{k-1}$, has been considered in [12], it follows:

Lemma 3 ([12]). *When* $R_k \leq F_{k-1}$, *for any* $k \in \{2 \ldots K - 1\}$, *we have*

$$\pi(m,k) = \frac{1 - \rho_k^{m-R_{k-1}}}{1 - \rho_k} \pi(R_{k-1} + 1, k) \text{ if } R_{k-1} + 2 \leq m \leq R_k \tag{9}$$

$$\pi(m,k) = \frac{1 - \rho_k^{m-R_{k-1}}}{1 - \rho_k} \pi(R_{k-1} + 1, k) \tag{10}$$

$$- \frac{\rho_k}{\rho_{k+1}} \frac{1 - \rho_k^{m-R_k}}{1 - \rho_k} \pi(R_{k+1,k+1}) \text{ if } R_{k+1} \leq m \leq F_{k-1} + 1,$$

$$\pi(m,k) = \rho_k^{m-F_{k-1}-1} \frac{1 - \rho_k^{F_{k-1}-R_{k-1}+1}}{1 - \rho_k} \pi(R_{k-1} + 1, k) \tag{11}$$

$$- \frac{\rho_k}{\rho_{k+1}} \frac{1 - \rho^{m-R_k}}{1 - \rho_k} \pi(R_k + 1, k + 1 \text{ if } F_{k-1} + 2 \leq m \leq F_k.$$

Proofs are given in [12].
Level K. Let us consider newt the level $k = K$.

Lemma 4 ([12]). *The steady-state probabilities for the level K: $\pi(m, K)$ are equal to:*

$$\pi(m,K) = \left(\frac{1 - \rho_k^{m-R_{K-1}}}{1 - \rho_K} \right) \pi(R_{K-1} + 1, K) \text{ if } R_{K-1} + 2 \leq m \leq F_{K-1} + 1,$$

$$\pi(m,K) = \left(\frac{1 - \rho_k^{F_{K-1}+1-R_{K-1}}}{1 - \rho_K} \right) \rho_K^{(m-F_{K-1}-1)} \pi(R_{K-1} + 1, K) \text{ if } F_{K-1} + 2 \leq m \leq B.$$

It is proved in [12].

4   Performance Measures and Energy Cost Parameters

We propose now to calculate the expected cost in terms of performance and energy consumption for the model presented in this paper. Once the steady-state vector is calculated, we get various performance and energy consumption measures. Indeed the cost is expressed as an expected Markov reward function $\mathcal{R}$, where $\mathcal{R} = \sum_{m,k} \pi(m, k) r(m, k)$ and $r(m, k)$ be the reward of state (m, k). Metrics of interest are described hereafter.

First, we give the performance measures. These one are related to the Service Level Agreement (SLA) which defines several QoS (Quality of Service) constraints that the provider should guarantee. Losses, queue lengths and processing speed are the main parameters that are taken into account.

The *mean number of customers* in the system is denoted by $\overline{N_C}$ and is equal to: $\overline{N_C} = \sum_{(m,k)\in A} \pi(m, k) \cdot m$.

The *mean number of losses* due to full queue by time unit is denoted by $\overline{N_R}$ and is equal to: $\overline{N_R} = \lambda \cdot \pi(B, K)$.

The *mean response time* is denoted by $\overline{R}$ and is equal to: $\overline{R} = \overline{N_C}/(\lambda \cdot (1 - \pi(B, K)))$.

Energy consumption measures are defined now. The energy costs representation adopted here is mainly based on [10]. In this paper, the energy costs of a VM in use can be decomposed in two parts: static and dynamic costs. Static costs are mainly independent of the workload and comprise idle (or standby) consumption of the nodes, routers and consumption of the data center (cooling system, power distribution units,.....) which is evaluated by the industrial metrics of the Power Unit Efficiency (PUE). On the other hand, dynamic costs include the energy consumption part of servers, storage devices and network that is induced by the resource usage and then depends on the workload. The hysteresis approach considers only the dynamic costs but static costs should be added in order to get the whole consumption of a VM. Hence, energy consumption is depending on both mean number of active servers (dynamic part of the cost) and mean number of their activation and deactivations, which represent the energy cost of the start (or pausing) and the data migration of a VM.

The *mean number of active servers* in the system is denoted by $\overline{N_S}$ and is equal to: $\overline{N_S} = \sum_{(m,k) \in A} S_k \cdot \pi(m, k)$.

The *mean number of activations triggered* by time unit, is denoted by $\overline{N_A}$ and is given by: $\overline{N_A} = \lambda \sum_{(m,k) \in A} (S_{k+1} - S_k) \cdot \mathbb{1}_{\{m=F_k; 1 \leq k \leq K-1\}} \cdot \pi(m, k)$.

The *mean number of deactivations triggered* by time unit is denoted by $\overline{N_D}$ and is given by:

$$\overline{N_D} = \sum_{(m,k) \in A} \mu \min\{S_k, m\}(S_k - S_{k-1})\pi(m, k) \cdot \mathbb{1}_{\{m=R_{k-1}+1; 1 \leq k \leq K-1\}}$$

In order to consider both performance and energy consumption, then we define the overall expected cost by time unit for the underlying model as follows: $\overline{C} = C_H \cdot \overline{N_C} + C_S \cdot \overline{N_S} + C_A \cdot \overline{N_A} + C_D \cdot \overline{N_D} + C_R \cdot \overline{N_R}$. Where, C_H is the per capita cost of holding one customer in the system within one time unit, C_S is the per capita cost of using one working server within one time unit, C_A is the activating cost (cost of switching one server from deactivating mode to activating mode), C_D is the deactivating cost and C_R is the cost of job losses due to full queue.

5   Numerical Results

This section focuses on the analysis of the queueing model defined before (Sect. 2). We perform some numerical examples in order to show the interest of the model and the improvement of the resolution methods for the analysis of Cloud performance.

First, the three resolution approaches depicted in this work (SCA, LQBD and closed form solution) are compared and we observe which approach is the most relevant in terms of computational complexity and results accuracy. At last, using the most relevant resolution method, some use cases of cloud systems are analyzed and we observe some performance metrics. All evaluations were

implemented in Matlab and performed on laptop with 64-bit Windows 10, 8 GB
RAM and 2.00 GHZ Intel i7-4750HQ CPU.

5.1 Comparing the Resolution Approaches

Our objective here is to determine among the proposed resolution methods, the
most relevant one. The relevance criteria are defined in terms of computation
time and accuracy of the results. So, a set of experiments are performed for
this purpose. We consider a threshold queueing model with hysteresis where the
activation and deactivation of VMs are occurred one by one (i.e. $S_1 = 1$, $S_{k+1} =
S_k + 1$, $\forall k < K$, which means that $C = S_K = K$). This one by one activation
case is considered since it represents the worst case in terms of computational
complexity, and is thus the best way to compare the different proposed methods.
We assume here that each server provides a service following an exponential
distribution with rate 1. We generate several instances by increasing the size of
the model (i.e. the number of levels (K) and the size of buffer B). We illustrate in
Table 1, the computation times of each method for these instances. The forward
and reverse thresholds are set as follow: $F = \{50, 100, 150, \dots, B\}$ and $R =
\{10, 40, 90, 140, 190, \dots, B\}$. Threshold values have been taken arbitrarily and
additional studies with different choices of threshold sequences will be the subject
of future work.

Table 1. Computational times (in seconds) of proposed resolution methods.

	SCA with Power method	SCA with GTH method	LDQBD	Closed form
$\lambda = 2$, K = 10, B = 750, (1271 states)	2.933	0.0406	0.0121	0.00905
$\lambda = 10$, K = 100, B = 7500, (13421 states)	4117.59	0.9587	0.9889	0.0823
$\lambda = 10$, K = 500, B = 37500, (67421 states)	+3600	41.573	88.01	0.4979
$\lambda = 10$, K = 1000, B = 75000, (134921 states)	+3600	307.54	1330.27	1.0561

Through this table, one can clearly see that the closed form solution is the
fastest one. This method is more than 1000 times faster than LQBD and 100
times than the SCA with GTH method. This result is expected because the
closed form solution is based on a set of formulas containing basic operators
contrary to LQBD method based on matrix inversion or SCA method with GTH
approach where the numerical resolution approach GTH has a cubic complexity.
It should be precised that since the SCA approach is a combination of state
aggregation technique and numerical solution of Markov chain, then we propose
to distinguish two numerical methods commonly used: the GTH [18] and power
methods [17].

Considering the precision of the results, it could be noticed that even if the SCA and LQBD methods are numerical resolution approach, their precisions are not so far from the closed form method. Indeed, the gap on the stationary distribution vector between the different methods is smaller than 10^{-12}.

To confirm our conclusions, we propose to observe the relevance of the resolution methods according to the variation of the arrival rate λ. For this example, previously cited in the Table 1, parameters are $C = K = 100$, $B = 7500$ and $\mu = 1$. Then, we let vary the arrival rate from $\lambda = 1$ to $\lambda = 100$, and assess the computational resolution times of the closed form, LDQBD and SCA+GTH methods. In view of the computation times of the SCA + Power method (rather longer) this method is not considered in this comparative study. The obtained results are illustrated in the following Fig. 2.

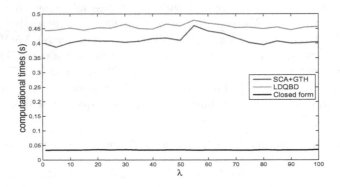

Fig. 2. Computational times (in seconds) versus arrival rate (λ).

In view of these results, it is clear that the closed form method is the most relevant resolution approach. However, for large ρ numerical methods could be more precise than the formal one due to the limits of the computer. This point should receive further investigations.

5.2   Performance and Energy Consumption Measures

In this part, all computations are made with the closed form formula. We assess here the performance of a large cloud system and illustrate the trade-off between performance and energy consumption. We consider a multi server queue model driven by a hysteresis policy. We want to see the impact of the number of servers on the performance and the energy-efficiency of a cloud henceforth, the metrics defined Sect. 4 are used. We exhibit several cases in which our data center is composed by a pool of C virtual machines. It is assumed that the number C ranges from 50 to 10000 VMs, this last number being the size of a small data center. The buffer size is set to $B = 1000$ jobs, the service rate of each VM is set to $\mu = 10$ and we let vary the arrival rate varying between 100 and

1000 jobs/min. We assume that for each considered model there are fifty levels ($K = 50$). The forward thresholds and reverse thresholds are set respectively to $F = \{20, 40, 60, \ldots, 1000\}$ and $R = F - 10$. The sequence of service levels is taken as follows: $S = \{s \mid s = i \times \lfloor \frac{C}{50} \rfloor, \forall i = 0 \ldots 50\}$. Concerning the energy consumption parameters, we set the costs to 1: the energy consumption of one working server within one time unit is $C_S = 1$, the cost of holding one job in the system within one time unit is $C_H = 1$, the cost of activating or deactivating one server are respectively $C_A = 1$ and $C_D = 1$, and the cost of job losses due to full queue is $C_R = 1$.

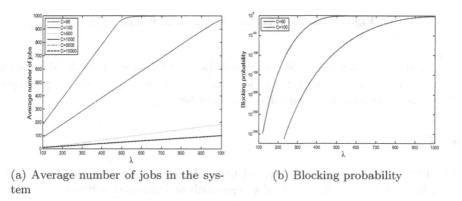

(a) Average number of jobs in the system

(b) Blocking probability

Fig. 3. Performance metrics versus arrival rate (λ).

The performance results are illustrated in Fig. 3(a) and (b). In Fig. 3(b), one illustrates only the curves for $C = 50$ and $C = 100$, since the other models have a zero blocking probability. From these figures, we can obviously observe that the number of servers increase improves the performance. This is shown, in Fig. 3, by the decrease of the number of jobs in the system and the blocking probability.

However, in terms of cost, one obviously observes the opposite when the system is moderately loaded. Hence, when the system is weakly/moderately loaded, the models that have a significant number of active VMs underperform comparatively to models with relatively few active VMs. Indeed, some VMs consume energy without performing any service. This can be seen on Fig. 4 for $C = 10000$. On the other hand, when the system is overloaded, the cost of losses increases and affects the global cost. This can be clearly seen Fig. 4 for the model with $C = 50$ when $\lambda > 450$ and for the one with $C = 100$ when $\lambda > 900$. This is consistent with intuition. The oscillation phenomenon observed for large C remains unclear and deserves to be studied in more detail.

It could be noticed that the closed form resolution allows to compute the performance measures of all the instances in very short times (smaller than 2 seconds) even in cases where the number of VMs is 10000. Since concrete small

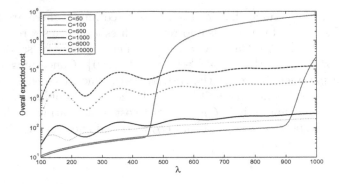

Fig. 4. Overall expected costs versus arrival rate (λ).

cloud systems or cloud modules have a number of VMs around 10000, this shows the practical value of our method for answering rather quickly the questions about energy consumption and network dimensioning.

6   Conclusion

We develop numerical and analytical methods for the analysis of a hysteresis queueing system modelling a cloud system with activation/deactivation by block of VMs. One important contribution of this paper is to suppose few constraints on the thresholds. We give numerical values of the performance even in the case of large Markov chains, and show that our methods are hugely faster than the classical ones. We define a global cost for performance and energy consumption in order to propose a trade off between performance and energy consumption, and we analyse the impact of the thresholds on it. For the future, we need to analyze real cloud architectures with concrete energy consumptions for the VMs in order to compute relevant cost values. We also want to develop optimization algorithms to obtain the thresholds which minimize the overall cost.

Acknowledgement. This work was supported by grant ANR MARMOTE (ANR-12-MONU-0019).

References

1. Aït-Salaht, F., Castel-Taleb, H.: Bounding aggregations on phase-type arrivals for performance analysis of clouds. In: 24th IEEE International Symposium on Modeling, Analysis and Simulation of Computer and Telecommunication Systems, MAS-COTS 2016, pp. 319–324. IEEE (2016)
2. Ardagna, D., Casale, G., Ciavotta, M., Pérez, J.F., Wang, W.: Quality-of-service in cloud computing: modeling techniques and their applications. J. Internet Serv. Appl. **5**, 11 (2014)

3. Artalejo, J.R., Economou, A., Lopez-Herrero, M.J.: Analysis of a multiserver queue with setup times. Queueing Syst. **51**(1–2), 53–76 (2005)
4. Asghari, N.M., Mandjes, M., Walid, A.: Energy-efficient scheduling in multi-core servers. Comput. Netw. **59**(11), 33–43 (2014)
5. Baumann, H., Sandmann, W.: Numerical solution of level dependent quasi-birth-and-death processes. Procedia Comput. Sci. **1**(1), 1561–1569 (2010)
6. Gandhi, A., Harchol-Balter, M., Adan, I.: Server farms with setup costs. Perform. Eval. **67**(11), 1123–1138 (2010)
7. Gaujal, B., Hyon, E., Jean-Marie, A.: Optimal routing in two parallel queues with exponential service times. Discrete Event Dyn. Syst. **16**(1), 71–107 (2006)
8. Kandi, M., Aït-Salaht, F., Castel-Taleb, H., Hyon, E.: Mathematical methods for analyzing performance and energy consumption in the cloud. Technical report, Institut Mines-Telecom Telecom SudParis (2017)
9. Kitaev, M.Y., Serfozo, R.F.: M/M/1 queues with switching costs and hysteretic optimal control. Oper. Res. **47**, 310–312 (1999)
10. Kurpicz, M., Orgerie, A.-C., Sobe, A.: How much does a vm cost? energy- proportional accounting in VM-based environments. In: PDP: Euromicro International Conference on Parallel, Distributed, and Network-Based Processing, pp. 651–658 (2016)
11. Latouche, G., Ramaswami, V.: A logarithmic reduction algoritm for quasi-birth-death processes. J. Appl. Prob. **30**, 650–674 (1993)
12. Le Ny, L.-M., Tuffin, B.: A simple analysis of heterogeneous multi-server threshold queues with hysteresis. In: Applied Telecommunication Symposium (ATS) (2002)
13. Lui, J.C.S., Golubchik, L.: Stochastic complement analysis of multi-server threshold queues with hysteresis. Perform. Eval. **35**(1), 19–48 (1999)
14. Mitrani, I.: Service center trade-offs between customer impatience and power consumption. Perform. Eval. **68**(11), 1222–1231 (2011)
15. Mitrani, I.: Managing performance and power consumption in a server farm. Ann. Oper. Res. **202**(1), 121–134 (2013)
16. Neuts, M.F.: Matrix-Geometric Solutions in Stochastic Models: An Algorithmic Approach. John Hopkins University Press, Baltimore (1981)
17. Philippe, B., Saad, Y., Stewart, W.J.: Numerical methods in markov chain modeling. Oper. Res. **40**(6), 1156–1179 (1992)
18. Stewart, W.J.: Introduction to the numerical Solution of Markov Chains. Princeton University Press, New Jersey (1995)

Deriving Power Models for Architecture-Level Energy Efficiency Analyses

Christian Stier[1(✉)], Dominik Werle[2], and Anne Koziolek[2]

[1] FZI Research Center for Information Technology, Karlsruhe, Germany
stier@fzi.de
[2] Karlsruhe Institute of Technology, Karlsruhe, Germany
{dominik.werle,koziolek}@kit.edu

Abstract. In early design phases and during software evolution, design-time energy efficiency analyses enable software architects to reason on the effect of design decisions on energy efficiency. Energy efficiency analyses rely on accurate power models to estimate power consumption. Deriving power models that are both accurate and usable for design time predictions requires extensive measurements and manual analysis. Existing approaches that aim to automate the extraction of power models focus on the construction of models for runtime estimation of power consumption. Power models constructed by these approaches do not allow users to identify the central set of system metrics that impact energy efficiency prediction accuracy. The identification of these central metrics is important for design time analyses, as an accurate prediction of each metric incurs modeling effort. We propose a methodology for the automated construction of multi-metric power models using systematic experimentation. Our approach enables the automated training and selection of power models for the design time prediction of power consumption. We validate our approach by evaluating the prediction accuracy of derived power models for a set of enterprise and data-intensive application benchmarks.

1 Introduction

Design-time quality analyses allow software architects to estimate quality characteristics of a designed system in early design phases and during software evolution. In the context of software systems, energy efficiency refers to the ratio of useful work the system performs and the energy it consumes, as Barroso et al. [1] outline. Energy efficiency is an essential quality characteristic as it determines a large portion of the deployed systems' operational cost. Power consumption accounts for over 15% of the Total Cost of Ownership (TCO) [2]. The usage profile of software determines the power consumption of servers on which it is deployed [3–6]. Meaningful reasoning on the energy efficiency of software hence requires the consideration of both design and deployment of software architectures [7].

The consideration of energy efficiency at design time enables software architects to reason on the implications of design decisions on infrastructure sizing and operational cost. The energy efficiency of software architectures can be predicted using approaches as proposed by Brunnert et al. [5] and in our previous

© Springer International Publishing AG 2017
P. Reinecke and A. Di Marco (Eds.): EPEW 2017, LNCS 10497, pp. 214–229, 2017.
DOI: 10.1007/978-3-319-66583-2_14

work [6]. The approaches utilize software performance models and power models to predict a system's power consumption at design time. Performance models predict performance and system metrics of a software system under a given workload. Power models then correlate the predicted system metric with power consumption of servers or individual hardware components to estimate power consumption. Using power models, previous work accurately predicts the effect of varying user workload [5] and architectural design decisions [6] on energy efficiency.

When extracting power models to evaluate the energy efficiency of a software system at design time, the implementation of the system is not yet fully available. Hence, the power models need to be trained on workloads for systems other than the system under design. Collecting representative measurements as training data is challenging as the relation between issued workload and values of the observed metrics is non-linear. A set of measurements hereby is representative if it allows to correlate the variance of power consumption with variances of system metrics. Individual workloads might not stress all the resources of the system under evaluation that impact its power consumption. In this case, individual workloads do not produce representative sets of measurements.

In early design phases and during software evolution, metrics such as throughput and utilization of processing units can be predicted with reasonable accuracy and modeling effort. Fine-grained system metrics such as the number of page faults per second are difficult to predict or require significant effort in refining the models. The effort in constructing fine-grained models should only be invested if it results in a significant increase in accuracy.

Existing approaches [5,6] for the design time prediction of energy efficiency use a manual process for selecting a suitable power model for a system under investigation. The authors assume that the utilization of certain server components significantly correlate with power consumption. They do not systematically select these metrics for each server under investigation. Previous work on the automated construction of power models [3,4] for run time estimation allow for an automated selection of metrics that correlate with power consumption. The approaches outlined in [3,4] do not consider the tradeoff between accuracy and effort that is essential to the extraction of power models for design time consumption predictions. Rather, they produce power models that rely on low-level system metrics and hardware performance counters.

We propose a methodology for the automated extraction of power models for design time energy efficiency analyses. Our approach enables software architects to evaluate which system metrics are worth considering for a specific server type based on their expected impact on power consumption prediction accuracy. Our Contributions are as follows:

C1: We define a profiling approach for automatically deriving power and system metric measurements based on representative workload combinations.
C2: We train a set of power models to identify the power models that most accurately predict our systems' power consumption.

C3: We outline a methodology for evaluating the effect of considering additional system metrics in the energy efficiency analysis.

We evaluate our approach with application workloads different from the workloads used in profiling. The evaluation workloads cover an enterprise application workload, SPECjbb2015 [8], and a set of diverse Big Data application workloads contained in the HiBench benchmarking suite [9]. To evaluate the benefit gained by applying our profiling approach (C1) we compare the prediction accuracy of power models derived from measurements extracted using our approach with a baseline approach. The baseline approach subsumes a set of profiling approaches found in related work [3,4]. We investigate the accuracy of power models constructed using our approach. The power models predict power consumption with an error of less than 4.9% for 19 of 27 considered models (C2). This confirms that we are able to construct power models that accurately predict the power consumption for application workloads not available at the time of profiling. We show that we correctly predict the accuracy gained by considering additional system metrics (C3).

This paper is structured as follows. Section 2 outlines our methodology. Section 3 outlines evaluation experiments and discusses their results. Section 4 discusses related work. Section 5 concludes and provides an outlook on future work.

2   Methodology

Figure 1 provides an overview of our methodology for deriving power models for architecture-level energy efficiency analyses. Our methodology consists of

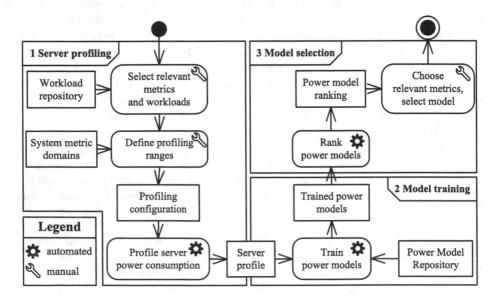

Fig. 1. Activity diagram overview of our power model extraction methodology

the three main steps *server profiling*, *model training*, and *model selection*. In server profiling, we automatically profile the power consumption for a set of system metrics to derive a representative server profile. We hereby consider a profile to be representative if it covers the typical values of system metrics of the system under the expected load. In model training, we construct a set of power models based on the server profile extracted in the first step. The final step *model selection* enables users to compare different power models and reason on the effect of system metrics on prediction accuracy. The following sections further elaborate on each of the three steps.

2.1 Server Profiling

In order to learn accurate power models for a server, we need representative measurements of the power consumption and relevant system metrics under different levels of utilization. A set of measurements is representative if it covers the typical behavior of the system under its expected workload. Using a single workload type to stress the server produces measurements that match only similar workload types. Thus, it is not sufficient to use an individual workload type as the foundation for learning power models. Different workload types need to be considered when learning power models. The measurements used to learn the models also need to cover different utilization levels for the model to be representative for possible workload mixes.

To the best of our knowledge, there does not exist an approach for targeting specific utilization levels for multiple resources using representative workloads.

We designed an approach for profiling the power consumption of a server under specific load levels. Our approach collects representative server profiles using workload mixes that use multiple resources. Our profiling approach controls the load intensity of a set of workloads to reach target values for a set of system metrics. This allows us to train power models that are representative of a large range of workloads and workload mixes. In order to validate our approach we implemented it upon the technical foundation of the Server Efficiency Rating Tool (SERT) [10,11] framework. This enabled us to reuse industry-proven workloads for classifying server energy efficiency. The ENERGY STAR program of the U.S. Environmental Protection Agency (EPA) uses SERT and its workloads to classify server energy efficiency [12].

The following elaborates our approach. First, we provide an overview of implementation and prerequisites of our approach. Based on a running example, we discuss the workload intensity calibration and measurement performed as part of the approach.

Implementation. We implemented our profiling approach atop the technical framework of SERT [10,11]. SERT evaluates the energy efficiency of servers for a set of transactional workloads. In order to reach different throughput levels, SERT linearly scales the rate of transactions in the system based on the maximum transaction rate the system can process. SERT applies representative workloads for different resources, such as CPU and storage I/O, and scales them from idle to maximum utilization. However, SERT does not allow the parallel execution of workloads that each stress different resources. Consequently, SERT does not produce a sample representative of the full combined domain of system metrics.

We implemented our profiling approach in a custom load driver. Our load driver controls the throughput by varying the delay time. It allows for the simultaneous execution of multiple workloads with different mean delay times.

Prerequisites. The prerequisites subsume all activities highlighted as manual in Fig. 1. Our approach requires the user to specify a set of target system metrics $M_{\text{profile}} = \{m_1, \ldots, m_n\} \subseteq M$. M_{profile} is the set of system metrics targeted by the profiling. M is the domain of measurable system metrics. Example metrics are the average utilization of all CPU cores u_{cpu}, storage write throughput tp_{write} and storage read throughput tp_{read} in kilobytes per second.

The user defines a set of workload mixes used in the server profiling. A workload mix is a tuple $(w_1, \ldots, w_n)$. w_j is a workload with a controllable load intensity parameter l, where there exists a monotonic relationship between l and measurements of m_j. An example element workload for u_{cpu} is the AES encryption workload $w_{\text{aes}} \in W_{u_{\text{cpu}}}$. The user defines a workload mix by selecting a workload w_j from a predefined set W_{m_i} for each $m_i \in M_{\text{profile}}$.

An individual user of our approach does not need to determine W_{m_i}. Rather, W_{m_i} has the role of a reusable repository. Once a monotonic relationship between l and the measurements m_i of a workload have been established for a workload w_{new}, any user can select from $W_{m_i}^{\text{new}} = W_{m_i} \cup \{w_{\text{new}}\}$.

Running Example. In the following, we will explain the profiling process with reference to the example workload mix $(w_{\text{aes}}, w_{\text{rwrite}})$. w_{rwrite} is a workload executing random disk writes. We outline the profiling process using one of the target level tuples we used in our experiments. A target level describes the utilization level the profiling aims to observe for a workload run. The target level tuple we selected is $(u_{\text{cpu}}, tp_{\text{write}}) = (0.55, 24\,000)$. Figure 2 illustrates the different steps involved in the profiling for this target level tuple. It shows measurements of tp_{write} and the load intensity of w_{rwrite} over time. The figure depicts the measured values in the upper graph. The lower graph shows the load intensity as the delay between two workload transactions. The figure shows the three phases of the calibration step (phases 1–3) and the three phases of the measurement step (phase 4–6) for $(u_{\text{cpu}}, tp_{\text{write}}) = (0.55, 24\,000)$.

Workload Intensity Calibration. The calibration phase has the goal of determining a suitable mean delay value for the transaction execution of every workload. We determine the mean value for each of the workloads in a workload mix in parallel. The calibrated mean delay value should result in a rate of transaction executions that induces the specified target level metric values.

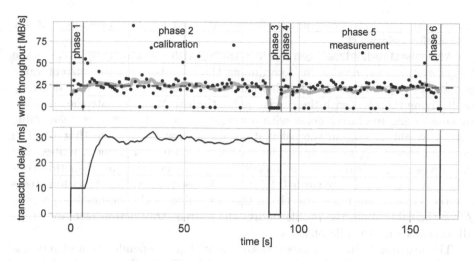

Fig. 2. Top: tp_{write} of exemplary run for target level $(u_{\text{cpu}}, tp_{\text{write}}) = (0.55, 24\,000)$. In gray: smoothed average of the measurements, and target value 24 MB/s. Bottom: Transaction delays for the storage intensive workload.

Figure 2 shows the transaction delay for workload w_{rwrite}, together with metric values of tp_{write}. The depicted workload calibration for the storage intensive workload runs in parallel with the calibration for the CPU intensive workload. In a first step, our profiling framework initializes the workload and starts transactions at an initial rate (phase 1). Subsequently, the actual calibration process starts, in which the transaction rate is varied (phase 2). Algorithm 1 lists the algorithm used during calibration. The profiling framework executes the algorithm in every measurement interval. The algorithm tries to reach a sensible starting value for the system metric, e.g., 10 MB/s for HDD write throughput

> **state**   : *thresholdReached* ← false
> **input**   : Current system metric value u, Target metric value u_t,
>   Threshold metric value u_{thold}, Metric-specific alpha α_m,
>   Initial delay *currentDelay*
> **output** : Delay to throttle workload *currentDelay*
> 1 **if** ¬*thresholdReached* **then**
> 2 $\quad$ **if** $u < u_{\text{thold}}$ **then** *thresholdReached* ← *true* ;
> 3 $\quad$ **else** *currentDelay* ← $2 \cdot$ *currentDelay* ;
> 4 **else**
> 5 $\quad$ *targetDelay* ← *currentDelay* $\cdot \frac{u}{u_t}$;
> 6 $\quad$ *currentDelay* ← *currentDelay* $\cdot (1 - \alpha_m) +$ *targetDelay* $\cdot \alpha_m$;
> 7 $\quad$ **if** $\alpha_m > 0.1$ **then** $\alpha_m \leftarrow 0.9 \cdot \alpha_m + 0.01$;

Algorithm 1. Adaptive calibration policy for controlling the workload intensity.

(lines 1–3). This avoids contention effects that occur for shared resources at low transaction delays.

After the threshold has been reached, the algorithm gradually approaches the target system metric value by determining the ratio between the current value u and the target value u_t. The algorithm determines the new target delay by multiplying this ratio with the current delay (line 5). We attenuate the adaption by considering the target delay with a weight α_m and the previous delay value with a weight $1 - \alpha_m$ in the calculation of the new delay value (line 6). The user can choose α_m for each metric. We set α_m to 0.2 for random writes, and 0.05 for sequential writes for the metric tp_{write}. In each run the algorithm continuously decreases α_m towards 0.1 (line 7). Consequently, the algorithm steers the transaction rate more directly in the beginning of the calibration process. After the calibration, the profiler stores the current transaction rate and stops all workloads for the idle phase 3.

The profiling framework executes the algorithm independently and simultaneously for each workload in the workload tuple. This enables the algorithm to adjust the load intensity based on interferences between the workloads. In the case of combining I/O-intensive with CPU-intensive workloads, the adjustment is necessary since most I/O-intensive workloads still utilize the CPU to perform operations on the read data, even though the CPU is not a potential bottleneck. The measured load would not match the target load for $(w_{\mathrm{cpu}}, w_{\mathrm{rwrite}})$ if we were to determine the delay time t_{cpu} that achieves the targeted average CPU utilization for (w_{cpu}) independently of the delay time t_{rwrite} for w_{rwrite}. Hence we need to determine delay times for the workload mix. The combined calibration allows us to achieve the utilization targets of both metrics with a parallel execution of the workload mix $(w_{\mathrm{cpu}}, w_{\mathrm{rwrite}})$.

Measurement. Throughout the calibration and measurement phases our profiling framework takes equidistant measurements of relevant system metrics. Idle phase 1 reduces instabilities between the measurement of two target level tuples. Idle phase 3 and warmup phase 4 aim to avoid instabilities when transitioning between calibration and measurement. We consider measurements of system metrics and power consumption taken during the measurement phase to be representative values of a system in a stable state under the used workload.

In the pre-measurement step (phase 4), our framework starts all workloads in the workload mix using the calibrated transaction rate. The pre-measurement phase allows the system to stabilize and mitigates warm-up effects. Our load driver runs the system with the stable transaction rate for the measurement phase (phase 5). For technical reasons, the load driver continues to run the workload mix in a short post-measurement phase (phase 6). It then stops all workloads.

2.2   Model Training

We use power models to reason on the power consumption of the profiled servers. We construct the models by means of statistical learning techniques. The power models are trained using the power consumption profile extracted using our profiling approach discussed in Sect. 2.1. We utilize a *Power Model Repository* [6] to persist a set of recurring power model types. Each power model type is associated with a regression model formula. The power model type references the system metrics it requires as input. To apply a power model type to a profiled system, we instantiate it by training its non-parametrized regression model. This produces a regression model we can use to predict the power consumption of a software system deployed on the server.

In the scope of this paper we use an iterated reweighted least squares algorithm based on a robust M-estimator as implemented by Rousseeuw et al. [13] to train the regression models. The central advantage of robust regression techniques is their robustness towards outliers and anomalous measurements. While techniques for non-parametric regression have been applied to power modeling [4], we did not find conclusive evidence that they are more accurate than parametrized learning.

2.3 Model Selection

Power models can be used to reason on the power consumption at runtime and design time. Over the years, different power models have been proposed to model the relation between system metrics and power consumption of servers [14]. The accuracy of power models depends on the server under investigation and the workload executed on the system. When training power models for runtime use, the target workload for which we want to analyze the power consumption may already be fully known. In this case, we can measure the accuracy of trained power models under the expected workload mix. Based on the measured accuracy, we can select a suited power model.

At design time, the implementation of the target workload is not yet fully available. We can not select the most accurate power model based on measurements for the target workload. Still, we need to make an informed trade-off decision between the accuracy of a candidate power model and the effort required to predict its input metrics. As we cannot measure the power consumption of the target application, we need to reason on power model accuracy independent of the final implementation of the designed application.

There exist different model selection techniques based on statistical methods such as residual sum of squares, k-fold cross-validation and Akaike's Information Criterion (AIC). k-fold cross validation is commonly used in software performance engineering to evaluate the predictive quality of models. AIC is an information-theoretic measure that quantifies the information loss between the evaluated model versus the "unknown true mechanism" [15] that actually produced the data which the model was trained on. Stone [16] has shown AIC and k-fold cross-validation to be asymptotically equivalent. We apply AIC to determine whether we can increase prediction accuracy by considering additional metrics. We opted for AIC over k-fold cross due to its simplicity.

We evaluate a set of candidate power models we maintain in a Power Model Repository to find the model that most accurately describes the power consumption of the profiled server. We determine the rank of each power model based on its difference to the minimal AIC as described by Burnham and Anderson [15]: $\Delta_{AIC} = AIC - AIC_{min}$. If all models considering a set of metrics M with $m \in M$ are dominated by any model with the metric set $M \backslash \{m\}$, we deduce that there is no benefit in considering m. Should the consideration of m increase accuracy, we compare the difference in ranking between the best-performing model with metrics M and $M \backslash \{m\}$.

3   Evaluation

In our evaluation we investigated four Evaluation Questions (EQs):

EQ1: Do the power models we derive from our server profile accurately predict power consumption across different types of workload?

EQ2: Does the simultaneous profiling of CPU and HDD profiles increase the accuracy over profiling CPU and HDD in isolation?

EQ3: Does our approach produce server profiles that are better suited for training power models than other approaches?

EQ4: Does the AIC-based selection of power models accurately predict the effect of considering system metrics on prediction accuracy?

We evaluate EQ1 by analyzing the accuracy of power models from literature, which we trained using the server profile produced by our profiling approach. We investigate EQ2 by comparing the accuracy of power models trained on a profile from simultaneous profiling, and a profile from isolated profiling. To evaluate EQ3, we compare the server profiles produced by our approach against a server profile produced by a commonly used alternative approach. To investigate EQ4, we compare our AIC-based ranking with the actual accuracy of the power models for a set of workloads.

3.1   Setup

We used a PowerEdge R815 with four Opteron 6174 CPUs and 256 GB RAM. The server utilized a built-in storage RAID with six 900 GB 10,000 RPM SAS. The server's resources were virtualized using XenServer 6.5. Profiling and power measurements were conducted within Ubuntu 14.04 VMs, with 48 virtual cores assigned to each VM. Only one VM was running at a single point in time. The SPECjbb2015 VM was assigned 32 GB RAM while the HiBench VM was allocated 16 GB RAM. Power monitoring was conducted using a ZES Zimmer LMG95 power meter connected to a dedicated notebook. The measurement data and analysis tooling used in our evaluation are available online[1].

We used the workloads *SequentialWrite*, *RandomWrite*, *XMLvalidate*, *CryptoAES* and *SOR* from the Server Efficiency Rating Tool (SERT) to profile our server under investigation. A detailed description of the used workloads can be

---

[1] https://sdqweb.ipd.kit.edu/wiki/Power_Consumption_Profiler.

found in the SERT design documents available for public review [11]. The profiling of each target level including warmup lasted around two and a half minutes. The framework collected around 60 power and system metric measurement samples per target level. The full profiling took approximately 38 h.

3.2 Considered Power Models

We collected power models based on system metrics from literature. Table 1 contains an overview of the considered power models. The models range from simple linear regression models (1), only parametrized by CPU utilization, to multi-factorial models with exponential components (3, 4). As explained in Sect. 2.2 we extracted the power models using robust non-linear regression.

Table 1. Overview of considered power models

No	Power model	Considered metrics
1	$P = c_0 + \sum_{m \in M} c_m u_m$	OS-level performance counters [3,5,17,18], or only CPU utilization [19,20]
2	$P = c_0 + \sum_{m \in M} (\sum_{l=1}^{l_{max}} c_l u_m{}^l)$	OS-level performance counters [18], or only CPU utilization [20]
3	$P = c_0 + \sum_{m \in M} \sum_{l=1}^{l_{max}} (e^{u_m} + c_l u_m{}^l)$	OS-level performance counters [18]
4	$P = c_1 \cdot e^{-(\frac{u_{cpu} - c_2}{\alpha_1})^2}$	CPU utilization [20]
5	$P = c_0 + c_1 u_{cpu} + c_2 u_{cpu}^{\alpha}$	CPU utilization [17,19]
6	$P = c_0 + c_1 u_{cpu}^{\alpha}$	CPU utilization

Previous work [17,18] has shown that the prediction accuracy of system metric based power models can be increased by considering additional metrics. To evaluate the impact of metric selection on prediction accuracy we instantiated each of the multi-metric power models 1, 2, and 3 with CPU and storage metrics. Models 2 and 3 contain a complexity parameter l that defines the polynomial degree of the function. We instantiate 2 and 3 for values of $l = \{1, 2, 3\}$.

3.3 Prediction Accuracy of Power Models

To investigate whether our profiling approach produces server profiles that are suited for training power models, we used it to train the power models described in the previous section. If the models produced by the robust regression are accurate, we can deduce that our approach produces server profiles representative of the power consumption of the system under investigation.

We used a diverse set of workloads from the HiBench benchmarking suite [9] and SPECjbb2015 [8] to evaluate the prediction error of the power models. From the considered workloads, *K-means*, *TeraSort*, *DFSIOe*, *Page Rank* and *Nutch Indexing* were I/O intensive. All other benchmarks mostly stressed the CPU, or no resources at all in the case of *Sleep*.

Surprisingly, the models had a smaller prediction error when trained via measurements from separate profiling. For the considered workloads and power models, the results thus negatively answer EQ2. One potential reason for this is the large number of measurements with high utilization for multiple metrics from simultaneous profiling. The used regression approach minimizes the prediction error for the training set. However, the application workloads considered in the evaluation rarely stress CPU or HDD at the same point in time.

To assess the total accuracy of the models learned with our approach, we calculated the Mean Absolute Error for each workload. Overall, robust regression was able to train all ty1pes of power models to reach low prediction errors. Power models of type 1 with $M = \{u_{cpu}\}$, 5 and 6 had a median prediction error below 2.3%. Models of types 1, 3 for $l = 1$, and 4 suffered from poor prediction accuracies for utilization levels close to idle as observed for the *Sleep* workload.

Aside from *Sleep*, all power models achieved an error of at most 5.9% across all other workloads. The power model of type 3 with $l = 2$ and $M = \{u_{cpu}, u_{read}, u_{write}\}$ reached a maximum error of 4.7%. The power model 5 meets this maximum error. In total, 19 of 27 considered power models have a maximum prediction error of 5.9% across all workloads. From this we conclude that our approach produces representative server profiles that are well suited for training power models with high accuracy (EQ1).

3.4   Comparison of Profiling Approach with State of the Art

To evaluate the benefit of our profiling approach we compared it to state of the art profiling approaches. We replicate the behavior of state of the art approaches [3,4] by monitoring the execution of SERT. As the SERT workloads individually stress the hardware components this matches the measurement procedure of state of the art approaches. We conducted a SERT run and collected measurements using the tooling described in Sect. 3.1.

The passive monitoring of SERT very rarely stressed storage to write more than 20 MB/s. Our profiling approach managed to reach write throughputs of up to 150 MB/s. This shows that the state of the art approach did not cover high write throughputs. Thus, the regression models trained on the resulting profile need to extrapolate for high write throughputs.

The power models built solely upon CPU utilization had high accuracy when trained using the profile from the SERT run. However, the models that consider both CPU utilization and storage throughput were significantly less accurate. Models 2 and 3 with $M = \{u_{cpu}, u_{read}, u_{write}\}$ deviated from the measured value by a factor of up to 70 for I/O-intensive workloads.

In conclusion, the profile obtained from monitoring SERT via a state of the art profiling approach can not be used to train multi metric power models. As our approach enabled us to train multi metric power models this confirms EQ3.

3.5   Impact of Metric Selection on Prediction Accuracy

We evaluated the impact of metric selection on prediction accuracy using the AIC-based ranking approach outlined in Sect. 2.3. Our intent was to evaluate whether the AIC-based ranking based on our server profile correctly predicted the effect of metric selection on prediction accuracy. For this, we compared the ranking with the prediction error of power models for our evaluation workloads. The ranking based on Δ_{AIC} indicated that the CPU-only models 6 followed by 5 had the highest likelihood of having the best prediction accuracy. Model 3 with $l = 3$ and $M = \{u_{cpu}, u_{read}, u_{write}\}$ followed third as the highest-placing model that considered storage metrics.

Since models 5 and 6 parametrized by both only CPU utilization outperformed all other models, we can deduce that considering storage metrics does not increase the prediction accuracy of trained power models for the models from Table 1. This was confirmed by the evaluation of error rates for the workloads outlined in Sect. 3.3. In the accuracy evaluation, model 6 had the lowest median prediction error. Considering storage write throughput did not reduce the average prediction error using our set of considered power models.

In conclusion, we were able to correctly predict the effect of considering additional metrics using the Δ_{AIC}-based ranking (EQ4). This indicates that the ranking is suited to the selection of a power model for consecutive use in design time predictions.

3.6   Threats to Validity

We conducted both profiling and measurements in a virtualized execution environment. This induces an overhead on the execution of both CPU and storage operations. We opted to perform the experiments in a virtualized environment as these environments are today's norm in the enterprise space. Benchmarks like SPECvirt [21] specifically target energy efficiency for virtualized environments. As with all models, power model abstract from system characteristics that can impact the power consumption. Examples for such system characteristics observed by Mccullough [18] are "hidden device states" and "significant variability" in power consumption of "identical components". Our approach does not consider these effects. Consequently, we can not quantify their significance to our findings.

Since we evaluated our approach for one specific server it cannot be guaranteed that our approach works for all server environments.

4   Related Work

Dayarathna et al. [14] provide an extensive overview of different power modeling techniques. The models covered by the survey range from manually created models to models trained using machine learning techniques. The following discusses a set of referenced modeling approaches that automate the creation or parametrization of their models.

Davis et al. [4] propose a methodology for automatically deriving power models based on OS-level performance counters. Their approach uses feature selection to identify the performance counters that strongly correlate with power consumption. Davis et al. use piecewise-defined regression power models. The profiling approach presented by the authors does not systematically vary load. Instead, it passively monitors the execution of a set of workloads to extract the measurement data needed to train the models. Davis et al. state that the workloads cover different load intensities and workload types which stress CPU, storage and network. However, their approach does not guarantee that measurements are collected for all relevant system metric levels and combinations.

Economou et al. [3] propose a profiling approach that individually stresses the hardware components of a server. Unlike our approach, it does not use hybrid workloads. Consequently, it does not support the investigation of interactions between multiple workloads on the measured system metrics. Section 3.4 had evaluated our approach against a profiling approach that replicated the behavior of the approaches by Davis et al. [4] and Economou et al. [3].

The PowerPack framework by Ge et al. [22] aims at profiling power consumption of distributed parallel applications. Like our work, Ge et al. investigate the effects of parallel job configurations on power consumption. In contrast to our work, PowerPack does not extract power models. Rather, it focuses on comparing the power consumption of the job configurations via measurements.

The Server Efficiency Rating Tool (SERT) [10] rates the energy efficiency of servers. It uses a set of workloads to stress the server under investigation. SERT varies the transaction rate of the workloads in order to assess the energy efficiency of the server at different load levels. Unlike our approach, SERT does not vary the workload to target system metric levels. SERT uses a hybrid workload based on the SSJ simulation library to assess efficiency for a mixed transactional workload. However, it does not assess the energy efficiency of workload combinations. This differs from our approach which simultaneously steers multiple workloads to reach target metric levels.

5   Conclusion

This paper presented an approach for the automated creation of power models using systematic experimentation. We outlined a methodology for deriving representative server profiles for training power models. Our approach allows for the creation of workload combinations from existing workloads. We presented an adaptive workload calibration policy that allows targeting system metric levels for combined workloads. We automatically parametrize a set of power models using the server profile produced by our profiling approach by means of robust nonlinear regression. To reason on the effect of considering additional system metrics on the power consumption prediction accuracy we rank the power models based on their Akaike's Information Criterion (AIC).

The evaluation investigated the applicability and accuracy of our approach by predicting the power consumption of a virtualized server system for a set of twelve benchmark applications, including the HiBench benchmarking suite [9] version 5.0 and SPECjbb2015 [8]. The evaluation showed that the power models parametrized by our approach accurately predicted the power consumption across all twelve applications (EQ1). We showed that separate profiling of CPU and HDD was sufficient to train accurate power models (EQ2). We compared a state of the art approach with our approach to determine whether our approach produced more representative server profiles for training power models (EQ3). The profile produced by our profiling was more representative of the system's power consumption. When we trained the power models based on the profiles collected using a state of the art approach, the model predictions deviated from the measured value by a factor of up to 70.

A comparison of our AIC-based ranking showed that we were able to estimate the effect of considering additional metric on prediction accuracy (EQ4). The most consistently accurate power model's prediction error ranged from 0.1% to 5.9%. Our AIC-based ranking had predicted this power model to have the highest likelihood of a high prediction accuracy. Four out of the six Pareto optimal power models from the evaluation had placed the highest in the ranking.

Our approach enables both software engineers to derive accurate power models of servers for design time predictions based on system metrics. It supports the combination of multiple workloads to create mixed system workloads. This enables engineers and operators to profile a server with workloads that more realistically match the behavior when hosting multiple collocated applications.

Our approach automates parametrization and ranking of power models. This reduces the effort for identifying a suitable power model for a given deployment environment. Engineers can choose from an extensible set of power models based on the system metrics they can predict. We ease reasoning on the effects of considering additional system metrics in power consumption analysis by ranking power models based on their estimated prediction accuracy.

In future work we will investigate how we can reduce the time needed for a profiling run. We plan to adaptively reduce the number of required measurement runs during profiling, reducing the total time required to create accurate power models for a server. To reason on the effect of adaptive server management policies we plan to include power consumption profiling for server reconfigurations. Examples for such reconfigurations are server shutdowns and bootups, as well as Virtual Machine migrations.

Acknowledgments. The research leading to these results has received funding from the European Union's Seventh Framework Programme under grant agreement 610711. This work was also supported by the German Research Foundation (DFG) as part of the Research Training Group GRK 2153: "Energy Status Data – Informatics Methods for its Collection, Analysis and Exploitation".

References

1. Barroso, L.A., Clidaras, J., Hölzle, U.: The Datacenter as a Computer: An Introduction to the Design of Warehouse-Scale Machines. Morgan & Claypool Publishers, California (2013)
2. Greenberg, A., Hamilton, J., Maltz, D.A., Patel, P.: The cost of a cloud: research problems in data center networks. ACM SIGCOMM Comput. Commun. Rev. **39**(1), 68–73 (2008)

3. Economou, D., Rivoire, S., Kozyrakis, C., Ranganathan, P.: Full-system power analysis and modeling for server environments. In: Workshop on Modeling Benchmarking and Simulation (MOBS) (2006)
4. Davis, J.D., Rivoire, S., Goldszmidt, M., Ardestani, E.K.: CHAOS: composable highly accurate OS-based power models. In: Proceedings of the 2012 IEEE International Symposium on Workload Characterization (IISWC), pp. 153–163, Washington, DC, USA (2012)
5. Brunnert, A., Wischer, K., Krcmar, H.: Using architecture-level performance models as resource profiles for enterprise applications. In: Proceedings of the 10th International ACM Sigsoft Conference on Quality of Software Architectures (QoSA 2014), pp. 53–62. ACM, Marcq-en-Bareul (2014)
6. Stier, C., Koziolek, A., Groenda, H., Reussner, R.: Model-based energy efficiency analysis of software architectures. In: Weyns, D., Mirandola, R., Crnkovic, I. (eds.) ECSA 2015. LNCS, vol. 9278, pp. 221–238. Springer, Cham (2015). doi:10.1007/978-3-319-23727-5_18
7. Jagroep, E.A., van der Werf, J.M., Brinkkemper, S., Procaccianti, G., Lago, P., Blom, L., van Vliet, R.: Software energy profiling: comparing releases of a software product. In: Proceedings of the 38th International Conference on Software Engineering Companion (ICSE 2016), pp. 523–532. ACM, Austin (2016)
8. SPECjbb2015 Benchmark Design Document. Technical report. Standard Performance Evaluation Corporation (SPEC), Gainesville (2015)
9. Huang, S., Huang, J., Dai, J., Xie, T., Huang, B.: The HiBench benchmark suite: characterization of the MapReduce-based data analysis. In: 2010 IEEE 26th International Conference on Data Engineering Workshops (ICDEW), pp. 41–51 (2010)
10. Block, H., Arnold, J.A., Beckett, J., Sharma, S., Tricker, M.G., Rogers, K.M.: Server efficiency rating tool (SERT) 1.0.2: an overview. In: Proceedings of the 5th ACM/SPEC International Conference on Performance Engineerin (ICPE 2014), pp. 229–230. ACM, Dublin (2014)
11. Server Efficiency Rating Tool (SERT) Design Document 1.1.1. Technical report. Standard Performance Evaluation Corporation (SPEC), Gainesville (2016)
12. ENERGY STAR Program Requirements for Computer Servers — Partner Commitments. U.S. Environmental Protection Agency. www.energystar.gov/ia/partners/proddevelopment/revisions/downloads/computerservers/ProgramRequirements V2.0.pdf
13. Rousseeuw, P., Croux, C., Todorov, V., Ruckstuhl, A., Salibian-Barrera, M., Verbeke, T., Koller, M., Maechler, M.: Robustbase: basic robust statistics. R package version 0.92-6 (2016). CRAN.R-project.org/package=robustbase
14. Dayarathna, M., Wen, Y., Fan, R.: Data center energy consumption modeling: a survey. IEEE Commun. Surv. Tutor. **18**(1), 732–794 (2016)
15. Burnham, K.P., Anderson, D.R.: Model Selection and Multimodel Inference: A Practical Information-Theoretic Approach. Springer, New York (2002)
16. Stone, M.: An asymptotic equivalence of choice of model by cross-validation and Akaike's criterion. J. R. Stat. Soc. Ser. B **39**, 44–47 (1977)
17. Rivoire, S., Ranganathan, P., Kozyrakis, C.: A comparison of high-level fullsystem power models. In: Proceedings of the 2008 Conference on Power Aware Computing and Systems (HotPower 2008), p. 3. USENIX Association (2008)
18. McCullough, J., Agarwal, Y., Chandrashekhar, J., Kuppuswamy, S., Snoeren, A.C., Gupta, R.: Evaluating the effectiveness of model-based power characterization. In: Proceedings of the USENIX Annual Technical Conference, Portland, OR (2011)
19. Fan, X., Weber, W.-D., Barroso, L.A.: Power provisioning for a warehouse-sized computer. SIGARCH Comput. Archit. News **35**(2), 13–23 (2007)

20. Zhang, X., Lu, J., Qin, X.: BFEPM: best fit energy prediction modeling based on CPU utilization. In: 2013 IEEE Eighth International Conference on Networking, Architecture and Storage (NAS), pp. 41–49 (2013)
21. SPECvirit_sc®2013, Standard Performance Evaluation Corporation (SPEC). www.spec.org/virtsc2013/. Accessed 26 Aug 2016
22. Ge, R., Feng, X., Song, S., Chang, H.C., Li, D., Cameron, K.W.: PowerPack: energy profiling and analysis of high-performance systems and applications. IEEE Trans. Parallel Distrib. Syst. **21**(5), 658–671 (2010)

ADaCS: A Tool for Analysing Data Collection Strategies

John C. Mace[✉], Nipun Thekkummal, Charles Morisset, and
Aad Van Moorsel

School of Computing Science, Newcastle University,
Newcastle upon Tyne NE1 7RU, UK
{john.mace,N.B.Thekkummal1,charles.morisset,
aad.vanmoorsel}@newcastle.ac.uk

Abstract. Given a model with multiple input parameters, and multiple
possible sources for collecting data for those parameters, a data collection strategy is a way of deciding from which sources to sample data,
in order to reduce the variance on the output of the model. Cain and
Van Moorsel have previously formulated the problem of optimal data
collection strategy, when each parameter can be associated with a prior
normal distribution, and when sampling is associated with a cost. In this
paper, we present ADaCS, a new tool built as an extension of PRISM,
which automatically analyses all possible data collection strategies for a
model, and selects the optimal one. We illustrate ADaCS on attack trees,
which are a structured approach to analyse the impact and the likelihood
of success of attacks and defenses on computer and socio-technical systems. Furthermore, we introduce a new strategy exploration heuristic
that significantly improves on a brute force approach.

Keywords: Security modeling · Risk management · Attack trees ·
Experiment design · Data collection

1 Introduction

To obtain model results that reflect realistic systems accurately, one collects data
from different sources and parameterises various variables in the model. Each
data collection source might have a different cost and a different accuracy. For
instance, a model for political forecast might take as input polls from population
samples, or aggregated data from social media. While opinions polls might bring
an option to ask specific questions, collecting data from social media might bring
a more global picture. Ideally, an organisation interested in political forecast
should collect data from both sources, but in practice, the combined cost might
be too high.

Cain and Van Moorsel proposed in [4] a general model for optimising data
collection strategies, based on the variance generated by these strategies and a
cost budget. We refine this approach here by presenting the tool ADaCS, which

© Springer International Publishing AG 2017
P. Reinecke and A. Di Marco (Eds.): EPEW 2017, LNCS 10497, pp. 230–245, 2017.
DOI: 10.1007/978-3-319-66583-2_15

takes a model as input, together with a cost model and a description of each source for each input parameter of the model, and returns automatically the best *data collection strategy*, i.e., the optimal way to spend a given budget on data collection. We illustrate ADaCS in the domain of attack trees, which take, among others, the likelihood of success of attacks and defensive mechanisms as input parameters. We believe that ADaCS could be useful to a security architect wanting to understand the best data collection strategy for a given attack tree.

The main contribution of this paper is therefore the tool ADaCS (Sect. 3), which is implemented as an extension of the probabilistic model-checker PRISM, together with the application of this tool on attack-trees (Sect. 4). We also present a heuristic for finding an approximation of the optimal data collection strategy reducing significantly the search space (Sect. 5).

2   Problem Formulation

In this section we articulate the data collection problem as a mathematical optimization problem, based on the earlier work of Cain and Van Moorsel [4]. Since we concentrate in this paper on tool support and practical applications, our presentation of the theory will be less general, but this reduces the complexity of the notation compared to [4]. We first present a general formulation of the problem, which we then refine on attack trees.

2.1   Optimal Data Collection Strategy

Let us consider a model, taking as inputs a set of parameters $P_1, \ldots, P_n$, and outputting a random variable Y. This model can be an attack tree, as we explain below, or any other discrete-event dynamic system model [11]. In order to compute the expected value of Y, which we write $E[Y]$, we consider a simulation-based approach: we repeat a number of M runs, such that at each run j, we instantiate each P_i with a value p_{ij}, compute the corresponding value y_j; the expected value of Y is calculated as $E[Y] = \sum_{j=1}^{M} y_i/M$, and the variance by $\sum_{j=i}^{M}(y_i - E[Y])^2/(M-1)$ [6].

The crucial point here is to instantiate each P_i with a value p_{ij}. Without any information, we could simply take a random value from the domain of P_i. However, many models can use data collection to determine the value of the parameters, for instance with the environmental metrics described above. For instance, in a socio-technical setting, such as the 'USB model' introduced in [3], the model describes the different security threats related to the use of USB sticks within an organisation, a parameter might be the typical behavior patterns of users of IT systems.

Hence, we now consider that each parameter P_i can be associated with a normal distribution $\mathcal{N}(\mu_i, \sigma_i/\sqrt{M_i})$ (this follows from the Central Limit Theorem, e.g., [6]), where μ_i represents the mean value for P_i, σ_i^2 the variance, and M_i the number of samples used to compute these values. In the USB model example, we could for instance ask 30 employees about how often they carry a USB stick to their customer, and we could go through 100 travel bookings to identify the duration of travel.

One of the key points of [4] is that a small number of data source samples M_i for P_i implies a wide normal distribution, and therefore implies a large variance in the values p_i drawn, which then implies a large variance in the output $E[Y]$ (which is not desirable). Their key argument is therefore to use the Central Limit Theorem to reflect that when one increases the number of samples M_i it will narrow down the normal distribution for P_i. Given an additional number of samples $N_i > 0$, one would run simulations drawing values p_{ij} from $\mathcal{N}(\mu_i, \sigma_i^2/\sqrt{M_i + N_i})$.

In order to add samples, given a set of parameters $X = \{P_1, \ldots, P_n\}$, we assume the existence of a set of sources $D = \{d_1, \ldots, d_m\}$, such that each source is a predicate $d_i : X \to \mathbb{B}$ (a source can sample multiple parameters at the same time), indicating which parameters it can sample. Each source d_i is associated with a cost c_i, indicating the cost of one sample of d_i (the cost is the same regardless of the number of parameters d_i can sample). A cost here represents an abstract notion, which could for instance correspond to a monetary value or the time required to sample.

A *data collection strategy* s is a set $\{N_1, \ldots, N_m\}$, indicating a number of samples $N_j(s)$ for each source d_j. Given a parameter X_i, we write $N_i(s)$ for the number of samples collected from all the sources, i.e., $N_i(s) = \sum_{d_j \in D | d_j(x_i)} N_j(s)$. The variance of Y using the strategy s, which we write $Var[E[Y] \mid s]$, is calculated by drawing for each parameter P_i the value p_i from $\mathcal{N}(\mu_i, \sigma_i^2/\sqrt{M_i + N_i(s)})$ in the simulation-based approach described above. Note that $Var[E[Y] \mid s]$ can be calculated using the equations in the first paragraph of this section, but that now the randomness is not only associated with the model but also with the parameter uncertainty under strategy s.

It is worth emphasising here that a strategy s decides which parameter distribution to narrow. Trivially, the best strategy would therefore be to add as many samples as possible for each source. However, in practice, a data collection strategy is bound by a budget, which leads us to the definition of the optimal collection strategy, simplified from [4].

Definition 1 (Optimal Strategy). *Given a budget B, a cost c_i for each sample provided by data source d_i, and a set S of all possible strategies, the optimal strategy is defined as:*

$$\arg\min_{s \in S} Var[E[Y] \mid s] \text{ subject to: } \sum_{i=1}^{m} N_i(s) \cdot c_i \leq B.$$

2.2   Attack Defence Trees

In an attack tree (AT) [17] each leaf corresponds to a basic action on the system, and each node is a composition of sub-trees using the logical operators $\wedge$ and $\vee$. For the sake of exposition, we might associate a node with a label, representing the attack corresponding to that node. For instance, a brute force attack on a password usually requires two sub-attacks to be successful: having the hash for the password, and being able to crack it. If we represent these two attacks by the atomic actions gethash and crack, respectively, then we can define the attack tree for a brute force attack as[1] bf $= \wedge($gethash, crack$)$. Similarly, we can represent the attack of getting the password for an account by either brute-forcing it, or by stealing it: getpw $= \vee($bf, steal$)$. In the remainder of this paper, we assume it is clear from the context whether a label corresponds to an atomic action (such as crack) or a composite attack (such as bf).

An attack-defence tree (ADT) [13] is an attack tree with the addition of the logical operator $\neg$, corresponding to the negation. Intuitively speaking, basic actions can either be attacks or defensive mechanisms, and the negation of a defensive mechanism corresponds to an attack. For instance, a typical defensive mechanism against someone getting the password is to two-factor authentication (TFA). Hence, getting the account can be defined as the conjunction of getting the password and not using TFA: getaccount $= \wedge($getpw, $\neg$tfa$)$.

ADT are useful to calculate the likelihood of an attack to succeed, by breaking it down to the likelihood of all atomic actions to succeed [2]. More precisely, given an ADT t with a set $A = \{a_1, \ldots, a_n\}$ of atomic actions, such that $p_i \in [0, 1]$ represents the likelihood of action a_i to succeed, we can define the function p_s which returns the likelihood of t to succeed as:

$$p_s(t) = \begin{cases} \max(p_s(t_1), \cdots, p_s(t_k)) & \text{if } t = \vee(t_1, \ldots, t_k), \\ p_s(t_1) * \cdots * p_s(t_k) & \text{if } t = \wedge(t_1, \ldots, t_k), \\ 1 - p_s(t') & \text{if } t = \neg t', \\ p_i & \text{if } t = a_i \end{cases}$$

In general, the probabilities p_i correspond to a parameter P_i which can be collected from the environment: the probability of a hashed password to be cracked depends on the time spent by the attacker and the entropy of the password domain; the probability of a user using two-factor authentication (assuming the user has such a choice) can be statistically computed; etc. The variance of these probabilities will depend on the source for data collection: an in-depth analysis of the system (e.g., using penetration testing techniques) will provide an accurate but costly view of the system, while relying on general statistics might be cheaper but with more variance. In this paper, we formulate the problem of data collection in the context of attack-defence trees: given a tree t, the parameters we consider are the likelihood of success $X = \{P_1, \ldots, P_n\}$ for all atomic actions, and the variable Y corresponds to the probability $p_s(t)$ to be higher than a given threshold.

3   ADaCS

In this section we introduce the tool ADaCS (*Analysing Data Collection Strategies*). ADaCS has been created by extending the probabilistic model checking

---

[1] We use here a prefix notation to avoid any ambiguity.

tool PRISM [14] with the functionality to analyse data collection strategies and compute the best strategy within a given budget. PRISM is used to verify the existence of different model properties such as the probability of reaching a specific system state. We begin this section by describing the process ADaCS takes to find the best data collection strategy. We then describe the main components of ADaCS in more detail.

3.1 Process of Finding Best Strategy

Figure 1 shows the main components of ADaCS and the process of computing an optimal data collection strategy for a given parameterised model: (1) files encoding the parameterised *PRISM model*, model *verification property*, and data collection *strategy configuration* are inputted to ADaCS; (2) the *strategy generator* analyses the inputted strategy configuration and generates all valid data collection strategies for the given model; (3) each valid strategy is tested M times in order to compute the model's mean output value (i.e., probability of holding verification property), and variance under that strategy. Each test run of a single data collection strategy is as follows: (3a) the *sample generator* generates the correct number of data samples per input parameter in accordance to the strategy and whose values respect the data source normal distributions encoded in the given strategy configuration; (3b) the *parameter generator* generates input parameter values for the model, where each input value represents a mean of the data source samples generated for that parameter; (3c) the *PRISM model checker* computes the maximum probability of the verification property existing in the given model under the generated input parameters; (3d) the probability value computed by PRISM is placed in the *result storage*. (4) the *strategy analyser* checks the M results generated under each strategy and computes the mean output and variance of the model under that strategy. (5) the *optimal strategy*, that is the strategy providing the minimum output variance, is outputted to a text file together with the strategy's variance and cost. We now describe the main components of ADaCS in more detail.

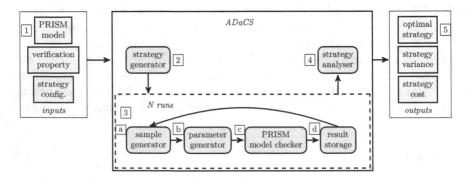

Fig. 1. Optimal strategy generation process

3.2   Strategy Configuration

Information stating how data may be collected is inputted to ADaCS by way of a *strategy configuration file* encoded in the Yet Another Markup Language (.yaml). YAML is a human-readable data serialisation language commonly used for configuration files such as the ones ADaCS requires. We will illustrate this by example, see Fig. 2.

```
            data_sources: [d1,d2,d3]
         input_parameters: [P1,P2,P3]
           input_mapping: [[1,0,0],[0,1,0],[0,0,1]]
             sample_mean: [[0.5,0,0],[0,0.5,0],[0,0,0.5]]
          sample_variance: [[0.01,0,0],[0,0.01,0],[0,0,0.01]]
  pre_collected_sample_count: [0,0,0]
             sample_cost: [1,1,1]
         sample_increment: [50,50,50]
       sample_startup_cost: [0,0,0]
              min_values: [0,0,0]
              max_values: [1,1,1]
                  budget: 200
```

Fig. 2. Example ADaCS strategy configuration file encoded in YAML

Figure 2 shows an example strategy configuration file for a model m_1 with three input parameters P_1, P_2, and P_3 mapped to three data sources d_1, d_2, and d_3. The internal structure of the model m_1 is not particularly relevant here, and is omitted for the sake of exposition. We can assume however that the input parameters of m_1 are probabilities therefore the min and max values are 0 and 1 respectively. The sample increment is 50 for each data source meaning samples can be drawn from each source in sets of 0, 50, 100, 150, and so on. The cost to draw each sample is 1 for each data source and the total maximum data collection budget is 200. A strategy collecting 50, 100, and 50 samples from d_1, d_2, and d_3, respectively, is valid (i.e., respecting the budget) whereas a strategy collecting 50, 100, and 100 samples from d_1, d_2, and d_3, respectively, would be invalid.

3.3   Extending PRISM Model-Checker

ADaCS is an extension of the probabilistic model checking tool PRISM, version 4.3 [14]. PRISM is an intuitive choice as it enables the specification, construction and analysis of parameterised probabilistic models, encoded as Markov chains and Markov decision processes for example. PRISM also comes with command line functionality and an open-source Java code base which is easily adaptable for our purposes. At runtime ADaCS inputs the model being analysed into PRISM which verifies the property expressed in the inputted verification property file.

Algorithm 1. finding optimal data collection strategy

1: **Inputs:**
  files: *config,model,property*
2: **Initialize:**
  opt_strategy = null, *runs* = 500
3: *strategies* = generateAllStrategies(*config*)
4: *results_data* = emptylist
5: **for** $j = 0 \rightarrow strategies.length - 1$ **do**
6: **for** $k = 1 \rightarrow runs$ **do**
7:   *params* = getParameters(*config*)
8:   *output* = solveModel(*model,property,params*)
9:   *results_data*.add(*output*)
10: *opt_strategy* = getOptimalStrategy(*results_data*)
11: *variance* = getVariance(*opt_strategy*)
12: *cost* = getCost(*opt_strategy*)
13: **return** *opt_strategy, variance, cost*

For instance, assume we wish PRISM to verify the maximum probability that a state can eventually be reached in a model m_1. This property is expressed as Pmax=? [F state], where F is the eventually operator and state is the model state we are interested in reaching.

ADaCS extends PRISM with the a new -adacs command line switch such that the following command can be executed:

```
$ prism m1.prism m1.props -adacs m1_config.yaml -exportresults
m1.adacs
```

The command tells PRISM to find the optimal data collection strategy under strategy configuration m1.yaml, given m1.prism and model.props, and output the strategy to m1.adacs. Algorithm 1 has been implemented in ADaCS which finds the optimal strategy, its variance and cost, by exploring all possible strategies. Example output written to m1.adacs is of the form:

```
optimal strategy: [100,50,50]
strategy variance: 0.02
    strategy cost: 200
```

The optimal strategy [100,50,50] states that 100 budgetary units should be invested in collecting data for parameter P_1 from data source d_1, 50 units for P_2 from d_2, and 50 units for P_3 from d_3.

4   Computer Virus Attack

In this section we demonstrate ADaCS by modelling a probabilistic computer virus attack scenario in PRISM and use ADaCS to calculate the optimal data collection strategy. Such analyses could be useful to security architects by informing them where best to invest limited time or money in analysing the likelihood its systems will fall prone to attacks. We describe an attack-defence tree of a virus attack, how the virus ADT is analysed with ADaCS, and present data collection strategy analysis results and tool performance.

4.1   Attack-Defence Tree

We consider the scenario of an attacker trying to infect a computer with a virus presented by Aslanyan, Neilson, and Parker in [2]. It is assumed an attacker will try to infect a computer in two phases. Attack Phase 1 (AP1) involves trying to put the virus file on the computer system followed by Attack Phase 2 (AP2), which involves executing the virus file. Two defence mechanisms exist on the computer system to prevent such an attack. First, an anti-virus mechanism aims to prevent the success of AP1, and second, a system rollback mechanism aims to prevent the success of AP2.

Figure 3 depicts the virus attack as an attack-defence tree, where each leaf corresponds to a basic action on the computer system. More precisely, there are three attack actions: (1) email, attacker sends virus as an email attachment; (2) usb, attacker distributes virus on a usb stick; (3) exefile, attacker executes virus; and two defence actions: (1) antivirus, anti-virus detects and removes virus; (2) rollback, system rollback restores system to secure state.

In order for the virus attack to succeed both attack phases AP1 and AP2 must succeed, in other words *attack success* $= \overrightarrow{\wedge}(AP1, AP2)$. The right arrow indicates AP1 must take place before AP2. For AP1 to be successful the virus must be uploaded and has not been subsequently detected or stopped by the anti-virus mechanism, such that $AP1 = \overrightarrow{\wedge}(virus\ upload, \neg antivirus)$. It follows that *virus upload* $= \vee(email, usb)$ meaning the attacker must send an email and/or distribute a usb stick containing the virus in order to upload it to the system. For AP2 to be successful the virus must be executed and the system has not been restored by the rollback mechanism, such that $AP2 = \wedge(exefile, \neg rollback)$.

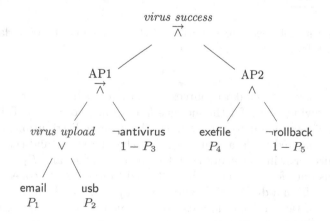

Fig. 3. Attack-defence tree representing virus attack on a computer system

4.2   Analysing Attack-Defence Tree with ADaCS

We model the virus ADT as a Markov decision process shown in Fig. 4 where the darker nodes represent virus attack failure states. The virus attack is successful if the state [¬rollback, exefile] can be reached. We assume the attacker's choice of virus deployment method is a non-deterministic attack action deployvirus whilst all other actions, email, usb, antivirus, rollback, and exefile, are probabilistic. The model is encoded in the PRISM language and has five input parameters $P_1, \ldots, P_5$ representing the probability of each of the five probabilistic actions succeeding. Each parameter is mapped to an action as follows: (email:P_1), (usb:P_2), (antivirus:P_3), (exefile:P_4), (rollback:P_5) as shown in Fig. 3. For instance, if the state [*virus upload*] is reached, the virus has been placed on the computer system. In this state, anti-virus removes the virus with probability P_3 to reach failure state [antivirus], or does not remove the virus with probability $1 - P_3$ to reach state [¬antivirus]. The model property we verify using the PRISM model checker is Pmax=? [F success], that is the maximum probability of eventually (F is the eventually operator) reaching the state [¬rollback, exefile], labelled as *virus success* in this case.

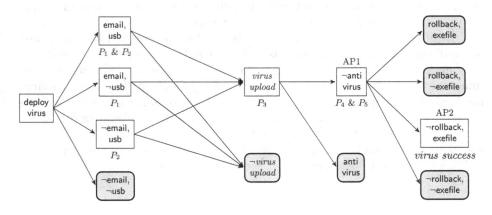

Fig. 4. Markov model style representation of the attack defence tree; darker nodes represent attack failure states.

Next we consider three data sources d_1, d_2 and d_3 from where data can be collected to provide values for the model's five input parameters. Table 1 shows the mapping of parameters to data sources, the normal distribution $\mathcal{N}(\mu, \sigma^2)$ for parameter P_i mapped to data source d_i, and the increment and cost to sample each data source. For instance, data can be collected for P_1 and P_2 from either d_1 or d_2, whereas data for P_5 can only be collected from d_3. Data collected from d_3 for P_5 is drawn from a distribution whose mean $\mu = 0.7$ and variance $\sigma^2 = 0.01$.

We assume attack method data can be sampled from a wide range of literature encompassing scientific and anecdotal evidence of computer virus attacks. Such data is likely to be easy to access but may not be highly accurate, and

Table 1. Virus attack-defence tree data collection configuration

	d_1	d2	d3
email:P_1	(0.5,0.2500)	(0.5,0.0025)	()
usb:P_2	(0.5,0.2500)	(0.5,0.0025)	()
antivirus:P_3	()	(0.7,0.0025)	()
exefile:P_4	(0.5,0.2500)	()	(0.5,0.0100)
rollback:P_5	()	()	(0.7,0.0100)
Sample increment	5	5	5
Sample cost	1	5	3

therefore comes with low cost and high variance. We represent literature-based data as data source $d1$. The values for parameters $P1$, P_2, and P_4, representing the probabilities of attack actions email, usb, and exefile being successful, can be drawn from d_1 in this instance. Next we assume attack data related to attack actions email and usb, and defence action antivirus can be sampled from the results of penetration testing. Such data is likely to be expensive and take a large amount of effort to obtain, but gives accurate scientific evidence and therefore comes with high cost and low variance. We represent penetration test results as data source d_2 from which values for parameters $P1$, P_2, and $P3$ can be derived. Lastly, data relating to the actions exefile and rollback can be sampled from the results of system testing, coming with medium cost and variance. Data source d_3 represents the results of such system tests, and can provide values for parameters P_4 and P_5. We further assume data may be collected from each data source in increments of 5 samples (e.g. 5, 10, 15, . . .), no samples have been pre-collected, there is no data source startup cost, and the maximum data collection budget is 100.

4.3 Data Collection Strategy Analysis

We analyse data collection strategies by executing ADaCS on a MacBook Pro with 2.7 GHz Intel Core i5 processor and 16 GB RAM. For each strategy we conduct M runs, as explained in Sect. 2, and we choose $M = 500$ based on trial runs, from which we concluded that $M = 500$ is large enough to compute the output variance sufficiently accurate. The choice for M is not the subject of study in this paper, but one potentially could further improve our approach by selecting a different number of runs M for different strategies, as long as for each strategy the calculated variance $Var[E[Y]|s]$ has a tight enough confidence interval. In our case, the virus ADT model is solved by PRISM 500 times for each strategy using input parameter values generated from samples drawn from the Normal distributions corresponding to the relevant data sources which come with a known mean μ and standard deviation σ^2. The variance of a strategy is calculated from the M output values computed by PRISM. Results generated

Table 2. ADaCS data collection strategy analysis results and performance

Strategy type	Samples [d1, d2, d3]	Strategy variance	Strategy cost	Strategies analysed	Runtime (m:s:ms)
Min. cost	[0,0,0]	1.81479E–4	0	1	00:22:067
Max. σ^2/max. cost	[100,0,0]	2.38172E–4	100	540	10:44:674
Min. σ^2	[0,5,25]	1.165047E–5	100	4617	58:16:819

by ADaCS and its performance are given in Table 2. In particular we focus on 3 analysis scenarios in order to illustrate the potential benefit of using ADaCS.

The first scenario highlights the case where no further data is collected to that given in Table 1. Analysis is carried out using ADaCS on the single data collection strategy [0,0,0] which indicates no samples are to be collected from any of the 3 data sources. In order to do so the total collection budget is simply set to 0. ADaCS runs the analysis in 22.067 s and computes the output variance of the model using this strategy to be 1.81479E–4.

The second strategy highlights the case where the total budget is spent without optimising the data collection strategy. We use ADaCS to identify the worst case strategy, that is the strategy that maximises the output variance of the virus ADT model with maximum cost (i.e., cost = budget). ADaCS analyses 540 strategies whose cost equals budget in 10 m 45s and returns the worst case strategy to be [100,0,0]. This strategy indicates the entire budget to be invested in collected data from data source d_1, that is literature based data regarding attack methods. Note, if this strategy was implemented, the output variance is roughly as bad as when no data would have been collected. In fact, the table shows a small increase in variance, but this difference is caused by the inherent uncertainty introduced by simulation.

The third scenario highlights ADaCS carrying out a full analysis of data collection strategies in order to compute the optimal, that is the strategy within budget which minimises the variance $Var[E[Y]|s]$. To find the optimal strategy, ADaCS analyses all strategies within budget, a total of 4617 strategies in this case. Taking this brute force approach comes with certain computational costs, for instance the runtime of ADaCS is 58 m 17s. The strategy computed as the optimal is [0,5,25] indicating 0 samples to be collected from d_1, 5 samples from d_2, and 25 samples from d_3. This is equivalent to 75 budgetary units, or 75% of time invested on the value for P_3 and 25 units, or 25% of time on the value for P_2. The strategy would indicate that system testing of the rollback defence mechanism is of most importance followed by some penetration testing of the antivirus mechanism. No more literature based data on attack methods need be collected which is in complete conflict with the worst case strategy highlighted in the second scenario above. The output variance of the virus ADT model, under the optimal strategy, reduces from 1.81479E–4 to 1.165047E–5.

5   Heuristic

In Sect. 4 we explained that by default ADaCS takes a brute force approach by analysing all data collection strategies within budget in order to find the optimal. We showed calculating the optimal strategy can be computationally expensive, even for models with a relatively small number of parameters and data sources. In this section we introduce a new strategy analysis heuristic that significantly reduces the strategy exploration space in order to find the best strategy.

We introduce the heuristic by example first. Figure 5 illustrates the heuristic for a model m_1, which can be parameterized with values coming from 3 data sources d_1, d_2, and d_3. Each data source has a sample set increment size of 50, cost per sample is 1 and total collection budget is 200. Rather than generating all valid strategies and analysing them, the heuristic generates strategies to analyse as necessary in a series of rounds; 5 rounds in the case of the example. In round 1 the base strategy is generated and analysed, this is the strategy [0,0,0] where no data is collected from any data source. Imagine the model output variance is computed to be 0.0676, this is set as the best strategy. In round 2 the next possible strategies are generated. As the sample increment size is 50, the next strategies are [50,0,0], [0,50,0], and [0,0,50] which are analysed. If a strategy s in round 2 has a lower output variance than the current best strategy then s is set as the best strategy, [50,0,0] in the case of the example. The heuristic moves on to the next round and keeps doing so until no strategy in round i has a lower output variance than the current best strategy, or the cost of the current best strategy equals the budget. The heuristic in this example cuts the brute force strategy space of 35 strategies to 13 strategies.

The general heuristic is presented in Algorithm 2. The heuristic has been implemented in ADaCS and results for 6 strategy analyses of the virus ADT model are shown in Table 3. Note the significant reduction in strategies analysed, 4617 in the brute force approach, to between 40 and 60 and a computation runtime of under 1 min. Note also the best strategy returned in each case indicates that most investment should be made in collecting data from source d_3, followed

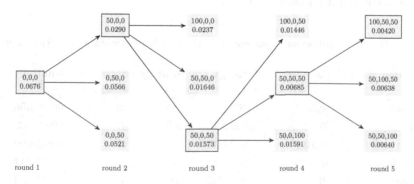

Fig. 5. Illustration of heuristic to find optimal data collection strategy for example model

Algorithm 2. Heuristic for finding best strategy
_____

1: **Inputs:**
  files: $config, model, property$
2: **Initialize:**
  $best\_strategy$ = null, $runs$ = 500
3: **while** true **do**
4:   $next\_strategies$ = emptylist
5:   **if** $best\_strategy$ == null **then**
6:    $base\_strategy$ = generateBaseStrategy()
7:    $next\_strategies$.add($base\_strategy$)
8:   **else**
9:    $next\_strategies$ = generateNextStrategies($best\_strategy, config$)
10:   $results\_data$ = emptylist
11:   **for** $j = 0 \rightarrow next\_strategies.length - 1$ **do**
12:    **for** $k = 1 \rightarrow runs$ **do**
13:     $params$ = getParameters($config$)
14:     $output$ = solveModel($model, property, params$)
15:     $results\_data$.add($output$)
16:   $strategy$ = getBestStrategy($results\_data$)
17:   **if** getVariance($strategy$) < getVariance($best\_strategy$) &
18: getCost($strategy$) ≤ getBudget($config$) **then**
19:    $best\_strategy$ = $strategy$
20:   **else**
21:    break
22: $variance$ = getVariance($best\_strategy$)
23: $cost$ = getCost($best\_strategy$)
24: **return** $best\_strategy, variance, cost$
_____

by d_2, and none from d_0 which matches the result of the brute force approach in Sect. 4.3. The heuristic does however only find a best strategy, and not the optimal one as shown by the increase in model output variance in Table 3. This would indicate the existence of a trade-off between strategy optimality (brute force) and computation time (heuristic).

Table 3. Best data collection strategies found using heuristic data collection method

Collection method	Samples [d1, d2, d3]	Strategy variance	Strategy cost	Strategies analysed	Runtime (m:s:ms)
Heuristic$_1$	[0,5,20]	2.56808E-5	85	49	00:46:856
Heuristic$_2$	[0,5,25]	2.19972E-5	100	57	00:52:432
Heuristic$_3$	[0,10,15]	2.65431E-5	95	49	00:45:999
Heuristic$_4$	[0,5,15]	2.68426E-5	70	41	00:39:711
Heuristic$_5$	[0,5,25]	1.99201E-5	100	57	00:51:534
Heuristic$_6$	[0,10,15]	2.61067E-5	95	49	00:46:047

6   Related Work

The data collection strategy optimization approach in this paper combines aspects of two main strands of analysis, namely sensitivity and uncertainty analysis [1,5,16], and adds to that specific detailed techniques based on statistics (the use of the central limit theorem) and optimization. In [4] we presented the related work, and explained that compared to known literature, the problem formulation in this paper is different because of its focus on strategies for deciding on data sources. This leads to a specific optimization problem not found in literature. For more details, please refer to [4]. The contributions in this paper focus on the practical application of and tool support for identifying data collection strategies, and this will also be the focus of this related work section.

The performance and dependability community has developed an important set of software tools that support the quantitative analysis of computer systems and networks, e.g., [15]. The specific problem formulation of the optimization problem in fact assumes that a performance or dependability model is built, and tool support for optimizing data collection would therefore naturally fit with tools such as Möbius [7], PRISM [14] or the PEPA workbench [9]. The data collection problem presented in this paper is related to sensitivity analysis. Sensitivity analysis aims at identifying the most important parameters, using techniques such as those within the classes of screening methods [12] and variance-based methods [5]. Augmenting the mentioned class of quantitative analysis tools for sensitivity analysis was first pursued by [10], which also articulates the computational challenge of exploring the parameter space when conducting sensitivity analysis.

The heuristics identified in this paper to improve the efficiency of exploring the 'space' of strategies is of importance in any of the related approaches in uncertainty and sensitivity analysis. For closed-form mathematical and simulation models with specific structure, advanced methods can be devised for the specific problem, see [12] and other references in [4]. Exploration of the strategy 'space' relates to exploration techniques in very different settings, such as codesign [8], and it may be mutually beneficial for the evaluation and codesign communities to jointly explore ideas and algorithms.

7   Conclusion

In this paper we presented the software tool ADaCS, which, to the best of our knowledge, is the first tool facilitating the calculation for optimal data collection strategies for model parameterization. In particular, ADaCS extends the probabilistic model-checker PRISM and includes both graphical and command line elements, and should therefore be easily usable by a model developer using PRISM. We also illustrated ADaCS on attack trees, which is a standard way to model security attacks and defenses in a system.

Since the bottleneck in determining data collection strategies is in traversing the space of all possible strategies, we elaborated on heuristics that limited the number of strategies for which the output variance needs to be computed. The algorithms can reduce the computation time by an order of magnitude, which may mean the difference between a feasible and intractable optimization.

Acknowledgements. The research in this paper was supported in part by UK EPSRC through 'Choice Architecture for Information Security' (EP/K006568/1) within the Research Institute for the Science of Cyber Security and by US NSA through the University of Illinois at Urbana-Champaign lablet in 'Science of Security Systems' (H98230-14-C-0141).

References

1. Ascough, J., Green, T., Ma, L., Ahuja, L.: Key criteria and selection of sensitivity analysis methods applied to natural resource models. In: Proceedings of the International Congress on Modeling and Simulation (2005)
2. Aslanyan, Z., Nielson, F., Parker, D.: Quantitative verification and synthesis of attack-defence scenarios. In: Proceedings of the 29th IEEE Computer Security Foundations Symposium (CSF 2016), pp. 105–119, June 2016
3. Beautement, A., et al.: Modelling the human and technological costs and benefits of USB memory stick security. In: Johnson, M.E. (ed.) Managing Information Risk and the Economics of Security, pp. 141–163. Springer, Boston (2009). doi:10.1007/978-0-387-09762-6_7
4. Cain, R., Van Moorsel, A.: Optimization of data collection strategies for model-based evaluation and decision-making. In: Proceedings of the 42nd Annual IEEE/IFIP International Conference on Dependable Systems and Networks (DSN 2012), pp. 1–10 (2012)
5. Chan, K., Saltelli, A., Tarantola, S.: Sensitivity analysis of model output: variance-based methods make the difference. In: Proceedings of the 29th Conference on Winter Simulation, pp. 261–268 (1997)
6. Cochran, W.: Sampling Techniques, 3rd edn. Wiley, New York (1977)
7. Deavours, D., Clark, G., Courtney, T., Daly, D., Derisavi, S., Doyle, J., Sanders, W., Webster, P.: The Möbius framework and its implementation. IEEE Trans. Softw. Eng. **28**(10), 956–969 (2002)
8. Gamble, C., Pierce, K.: Design space exploration for embedded systems using co-simulation. In: Fitzgerald, J., Larsen, P.G., Verhoef, M. (eds.) Collaborative Design for Embedded Systems, pp. 199–222. Springer, Heidelberg (2014). doi:10.1007/978-3-642-54118-6_10
9. Gilmore, S., Hillston, J.: The PEPA workbench: a tool to support a process algebra-based approach to performance modelling. In: Haring, G., Kotsis, G. (eds.) TOOLS 1994. LNCS, vol. 794, pp. 353–368. Springer, Heidelberg (1994). doi:10.1007/3-540-58021-2_20
10. Haverkort, B., Meeuwissen, A.: Sensitivity and uncertainty analysis of Markov-reward models. IEEE Trans. Reliab. **44**(1), 147–154 (1995)
11. Ho, Y.-C.: Introduction to special issue on dynamics of discrete event systems. Proc. IEEE **77**(1), 3–6 (1989)
12. Kleijnen, J.P.: Sensitivity analysis and related analyses: a review of some statistical techniques. J. Stat. Comput. Simul. **57**(1), 111–142 (1997)
13. Kordy, B., Mauw, S., Radomirović, S., Schweitzer, P.: Foundations of attack–defense trees. In: Degano, P., Etalle, S., Guttman, J. (eds.) FAST 2010. LNCS, vol. 6561, pp. 80–95. Springer, Heidelberg (2011). doi:10.1007/978-3-642-19751-2_6

14. Kwiatkowska, M., Norman, G., Parker, D.: PRISM 4.0: verification of probabilistic real-time systems. In: Gopalakrishnan, G., Qadeer, S. (eds.) CAV 2011. LNCS, vol. 6806, pp. 585–591. Springer, Heidelberg (2011). doi:10.1007/978-3-642-22110-1_47
15. Sahner, R., Trivedi, K., Puliafito, A.: Performance and Reliability Analysis of Computer Systems: An Example-Based Approach Using the SHARPE Software Package. Kluwer, Boston (1996)
16. Saltelli, A., Ratto, M., Andre, T., Campolongo, F., Cariboni, J., Gatelli, D., Saisana, M., Tarantola, S.: Global Sensitivity Analysis. The Primer. Wiley, West Sussex (2008)
17. Schneier, B.: Attack trees. Dr. Dobb's J. **24**(12), 21–29 (1999)

Case Studies

Improving ZooKeeper Atomic Broadcast Performance by Coin Tossing

Ibrahim EL-Sanosi[1,2]($\boxtimes$) and Paul Ezhilchelvan[2]($\boxtimes$)

[1] Faculty of Information Technology, Sebha University, Sebha, Libya
i.elsanosi@sebhau.edu.ly
[2] School of Computing Science, Newcastle University, Newcastle upon Tyne, UK
{i.s.el-sanosi,paul.ezhilchelvan}@ncl.ac.uk

Abstract. ZooKeeper atomic broadcast (Zab) is at the core of ZooKeeper system, enforcing a total order on service requests that seek to modify the replicated service state. Since it is a leader based protocol, its performance degrades when the leader server is made to handle an increased message traffic. We address this concern by having the other, non-leader server replicas toss a coin and broadcast their acknowledgement of a leader's proposal only if the toss results in an outcome of *Head*. We model the coin-tossing process and derive analytical expressions for estimating the coin's probability of *Head* for a given arrival rate of service requests such that the dual objectives of performance gains and traffic reduction can be accomplished. Experiments compare the performance of our coin-tossing Zab version (ZabCT) with Zab performance and confirm that the dual objectives are demonstrably met under heavy workloads. Moreover, ZabCT meets all requirements essential for crash-tolerance provisions within Zab which can therefore be adopted in any ZabCT implementation.

Keywords: ZooKeeper · Atomic broadcast · Implicit acknowledgements · Coin tossing · Probability estimation · Protocol latency · Throughput · Performance comparison

1 Introduction

Apache ZooKeeper [8] is a high-availability system offering coordination services to Internet-scale distributed applications. These services include: leader election (used by Apache Hadoop [15]), failure-detection and group membership configuration (by HBase [6]) and reliable information storage and update (by Storm in Twitter [16]). ZooKeeper itself is a replicated system made up of $N, N \geq 3$, servers that can crash at any moment and recover after an arbitrary downtime with pre-crash state in stable store. Server crashes may even be correlated and all servers may crash at the same time. Despite these failure possibilities, ZooKeeper is guaranteed to provide uninterrupted services, so long as at least $\lceil \frac{N+1}{2} \rceil$ servers are operative and connected.

© Springer International Publishing AG 2017
P. Reinecke and A. Di Marco (Eds.): EPEW 2017, LNCS 10497, pp. 249–265, 2017.
DOI: 10.1007/978-3-319-66583-2_16

At the heart of ZooKeeper is the ZooKeeper atomic broadcast protocol, Zab for short, to ensure that the service state is kept mutually consistent across all operational servers. Zab performance therefore impacts directly that of Zookeeper. Furthermore, efficient atomic broadcast protocols have far wider applications, e.g., in coordinating transactions particularly in large-scale in-memory database systems [5,14]. In such applications, the atomic broadcast protocol typically operates in heavy load conditions and is expected to offer low latencies even at such extreme loads. Thus, it remains a practical research problem to explore ways of improving Zab performance particularly under heavy loads.

Zab is a leader-based protocol and, like many other leader-based ones, it tends to offer worsening performance when the load on the leader increases. Experiments in [7] show that ZooKeeper throughput decreases gradually as write requests outnumber read requests in a cluster of *any* size. This is because read requests can be processed without involving Zab, while writes cannot start until Zab execution completes.

The aim of this paper is to improve Zab performance by reducing message traffic, both inbound and outbound, at the leader. This requires modifying the behaviour of non-leader servers, also known as *followers*, in two simple but important ways. In Zab, followers respond to the leader through unicast (1-to-1) communication which are changed to broadcasts. This allows followers to decide autonomously and relieves the leader from being the sole decision maker and, importantly, from having to broadcast its decisions to followers. This, in turn, reduces the leader's outbound traffic.

Secondly, a follower's broadcast is conditioned on the outcome of a *coin toss*: it is made if the outcome is *Head*; otherwise, not. This conditional broadcasting allows the inbound traffic at the leader to be reduced. It also introduces many design challenges. The principal one is in choosing the coin's probability p of a toss outcome being *Head* in such a way that enough followers broadcast for reaching decisions swiftly and thus keeping latencies small, but not to allow too many to broadcast at the same time. That is, determining p involves a trade-off between competing requirements. We model the coin-tossing process and derive analytical expressions for making this trade-off.

In extreme cases, p may not exist and servers must switch to Zab in a seamless manner. These aspects are addressed; the resulting protocol, termed as Zab with Coin Tossing or ZabCT for short, also maintains the well-understood and implementation-friendly structure of Zab itself. Moreover, ZabCT differs from Zab only in the latter's fail-free part and preserves all Zab invariants so that the crash-recovery part of Zab can be used unchanged. So, ZabCT can be implemented easily using Zab implementations.

The paper is structured as follows. Section 2 describes Zab for completeness. Section 3 presents ZabCT design objectives and challenges, together with a complete set of solutions and their correctness or rationale. Section 4 is devoted to comparing the performance of Zab and ZabCT using latency and throughput as metrics; for space reasons, we only consider heavy work load and crash-free settings so that p can always be found and switch to Zab is not warranted. All our experiments confirm our design objectives: smaller average ZabCT latencies and reduced traffic at the leader. Section 5 discusses the related work. Finally, Sect. 6 concludes the paper.

2   ZooKeeper Atomic Broadcast Protocol

ZooKeeper implements replicated services using an ensemble of N, $N \geq 3$, connected servers. N is typically an odd number and commonly 3 or 5.

Assumption A1 - Server Crashes: A server can crash at any time and recover after a downtime of arbitrary duration. It has a stable store or *log* and the log contents survive a crash. A server that is operative is also said to be *correct*.

Assumption A2 - Server Communication: Servers are connected by a reliable communication subsystem: messages sent by a correct server are never permanently lost and are received by all correct destinations in the order sent.

Servers are replicas of each other and maintain a copy of the application state. Zookeeper clients can submit their requests to any one of N servers. Requests may be broadly categorised as *read* or *write*; the latter seek state modification while the former do not and are serviced only by the server receiving it. Write requests are first subject to total ordering through an execution of ZooKeeper atomic broadcast (*Zab*) protocol and then are carried out by all servers as per the order decided.

Let $\Pi = \{p_1, p_2, ..., p_N\}$ denote the set of Zab processes, one in each server. One Zab process is designated as the *leader* and the rest as *followers*. As in 2-Phase commit protocol, only the leader initiates atomic broadcasting of m, *abcast(m)* for short, and the followers respond to what they receive. So, when a follower receives a write request m for ordering, it forwards m to the leader for initiating *abcast(m)*. When Zab execution for m terminates, both the leader and followers deliver m locally for ordered processing, and this event is denoted as *abdeliver(m)*.

Since the leader can crash any moment, Zab, like Paxos [10], exploits the notion of *quorums*: a quorum Q is any majority subset of Π and any two quorums must intersect. Let $\boldsymbol{Q}$ be the set of all quorums in Π: $\boldsymbol{Q} = \{Q : Q \subseteq \Pi \wedge |Q| \geq \lceil \frac{N+1}{2} \rceil\}$. $\forall Q, Q' \in \boldsymbol{Q} : Q \cap Q' \neq \{ \}$; e.g., when $N = 3$, $\boldsymbol{Q} = \{\{p_1, p_2\}, \{p_2, p_3\}, \{p_3, p_1\}, \{p_1, p_2, p_3\}\}$.

By the *liveness* arguments in [8] (see Claim 7), when the leader crashes, another process gets elected as the new leader so long as a quorum of processes are correct and can communicate in a timely manner. The new leader starts *abcasting* after it has synchronised its *abdelivered* message history with the quorum that elected it. We refer the reader to [8] for crash-tolerance details and order guarantees, and focus on aspects of the Zab protocol during crash-free runs.

2.1   Zab Protocol

It consists of the following steps.

- L1: Leader initiates *abcast(m)* by proposing a sequence number $m.c$ for m and by broadcasting its *proposal(m)* (to all processes, including itself);

- F1: A follower, on receiving *proposal(m)*, logs m and then sends an acknowledgement, *ack(m)*, to the leader;
- L2: Leader sends *ack(m)* to itself after logging m. On receiving *ack(m)* from a quorum, it broadcasts *commit(m)* before *commit(m': m'.c = m.c + 1)* is broadcast;
- F2: A follower, on receiving *commit(m)*, executes *abdeliver(m)*.
- L3: Leader, on receiving *commit(m)* (from itself), executes *abdeliver(m)*.

Note that processes receive *commit(m)* in the increasing order of $m.c$ and hence observe an identical *abdelivery* order. Also, the protocol steps need not be sequential: the leader can use concurrent threads to execute L1, L2 and L3, and so can followers to execute F1 and F2. The following **invariant** holds for every *abdelivery*:

If a process executes *abdeliver(m)*, then all processes in some $Q \in \boldsymbol{Q}$ have logged m.

This invariant is necessary and sufficient for correct replacement of a crashed leader: any m that might have been *abdelivered* under the old leader is guaranteed to be *abdelivered* by the new leader since (i) the latter synchronises itself with a quorum that elects it, and (ii) any two quorums ought to intersect. ZabCT is designed to preserve this invariant. Leader crash and subsequent replacement can therefore be dealt with using Zab mechanisms and hence are not addressed here.

3   Coin-Tossing Zab (ZabCT)

In presenting ZabCT design objectives and details, we will *initially* assume that no follower crashes, there are n, $\lceil \frac{N+1}{2} \rceil \leq n \leq N - 1$, operative followers, and that n is known to followers via a membership view management service such as JGroups [1]; also that the leader starts its *abcasting* epoch with initial sequence number $m.c_0$.

3.1   Design Objectives

They are primarily two-fold: to reduce inbound and outbound traffic at the leader with no overall performance loss and an increased outbound traffic at followers.

The leader reduces its outbound traffic by not broadcasting *commit* at all, but leaving it to the followers to decide locally when a given m is to be committed. The latter requires that (i) followers broadcast their *ack*s (not just unicast to the leader) and (ii) $n \geq \lceil \frac{N+1}{2} \rceil$ which makes ZabCT less crash-resilient than Zab; e.g., ZabCT is *not* viable if a follower crashes in a system of $N = 3$ processes. We later address this restriction by letting processes switch between ZabCT and Zab without stopping *abdelivery*.

Inbound traffic at the leader is reduced by the use of implicit acknowledgments and coin-tossing by followers. When a follower has logged m and is ready to broadcast *ack(m)*, it tosses a coin: if the outcome is *Head*, *ack(m)* is broadcast; if *Tail*, *ack(m)* is sent only to itself. Further, whenever *ack(m)* is broadcast, it indicates to recipients that the broadcaster has locally logged all $proposal(m')$, $m.c_0 \leq m'.c < m.c$, and every such $proposal(m')$ is thereby being *implicitly* acknowledged. Recall that the assumption A2 guarantees that $proposal(m')$, $m'.c < m.c$, is received before $proposal(m)$ and hence that m' is logged no later than m.

The lines of pseudo-code executed (possibly by concurrent threads) at the leader are as follows.

- L1: Leader initiates $abcast(m)$ by broadcasting $proposal(m)$ to all processes;
- L2: On receiving (from itself) $proposal(m)$: log m; send $ack(m)$ to itself;
- L3: Upon receiving either $ack(m)$ or implicit ack for m from a quorum: send $commit(m)$ to itself;
- L4: On receiving $commit(m)$: $abdeliver(m)$ before $abdeliver(m')$, $m'.c > m.c$;

Those executed at a follower are:

- F1: On receiving $proposal(m)$ from the leader: log m; send $ack(m)$ to itself; toss the coin; if (coin $=$ *Head*) then broadcast $ack(m)$ to leader and all other followers;
- F2: On receiving $ack(m)$ or an implicit ack for m from a quorum of followers, send $commit(m)$ to itself.
- F3: On receiving $commit(m)$: $abdeliver(m)$ before $abdeliver(m')$, $m'.c > m.c$;

Let $committable(m)$ be a stable predicate that becomes true at a process if the process has received either $ack(m)$ or implicit ack for m from a quorum. At any p_i, $committable(m) \Rightarrow committable(m')$, $\forall m' : m.c_0 \leq m'.c \leq m.c$; also, $\neg\, committable(m') \Rightarrow \neg\, committable(m)$. Note that these properties also hold true in Zab and the use of implicit acks does not invalidate them. Coin-tossing, however, brings in challenges not present in Zab.

3.2   Coin Toss Challenges

Let us focus on $committable(m)$ becoming true for a given m that the leader $abcast$ at, say, time t_0. Subsequent to m, let the leader $abcast$ m_i, $i \geq 1$, at time t_i, $t_{i-1} < t_i < t_{i+1}$. Assume for brevity that time taken for message processing and transmission is zero. Thus, followers toss their coins at t_0 for m and at t_i for m_i as shown in Fig. 1(a).

Let p denote the probability that a coin-toss results in *Head*; so, prob $(Tail) = 1 - p$. Let $N = 5$ and the leader of this 5-process system be p_1 also denoted as p_ℓ, $\ell = 1$.

In scenario 1, at time t_0, the followers p_2 and p_3 are assumed to get *Head* and others a *Tail* outcome. This outcome is abbreviated in Fig. 1(b) as (H_2, H_3) with subscripts indicating followers that got *Head* and *Tail* outcomes not explicitly shown. $committable(m)$ becomes true for $\{p_\ell, p_4, p_5\}$ and not for $\{p_2, p_3\}$ which have only two $ack(m)$. When the coin-toss outcome at t_1 is (H_4, H_5), p_4 and p_5 broadcast $ack(m_1)$ and thereby implicitly ack m. Thus, $committable(m)$ becomes

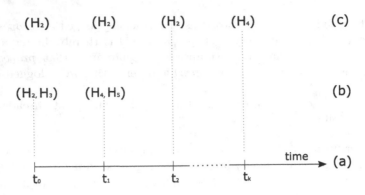

Fig. 1. (a) Coin toss instances; (b) Scenario 1; (c) Scenario 2

true for $\{p_2, p_3\}$ at t_1. Note that, also at t_1, *committable(m_1)* becomes true for $\{p_\ell, p_2, p_3\}$ (but not for $\{p_4, p_5\}$).

Thus, the system-wide coin-toss outcomes determine when individual processes can *abdeliver* a given m. We next claim that p_ℓ is always in the first wave of at least 2 processes that *abdeliver* any m.

Lemma 1: If *committable(m)* becomes true for a follower at t, then *committable(m)* becomes true for p_ℓ at t or earlier (when zero message transmission time is assumed).

Proof is simple and done by contradiction: if a follower can receive *ack(m)* or implicit *ack* for m from at least $\lceil \frac{N-1}{2} \rceil$ *other* followers, p_ℓ would also receive the same.

Lemma 2: If *committable(m)* becomes true for p_ℓ at t then there exists at least one follower for which *committable(m)* becomes true also at t.

Proof follows from the requirement $n \geq \lceil \frac{N+1}{2} \rceil = \lceil \frac{N-1}{2} \rceil + 1$: even if p_ℓ receives *ack(m)* or implicit *ack* for m from exactly $\lceil \frac{N-1}{2} \rceil$ followers, there must be one follower that receives the same from *other* followers at the same time.

Thus, the earliest time a follower can *abdeliver* m is when p_ℓ *abdelivers* m.

Scenario 2 in Fig. 1(c) is used to illustrate ZabCT reliance on subsequent *abcasts* for *abdelivering* m and hence the expected impact on performance if p is kept too small: only p_2 gets *Head* until t_{k-1}, for some $k > 1$, and at t_k only p_4 gets *Head*. Only at t_k, p_ℓ can *abdeliver* m together with followers p_3 and p_5; $\{p_\ell, p_3, p_5\}$ can also *abdeliver* $m_1, m_2, \ldots, m_{k-1}$ at t_k due to *acks* broadcast by p_2 and implicit *ack* from *ack(m_k)* broadcast by p_4.

Observe that p_ℓ requires 0 and k *abcasts* subsequent to m in order to *abdeliver* m in scenarios 1 and 2 respectively. For a given p, let $W(p)$ be the *expected* number of *abcasts* required subsequent to m for p_ℓ to *abdeliver* any m. (It is the average over all possible coin-toss outcomes for a given p.)

Note also that $W(p) = 0$ when $p = 1$ and $W(p) \to \infty$ as $p \to 0$. We thus observe that *abdelivery* latencies depend on $W(p)$ and the intervals between successive *abcasts*. Let λ be the average rate at which p_ℓ makes *abcasts*. Therefore, we have:

Challenge 1: p must be chosen by taking into account the prevailing value of λ if the average *abdelivery* latency by ZabCT is to be smaller than that by Zab.

It is possible that the value of λ drops suddenly; if that happens, $(t_{i+1} - t_i)$ for some $i < k$ in scenario 2, for example, can be too long and *abdelivery* of m is delayed considerably. In these circumstances, followers are *forced* to carry out coin-tossing.

Challenge 2: Enforce coin-tossing by followers, when necessary, so that the average *abdelivery* latency by ZabCT does not exceed that by Zab.

Suppose that followers are forced to coin-toss quite frequently. This obviously tends to reduce ZabCT latencies but also increases the rate at which followers generate *acks* (for any given $p > 0$). The latter has two implications: first, our design objective of reducing inbound traffic at the leader is undermined; secondly, a follower, due to an increased inbound traffic of *acks*, cannot speedily respond to read requests.

Challenge 3: The rate of *ack* arrivals at a follower is bounded by $\theta \leq \lambda$.

A follower receives *commit* messages in Zab at the rate of λ, i.e., one *commit(m)* for every *abcast(m)* and hence *commit* messages arrive at a follower at rate λ in steady state. There are no commit messages in ZabCT but followers' *acks* are broadcast. So, $\theta = \lambda$ ensures that followers handle the same inbound traffic in both protocols.

Let us note that when followers toss coins more frequently or use larger value of p, ZabCT latencies tend to be smaller and the rate at which *acks* are broadcast tends to be larger. This means that addressing the first two challenges can *at times* make addressing the third one impossible and *vice versa*. That is, it may not always be possible to have ZabCT out-performing Zab; this observation leads to:

Challenge 4: If ZabCT is judged not to offer performance benefits over Zab, processes should be able to switch autonomously to Zab.

We next address Challenge 2, then Challenges 1 and 2, and finally Challenge 4.

3.3   Enforced Coin Tossing

Since coin-tossing is done only by followers, enforcing it causes no change in the pseudo-code of the leader in Subsect. 3.1. For followers, F2 and F3 are unchanged, F1 is modified and F4 is added:

- F1: On receiving *proposal(m)* from the leader: log m; send *ack(m)* to itself; reset timer(D); toss coin; if (coin = *Head*) then broadcast *ack(m)*;
- F2: // As in Subsect. 3.1;
- F3: // As in Subsect. 3.1;

– F4: On timer(D) expiry: reset timer(D); if ($\exists m'$: not implicitly *acked* by this process $\wedge$ *ack(m')* not broadcast $\wedge \neg$ *committable(m')*) then {select $m, m.c \geq m'c$; toss coin; if (coin = *Head*) then broadcast *ack(m)*};

Every time a follower receives a *proposal*, it sets a timer for duration D (in F1). When the timer expires (in F4), the follower·resets it and looks for *proposal(m')* whose m' has been neither implicitly nor explicitly *acked* and not *committed* as well. If it has such a *proposal(m')*, then it selects the *proposal(m)* with the largest $m.c \geq m'.c$. Note that if $m \neq m'$, m would also not have been *committed* nor *acked* implicitly or explicitly. The follower broadcasts *ack(m)*, if the outcome of coin toss is *Head*. Thus, a follower's coin tossing rate is $maximum\{\lambda, \frac{1}{D}\}$ which is no smaller than $\frac{1}{D}$.

3.4 Computing Coin Toss Probabilities

We will continue to retain (for now) the simplifying assumptions that n is known, fixed and is at least $\lceil \frac{N+1}{2} \rceil$. Followers must meet two (competing) requirements:
R1: The average *abdelivery* latency is less than the average latency in Zab; and,
R2: The average rate at which followers broadcast *acks* is bounded by $\theta \leq \lambda$.

Let L denote the average leader latency in Zab and d the average transmission delay for *commit* messages to reach the followers. Thus, the average follower latency in Zab is $L + d$.

Suppose that ZabCT is run with $p = 1$; i.e., followers broadcast their *acks* (instead of unicasting them to the leader as in the equivalent Zab runs). If the broadcasting overheads are ignored, L is also the average ZabCT latency for all $p_i \in \Pi$. However, when $p < 1$, leader requires an average of $W(p)$ follow-up *abcasts* for *abdelivery*, and each of these *abcasts* is separated by an average duration of $\min\{\frac{1}{\lambda}, D\}$. So, the average leader latency in ZabCT is $L + W(p) \times \min\{\frac{1}{\lambda}, D\}$. By Lemmas 1 and 2, a follower can *abdeliver* only as early as the leader. So, a necessary condition for **R1** is:

$$W(p) \times \min\{\frac{1}{\lambda}, D\} < d \tag{1}$$

Followers toss coins at the average rate of $\max\{\lambda, \frac{1}{D}\}$ and the expected number of heads in each of these tosses is np. Thus, R2 requires $np \times \max\{\lambda, \frac{1}{D}\} < \theta$; so,

$$p < \left(\frac{\theta}{n}\right) \times \min\{\frac{1}{\lambda}, D\} \tag{2}$$

With D fixed at the start, each follower periodically measures λ and computes *prob(Head)* as follows. It estimates P_1 as the *smallest* probability that satisfies Eq. 1, and P_2 as the *largest* probability that satisfies Eq. 2. When $P_2 \geq P_1$, *prob(Head)* for ZabCT is chosen as some p, $P_1 \leq p \leq P_2$ with a default choice of $p = P_1$; when $P_2 > P_1$, ZabCT is not feasible and a switch to Zab is needed.

Computing $W(p) = \mathbf{E}$(No. of subsequent *abcasts* for *abdelivery* by Leader)

It is similar to computing the hitting times of specific states in a transient Markov chain. Let us assume that message transmission and processing times are zero and focus on *abdelivery* of m that was *abcast* at t_0 as depicted in Fig. (1a). Let S_i, $0 \le i \le n$, refers to the system state in which i followers have broadcast either $ack(m)$ or $ack(m' : m'.c > m.c)$. $f(h; g) = \binom{g}{h}p^h(1 - p)^{(g-h)}$ is the binomial probability that h heads occur when g followers toss their coins. Thus, S_i is reached at t_0 with probability $f(i; n)$, i.e., when i of n followers get *Head* in their coin-toss for m at t_0.

When $i \ge a = \lceil \frac{N-1}{2} \rceil$, the leader *abdelivers* m, and hence $S_a, S_{a+1}, \ldots S_n$ are called *absorption* states which, if reached, require no further *abcasts* for m to be *abdelivered* by the leader. Let $W_i(p)$ be the expected number of *abcasts* required for leader to *abdeliver* m, *given* that system is in S_i at t_0; note that $W_i(p) = 0, \forall i \ge a$.

Let q_{ij} be the probability that the system transits from S_i to $S_j, j \ge i$, when one more *abcast* is made. It is the probability that $(j - i)$ followers, out of those $(n - i)$ followers that have not yet got *Head* since receiving m, get *Head* for the latest *abcast*. So, $q_{ij} = \binom{n-i}{j-i}p^{(j-i)}(1 - p)^{(n-j)}$. $W_i(p) = (1 + W_j(p))$, given that S_i prevails at t_0 and S_j at t_1; any $S_j, i \le j \le n$, is possible at t_1 with probability q_{ij}. So, we have:

$$W_i(p) = \sum_{j=i}^{n} q_{ij}(1 + W_j(p)) = \sum_{j=i}^{n} q_{ij} + \sum_{j=i}^{a-1} q_{ij}W_j(p) = 1 + \sum_{j=i}^{a-1} q_{ij}W_j(p) \quad (3)$$

$$\therefore W(p) = \sum_{i=0}^{n} f(i; n)W_i(p) = \sum_{i=0}^{a-1} f(i; n)W_i(p).$$

For example, when $N = 5$ and $n = 4$, $W_4(p) = W_3(p) = W_2(p) = 0$; from Eq. 3, $W_1(p) = \frac{1}{1-q_{11}}$ and $W_0(p) = \frac{q_{01} \times W_1(p)}{1-q_{00}}$. $W(p) = f(0; 4)W_0(p) + f(1; 4)W_1(p)$.

3.5 Protocol Switching

A follower may wish to switch to executing Zab on two occasions: (i) p could not be computed as per Eqs. 1 and 2; and, (ii) another follower crashes, value of n changes and the membership service is yet to update the new membership. In the latter case, the value of p being used may be inappropriate and *abcasts* can remain uncommitted for too long. This is deduced by setting timer(C_m) on receiving *proposal(m)*.

Protocol switching is organised similar to 2-Phase commit: even one follower's *vote* to quit ZabCT is enough for all to switch to Zab, and all followers must vote for ZabCT for switching from Zab to ZabCT; moreover, the leader *decides* based on followers' votes and informs them of its decision. Followers use a message field *prot* in their *acks* to indicate their votes, and the leader uses *prot* in its *commit* messages to inform followers of its decision.

If a follower, while executing ZabCT, experiences timer(C_m) or cannot find p, it unicasts its *ack* (as in Zab) to the leader with *prot* set to Zab. Whenever the leader receives an *ack(m)* with *prot* = Zab, it broadcasts *commit(m)* with *prot* = Zab to all followers, when it sends, or if it has already sent, *commit(m)* to itself. When a follower executing ZabCT receives *commit(m)* with *prot* = Zab, it starts executing Zab.

A follower that executes Zab still measures λ and attempts to compute p; if p can be computed successfully on several consecutive iterations and membership remains unchanged for a prolonged period, a follower votes for ZabCT using *prot*. If the leader receives votes for ZabCT from all n, $n \geq \lceil \frac{N+1}{2} \rceil$, followers, it broadcasts its *commit* with *prot* = ZabCT and thus instructs the followers to switch to ZabCT.

4   Performance Comparison

In this section, we compare the performances of Zab and ZabCT. Atomic broadcast latency and throughput are the two metrics used for comparison.

We use 250 concurrent clients distributed equally on 10 identical machines; each machine thus hosts 25 clients. At most 9 machines were dedicated to running the protocols, thus covering $N = 3, 5, 7, 9$. Machines used in our experiments are commodity PCs of 2.80 GHz Intel Core i7 CPU and 8 GB of RAM, running Fedora 21 and communicating over 100 Mbps Switched Ethernet. Connections between machines were established at the beginning of the experiment.

The protocols were implemented in Java (JDK 1.8.0) on top of the JGroups framework. JGroups is a toolkit for reliable communication and also supports crash detection, joining of a recovered process and installation of group membership views [1]. Messages are transmitted using JGroups' FIFO reliable UDP, more precisely, by using UNICAST3 protocol in JGroups suite which is functionally identical to TCP.

Each client generates a read or write request with a payload of 1Kbyte and sends the request to one of N servers. If the request is of *read* type, then the server simply returns the request as the response; if the request is of *write* type, the server (if not the leader) forwards it for *abcasting*; when a server *abdelivers* a request it had received directly from a client, it sends the request back to the client as the response. Thus, no read/write operations actually occur since the aim is to measure and compare *abdelivery* latencies and throughput. On receiving the response, the client repeats its action after a specified *wait-time* and selects the destination server in a round-robin manner. If *wait-time* is zero, servers collectively handle at most 250 requests at any moment.

We use *write-ratio*, $WR, 0 < WR \leq 1$, for clients to vary the load they impose on servers. For every write request that a given client generates, it will generate $\frac{1-WR}{WR}$ read requests; in other words, $WR > 0$ is the probability that a request generated by a client is of *write* type. Experiments reported consider WR values ranging from 10% to 100% in steps of 10%.

In an experiment, where the protocol, WR and N are fixed, clients send, and receive responses for, a total of 10000 write requests after the warm-up phase. For example, if $WR = 50\%$, the server system will process $\frac{10000}{0.5} = 20000$ read/write requests, i.e., each of the 250 clients will issue 80 requests. Note that servers handle at most $250 \times WR$ *abcasts* at any moment when client *wait-time* is zero.

Let t_0 and t_1 be the instances when a server receives a request from a client and *abdelivers* that request respectively; $t_1 - t_0$ defines the *abdelivery* latency for that request. We compute the average of 10000 such latencies and repeat the experiment 20 times for a confidence interval of 95%. Throughput is defined as the number of *abdeliveries* (*abds*) made by all servers per unit time and is computed, like latencies, with a 95% confidence interval. For space reasons, we report latency/throughput improvements offered by ZabCT over Zab and are computed as follows. Let X and X_{ct} be metrics for Zab and ZabCT respectively; improvement in latency (L) is $\frac{L - L_{ct}}{L}$ and that in throughput (T) is $\frac{T_{ct} - T}{T}$. (Thus, a positive value implies ZabCT is better.)

Experiments are run in failure-free scenarios. Furthermore, servers do not log m in disk (as ideally required) but only record m in main-memory. Thus the performance figures we present here do not include disk write delays, but only network delays. This kind of evaluations corresponds to the 'Net-Only' category of the evaluations in [8] where several ways of logging have been considered. Since both protocols require logging of m exactly at the same point in the execution for every *abcast(m)*, ignoring delays due to disk writes cannot invalidate the integrity of observations made and conclusions drawn from the performance figures.

We consider two values for client *wait-time*: zero and a random value that is uniformly distributed (*u.d.* for short) on (25, 75) millisecond (ms), with the average of 50 ms. In the former, client does not wait between receiving response and issuing its next request, whereas in the latter client waits for an average of 50 ms. Thus, the arrival rate of *proposals*, λ, measured by followers every second, will be different for different values of *wait-time* and WR used.

Ideally, θ in Eq. 2 must satisfy $\theta \leq \lambda$ - see Challenge 3 in Subsect. 3.2. To avoid followers being unable to compute p and thereby having to switch to Zab in experiments, we set $\theta = \lambda$ when $WR = 1$ when Zab was run. That is, we measured the average value of λ encountered when Zab was run for $WR = 1$, and used that values to *fix* θ in ZabCT for *all* values of WR (including $WR = 1$).

Thus, with zero client *wait-time*, the θ values used in ZabCT are: 3967, 2351, 1639, 1332 when $N = 3, 5, 7, 9$ respectively; similarly, for *wait-time u.d.* on (25, 75), θ values for ZabCT are: 3597, 2236, 1597, 1302 when $N = 3, 5, 7, 9$ respectively.

Finally, each follower continually measures d as the communication delay (one-way transmission) from the leader to itself (see Subsect. 3.4), without clock synchronisation. This is done by a follower selectively timestamping its *ack* and the leader incorporating the duration elapsed between receiving a timestamped *ack* and broadcasting its next timestamped *proposal*.

4.1 Observations

Figure 2 presents the average latency and throughput comparison for $N = 5$ and zero client wait time. Let us first focus on latency comparison depicted in Fig. 2a. As we can observe, ZabCT offers lower latencies compared to Zab for all WR values.

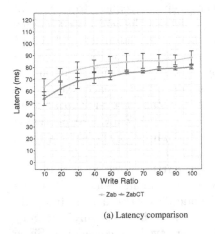

(a) Latency comparison

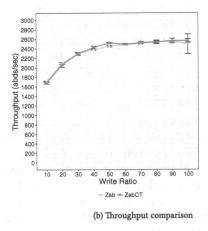

(b) Throughput comparison

Fig. 2. Performance comparison for $N = 5$ and zero client *wait-time*

Table 1. Zero client *wait-time*

(a) Number of acks per commit

WR	N=3	N=5	N=7	N=9
10	1.607	1.319	1.041	1.036
20	1.177	1.118	0.959	1.038
30	0.996	0.628	0.934	1.045
40	0.773	0.602	0.932	1.044
50	0.766	0.568	0.938	1.027
60	0.758	0.596	0.925	0.662
70	0.669	0.587	0.923	0.543
80	0.718	0.598	0.926	0.544
90	0.732	0.592	0.935	0.519
100	0.788	0.599	0.924	0.547

(b) Coin-toss probabilities

WR	N=3	N=5	N=7	N=9
10	0.800	0.338	0.171	0.132
20	0.594	0.289	0.154	0.129
30	0.505	0.160	0.149	0.126
40	0.388	0.148	0.150	0.128
50	0.387	0.137	0.154	0.123
60	0.380	0.156	0.152	0.085
70	0.353	0.156	0.156	0.073
80	0.359	0.155	0.156	0.072
90	0.373	0.154	0.158	0.071
100	0.403	0.156	0.151	0.071

The difference between Zab and ZabCT varies between 7 and 11 ms. This can be attributed to (i) absence of *commit* message transmissions in ZabCT and (ii) ZabCT leader receiving fewer *acks* compared to Zab leader, see column $N = 5$ in Table 1a. (Recall that number of *acks* received by the Zab leader per *commit* is $N - 1$.) Due to (i) and (ii), the leader and followers have fewer messages in their buffer, which results in messages being received faster at destinations and in reduced *abdelivery* latency.

Figure 2b compares throughput with zero client *wait-time*. The throughput of ZabCT is at least as good as, if not better than, Zab; when $WR = 100\%$, the difference is maximum at about 70 abds/sec.

Table 1 shows the number of *acks* received by the leader per *commit* and the coin-toss probabilities computed for experiments with zero client *wait-time*. An important observation to be drawn from the Table 1a that, in all N, the ZabCT leader receives less incoming traffic compared to the Zab leader. For example, when $N = 5$ and at $WR = 10\%, 100\%$, ZabCT leader receives 1.319 and 0.599 *acks* per commit respectively whereas in Zab, the leader would receive $N - 1 = 4$ *acks*. This reduction in *ack* messages for ZabCT leader corresponds to the small coin-toss probabilities of 0.338 chosen for $WR = 10$ and 0.156 for $WR = 100$. This is the main reason we observe lower latency and relatively high throughput as shown in Fig. 2.

Figure 3 shows latency and throughput comparison using an average of 50 ms client *wait-times* (*u.d.* on (25, 75)). An interesting finding is that in Fig. 3a at $WR = 10\%, 20\%, 30\%$, the latency becomes nearly equal for ZabCT and Zab. A possible explanation for these results may be λ is low (due to (1) non-zero client *wait-time* and (2) reads far out-numbering writes) which leads to high coin-toss probabilities 0.750, 0.561 and 0.381 respectively (see Table 2b column $N = 5$). This results in increasing incoming traffic for leader and followers (increasing the number of *acks* per commit) to 3.253, 2.688 and 2.046 respectively (see Table 2a column $N = 5$). However, as WR increases, λ becomes high. This leads to less incoming traffic on the leader and followers, resulting in ZabCT latency being smaller than Zab, with a maximum difference of 1 ms at $WR = 40\%$ and increasing to about 5 ms at $WR = 100\%$.

Figure 3b compares throughput for $N = 5$. It is obvious that ZabCT demonstrates high throughput. With $WR = 70\%, 80\%, 90\%, 100\%$ the difference is about 130 abds/sec.

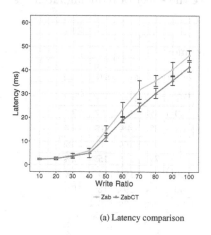

(a) Latency comparison

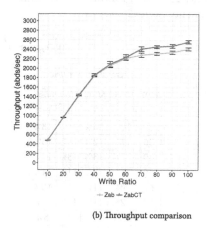

(b) Throughput comparison

Fig. 3. Performance comparison for $N = 5$ and client *wait-time* on (25, 75) ms

Table 2 indicates the number of *acks* received by ZabCT leader per commit and coin-toss probabilities for an average of 50 ms client *wait-time*. Consider Table 2a; it is significant to notice that the number of *acks* per commit is higher than that shown in Table 1a. This is explained by the fact that λ decreases for all N and WR (due to non-zero *wait-times*), resulted in probability, $Prob(Head)$ increases, hence the likelihood of sending an *ack* increases as well (see Table 2b).

Table 2. Client wait time in (25, 75) ms

(a) Number of acks per commit

WR	N=3	N=5	N=7	N=9
10	1.600	3.253	3.757	5.783
20	1.601	2.688	2.374	1.439
30	1.603	2.046	1.731	1.052
40	1.600	1.733	1.590	1.045
50	1.537	1.288	0.949	1.038
60	1.101	1.063	0.941	0.762
70	1.116	0.975	0.934	0.550
80	1.014	0.986	0.921	0.549
90	0.742	1.152	0.942	0.540
100	0.793	1.117	0.930	0.544

(b) Coin-toss probabilities

WR	N=3	N=5	N=7	N=9
10	0.800	0.750	0.552	0.667
20	0.800	0.561	0.275	0.400
30	0.800	0.381	0.180	0.129
40	0.801	0.297	0.165	0.130
50	0.771	0.197	0.155	0.126
60	0.553	0.209	0.155	0.090
70	0.564	0.181	0.160	0.067
80	0.534	0.176	0.154	0.071
90	0.402	0.180	0.158	0.066
100	0.429	0.190	0.159	0.071

Table 3 shows latency improvements for all N and WR, and for both zero and 50 ms client *wait-time* experiments. Overall, what is interesting to note is that the performance of ZabCT nearly outweighs that of Zab for all N and WR. Frequent *abcasting* leads to frequent coin-tosses which in turn reduce the delays due to the leader having to *commit* by receiving implicit *acks* from followers; moreover, the incoming traffic at the leader reduces remarkably (see Tables 1a and 2a) when followers toss coins which will have the effect of reducing latencies at the leader.

Table 3. Latency improvement

(a) Zero client *wait-time*

WR	N=3	N=5	N=7	N=9
10	25%	15%	18%	17%
20	10%	16%	14%	18%
30	11%	12%	16%	15%
40	13%	13%	13%	12%
50	9%	13%	11%	11%
60	8%	11%	11%	10%
70	12%	11%	11%	9%
80	12%	8%	9%	8%
90	13%	8%	8%	8%
100	10%	9%	8%	9%

(b) Client *wait-time* on (25, 75) ms

WR	N=3	N=5	N=7	N=9
10	4%	5%	6%	17%
20	2%	4%	4%	15%
30	3%	6%	4%	23%
40	7%	18%	15%	12%
50	8%	19%	14%	11%
60	3%	19%	12%	10%
70	10%	23%	10%	9%
80	14%	15%	10%	8%
90	27%	12%	8%	8%
100	50%	10%	8%	9%

5 Related Work

As per [4], Zab belongs to the group of fixed sequencer protocols and its intellectual ancestor is Paxos [3,10]. Unlike Paxos, Zab permits at most one leader at any moment and a new leader cannot commence its leadership role until a quorum of servers have disowned the old leader; thus, it avoids Paxos-style ballots, but it may omit *abdelivering* some *abcasts* during, and because of, leader change. Consequently, Zab cannot guarantee causal order delivery as traditionally understood [9].

Leader based protocols tend to overload the leader and several authors [2,11, 12,17] have sought to remedy this drawback. S-Paxos [2] relieves the leader from broadcasting client requests by separating the roles of request dissemination and request ordering. Each process directly broadcasts client requests to others and request ordering is done using only request identifiers.

Mencius [12] allows each process to act as a leader by numbering its own *abcasts* with unique and increasing $m.c$ such that all *abcasts* are uniquely and continuously numbered. It thus achieves a high throughput but *any* server crash could stop *abdelivery* until reconfiguration. Chain replication [17] reduces the leader load by distributing the role between two servers called the *head* and the *tail* but involves sequential transmission of m which tends to increase *abdelivery* latencies for large N.

Broadcasting an acknowledgement is common in symmetric (leaderless) atomic broadcast protocols such as [14]. That it helps to avoid the leader broadcasting *commit* messages has been hinted by Zab authors themselves (e.g., [8,13]). In this paper, we explored this idea with coin-tossing approach to reduce the number of acknowledgements being broadcast. Implicit acknowledgments and membership service which we have used here are not new. The former are commonly used in TCP implementations where they are also called cumulative acknowledgements. The latter is readily offered by the (open-source) JGroups framework [1].

6 Conclusions and Future Work

We have extended the well-known Zab protocol under its original fault assumptions. Extensions use *ack* broadcasting - not an unknown idea [8,13] - but subject to coin-toss outcomes to reduce network traffic and also the traffic at the leader. Coin-tossing is one instance of the general concept of using only a subset of randomly selected nodes to engage in communication at any given time in order to reduce traffic, particularly at bottleneck nodes. Examples are: controlling *ack* implosion at multicasting nodes and information dissemination through gossiping in large systems.

While coin-toss reduces leader traffic, it also delays *abdelivery* which requires future *abcasts* to be made or coin-tossing to be forced. This paper demonstrates that the effect of leader traffic reduction is so overwhelming that much smaller latencies can be obtained particularly at heavy loads when coin-toss probability is appropriately chosen. This is our principal contribution and, to the best of our knowledge, improving Zab performance through coin-toss guided *ack* broadcasting has not been investigated. Followers broadcasting their acks to eliminate commit phase in Zab has been deemed impractical in [13]; here, we demonstrate that it is indeed a practical approach to improve Zab performance when it is combined with coin-tossing.

Having established that coin-tossing is effective, irrespective of WR and N, when λ is relatively large, we plan to conduct more evaluations with θ varying realistically with λ and with a follower being allowed to crash which would force protocol switching (Subsect. 3.5). We also plan to investigate ZabCT performance at heavier loads that saturate the Zab leader to an extent that Zab throughput starts deteriorating.

References

1. Ban, B.: Jgroups, a toolkit for reliable multicast communication (2002). http://www.jgroups.org
2. Biely, M., Milosevic, Z., Santos, N., Schiper, A.: S-paxos: offloading the leader for high throughput state machine replication. In: IEEE 31st Symposium on Reliable Distributed Systems (SRDS), pp. 111–120 (2012)
3. Chandra, T.D., Griesemer, R., Redstone, J.: Paxos made live: an engineering perspective. In: Proceedings of the Twenty-Sixth Annual ACM Symposium on Principles of Distributed Computing, pp. 398–407 (2007)
4. Défago, X., Schiper, A., Urbán, P.: Total order broadcast and multicast algorithms: taxonomy and survey. ACM Comput. Surv. (CSUR) **36**(4), 372–421 (2004)
5. Emerson, R., Ezhilchelvan, P.: An atomic-multicast service for scalable in-memory transaction systems. In: 2014 IEEE 6th International Conference on Cloud Computing Technology and Science (CloudCom), pp. 743–746. IEEE (2014)
6. George, L.: HBase: The Definitive Guide. O'Reilly Media, Inc., Sebastopol (2011)
7. Hunt, P., Konar, M., Junqueira, F.P., Reed, B.: Zookeeper: wait-free coordination for internet-scale systems. In: USENIX Annual Technical Conference, vol. 8, p. 9 (2010)
8. Junqueira, F.P., Reed, B.C., Serafini, M.: Zab: high-performance broadcast for primary-backup systems. In: IEEE/IFIP 41th International Conference on Dependable Systems & Networks (DSN), pp. 245–256. IEEE (2011)
9. Lamport, L.: Time, clocks, and the ordering of events in a distributed system. Commun. ACM **21**(7), 558–565 (1978)
10. Lamport, L.: Paxos made simple. ACM Sigact News **32**(4), 18–25 (2001)
11. Lamport, L.: Fast paxos. Distrib. Comput. **19**(2), 79–103 (2006)
12. Mao, Y., Junqueira, F.P., Marzullo, K.: Mencius: building efficient replicated state machines for WANs. OSDI **8**, 369–384 (2008)
13. Reed, B., Junqueira, F.P.: A simple totally ordered broadcast protocol. In: proceedings of the 2nd Workshop on Large-Scale Distributed Systems and Middleware (LADIS), pp. 1–6. ACM (2008)
14. Ruivo, P., Couceiro, M., Romano, P., Rodrigues, L.: Exploiting total order multicast in weakly consistent transactional caches. In: IEEE 17th Pacific Rim International Symposium on Dependable Computing (PRDC), pp. 99–108 (2011)

15. Shvachko, K., Kuang, H., Radia, S., Chansler, R.: The hadoop distributed file system. In: IEEE 26th Symposium on Mass Storage Systems and Technologies (MSST), vol. 2, pp. 1–10 (2010)
16. Toshniwal, A., Taneja, S., Shukla, A., Ramasamy, K., Patel, J.M., Kulkarni, S., Jackson, J., Gade, K., Fu, M., Donham, J.: Storm@twitter. In: Proceedings of the 2014 ACM SIGMOD International Conference on Management of Data, pp. 147–156 (2014)
17. Van Renesse, R., Schneider, F.B.: Chain replication for supporting high throughput and availability. OSDI 4, 91–104 (2004)

Modelling and Analysis of Commit Protocols with PEPA

Said Naser Said Kamil[(⊠)] and Nigel Thomas[(⊠)]

School of Computing Science, Newcastle University, Newcastle upon Tyne, UK
{Said.Kamil,Nigel.Thomas}@ncl.ac.uk

Abstract. This paper introduces performance models of two phase and three phase commit protocols specified formally using the Markovian process algebra PEPA. We show how we can investigate the performance of such distributed commit protocols to get more insight into the system behaviour under different loads. The commit phases of the protocols are examined using discrete state space (CTMC) and fluid (ODE) analysis and then compared to better understand how performance is affected by the different protocol behaviours.

Keywords: Commit protocols · Distributed computing · Performance evaluation · Scalability · Stochastic process algebra

1 Introduction

In a distributed database system a transaction is logically an atomic operation, thus either it will be executed to completion (i.e. committed) or not executed (i.e. aborted) [15–17]. Nonetheless, a transaction physically is comprised of a sequence of sub-operations. This discrepancy causes a considerable problem in distributed systems implementations [15]. Commit protocols are used for preserving the atomicity of distributed transaction, where all participating servers either unanimously abort or unanimously commit a transaction. Examples of distributed database systems include banking applications, airline reservation systems, and stock-market transactions.

Two Phase Commit Protocol (2PC) [7] is a distributed algorithm used in distributed database systems. Also, it is a simple commit protocol [7,14] that allows unilateral abort. This characteristic preserves transaction atomicity in the absence of failures. Skeen's Three Phase Commit protocol (3PC) is an extension of the 2PC protocol [15], to cope with the blocking problem of the 2PC protocol. The 3PC protocol is a replicated data management protocol, which not only increases data availability but also, decreases the cost of data retrieval [6]. Consequently, it provides high performance and also, can tolerate performance failures. 2PC and 3PC protocols have long been widely used in transaction processing, computer networking and databases [3,7,20]. Additionally, these protocols are widely used in numerous distributed database environments,

© Springer International Publishing AG 2017
P. Reinecke and A. Di Marco (Eds.): EPEW 2017, LNCS 10497, pp. 266–281, 2017.
DOI: 10.1007/978-3-319-66583-2_17

for instance, real-time databases [8], Web databases [21], distributed database systems homogeneous [4] and heterogeneous [1].

According to [15] commit protocols are classified into two generic classes: The first, centralized class, designate a coordinator to control the transaction execution at other servers. The second class is completely decentralized, as in the models in this paper. The centralized model of the 2PC protocol, however, is a costly protocol due to required logs and communication between servers and potentially the blocking of a server in the case of failure; which makes the other participating servers wait until the failed server recovers.

This paper aims to model and analyse the performance of the decentralized 2PC and 3PC protocols in the commit phase. We do not consider the failure and recovery phases of these protocols. Our motivation is to better understand how the behaviour of these two protocols affects their performance. The outline of this paper is as follows. The background for decentralized 2PC protocol is given in the next section, and then its PEPA model is introduced. In Sect. 3 we present the decentralized 3PC protocol and in Sect. 4 we introduce our performance metrics. In Sect. 5 we discuss the experiments and results and we end the paper with Sect. 6 with the conclusion and our future work.

2 Decentralized Two Phase Commit Protocol

According to Skeen and Stonebraker [15], the decentralized model has the following characteristics: Servers are equally participating in the protocol and the same protocol is executed by all servers. Additionally, each server communicates with every other participating server. Successive rounds of message interchanges are one of the main characteristics of decentralized protocols, where each server will send the same message to all other servers during a message round interchange. Also, before beginning the next round of messages the sender will wait until receiving messages from all other servers. The decentralized 2PC protocol is the simplest decentralized commit protocol as illustrated in Fig. 1.

As reported by [15] the execution of a transaction at each server is modelled as finite state automaton (FSA), where the local states of $server_i$ are the states of the FSA for $server_i$. A state transition involves the server reading nonempty received messages, writing messages, and then proceeding to the next local state. The local state change is an instantaneous event, indicating the end of the transition. Basically, each FSA has four local states: an initial state (q_i), a wait state (w_i), an abort state (a_i), and a commit state (c_i). Commit and abort represent the final states, specifying that the transaction has been either committed or aborted. Furthermore, server state transitions are asynchronous with other servers.

The finite state automata of 2PC has the following properties [15,17]: First, they are nondeterministic and the order of the received messages by one server is arbitrary. Second, the commit states and abort states represent the final states of the FSA. Next, transitions are not allowed to non-abort states as long as a transition to an abort state has been made. Consequently, the same restrictions are applied to commit states. So, committing and aborting are irreversible

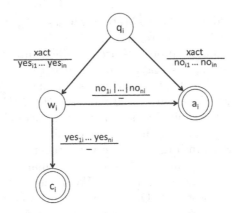

Fig. 1. Decentralized two phase commit protocol [15]

operations. The FSA state diagram is acyclic, thus the execution will eventually terminate at every server. Distributed commit protocols have at least two phases.

The global transaction state [15] consists of: a global state vector containing the local states of all participating FSA's, as well as the outstanding network messages. Moreover, the complete transaction processing state is defined by the global state. So, a global state is inconsistent in the case it has both local commit and local abort states. Hence, a protocol that maintains the atomicity of a transaction will not have any inconsistent global states. A global state will be either a final state or a terminal state. Final state if all its local states are final states, and, if there are no instantly reachable successors it is said to be a terminal state.

The concurrency set of a $server_i$ state consists of all local states concurrently occupied by other servers that can be derived from the reachable global state graph. A local state is said to be committable if all occupied servers voted to commit the transaction. On the other hand, a non-committable state is the state that a server does not know whether other servers voted to commit or not. All messages interchanged in a round have two subscripts; the first refers to the sender server and the other refers to the receiver server. Skeen [15] states that the decentralized 2PC protocol is synchronous in one state transition.

Let us assume that there are three servers obeying the decentralized 2PC protocol, and $server_1$ has received a request message. The interchange communications between these servers will be as follows: First phase, $server_1$ sends "*xact*" message to both $server_2$ and $server_3$, and moves to the waiting state. On receiving "*xact*" message, $server_2$ and $server_3$ decide whether to unilaterally abort this message, and they send that decision to each of their peers. Second phase, each server receives all the decisions and then moves to a final state (i.e. commit or abort).

2.1   Decentralized Two Phase Commit Protocol PEPA Model

In this section the decentralized 2PC protocol PEPA model will be presented. The created PEPA model will be used to simulate the performance of the decentralized 2PC protocol (commit phase). Performance Evaluation Process Algebra PEPA is a high-level quantitative modelling language developed by Hillston [12], and one of its essential uses is for modelling distributed systems. Models are built at a high level of abstraction using stochastic process algebra (SPA), for instance (PEPA [12], EMPA [2] and SPADES [9]), and stochastic Petri net (SPN) [5].

PEPA offers several significant features in performance modelling, such as, compositionality, formality and abstraction [13]. The PEPA Eclipse Plug-in tool [19] is a supporting tool which has been used for developing and analysing the performance of systems, offering a variety of analysis techniques, for example, continuous time Markov chain (CTMC), Stochastic Simulation and Ordinary Differential Equations (ODEs). Such approaches allow the observation of a system as it evolves from an initial state over a period of time [19]. Furthermore, each action within the PEPA model has a rate which is the reciprocal of the average duration, or delay, to undertake by the action.

The specified PEPA model consists of three main components: *Request, Client and Server*. The model is not only analysed by the Continuous Time Markov Chain (CTMC) steady state analysis, but also, the behaviour of the model is approximated using the Ordinary Differential Equations (ODEs) analysis for both throughput and population, which supports the numerical calculation of a large scale model with a large number of *Client* and *Request* components. ODE solvers are continuous and deterministic [19] and have been used by [10, 11] as a solution to the CTMC state space explosion problem. The number of clients and requests has been varied from 1 to 10. Each client generates 20 threads and sends requests to the *Request* component. Accordingly the number of requests that will be sent to a server equals the number of clients times 20 threads (e.g. 10 clients * 20 threads = 200 requests sent in parallel).

Several simplifications have been made on the 2PC protocol PEPA model. We have applied a model of abstraction on the 2PC protocol, because we are interested to look at the behaviour of one of the servers implementing the protocol. So, we are not only looking at one server but also, we are looking at only the actions that have impact on that server. These actions are identified as they have the biggest impact on a server, and the other actions are generally messaging actions which have a very low impact on the server. Actually, the messaging does not have much impact on the servers. For example, the time to generate or read a request is not the same as the server action (e.g. Commit), because most of that time is a transmission. But the processing part does have an impact on the servers. Nevertheless, these simplifications have not affected the overall performance of the model, and allow us to improve the model scalability.

The *Request* component has all actions that are processed by the protocol. For convenience of reference we have separated this component up into named derivatives $Request_i$, $1 \le i \le 10$. Furthermore, each of the components (*Client* and *Server*) cooperates with the main component *Request*. In other words, these components are simultaneously communicating with the *Request* component in order to correctly follow the sequence of actions in the protocol. Additionally, the component *Client* is used to represent a client who sends a write request or update request. Accordingly, the actions sequence is preserved in the model through the use of the *Request* component. Moreover, the model is cyclic, where it starts when a client sends a request message and then waits until it is processed. Thus the model is returned back to the initial state where a client will be able to send a new request. The following specifies the decentralized 2PC in PEPA:

$$Client \stackrel{def}{=} (sendRequest, r_1).(getRequest, r_1).Client$$
$$Request \stackrel{def}{=} (sendRequest, r_1).Request_1$$
$$Request_1 \stackrel{def}{=} (receiveRequest, r_1).Request_2$$
$$Request_2 \stackrel{def}{=} (cpu, c).Request_3$$
$$Request_3 \stackrel{def}{=} (snd\_xactToServer2, r_4).Request_4$$
$$Request_4 \stackrel{def}{=} (snd\_xactToServer3, r_4).Request_5$$
$$Request_5 \stackrel{def}{=} (processAckServer2, r_9).Request_{5a}$$
$$Request_{5a} \stackrel{def}{=} (processAckServer3, r_9).Request_6$$
$$Request_6 \stackrel{def}{=} (ackServer2, r_5).Request_{6a}$$
$$Request_{6a} \stackrel{def}{=} (ackServer3, r_5).Request_7$$
$$Request_7 \stackrel{def}{=} (cpu, c).Request_8$$
$$Request_8 \stackrel{def}{=} (commitServer2, r_7).Request_9$$
$$Request_9 \stackrel{def}{=} (commitServer3, r_7).Request_{10}$$
$$Request_{10} \stackrel{def}{=} (getRequest, r_1).Request$$
$$Server \stackrel{def}{=} (cpu, c).Server$$
$$System \stackrel{def}{=} Client[N] \underset{S_1}{\bowtie} Request[N] \underset{S_2}{\bowtie} Server$$

As shown above, the system equation representing the components of the model and the cooperation between these components over the sets S_1 and S_2. Whereas, N has been varied from 20 to 200, and the cooperation sets S_1 and S_2 have the following actions:

$$S_1 = \{sendRequest, getRequest\}$$
$$S_2 = \{cpu\}$$

Some local actions of the *Server*, such as *addRequest* and *processingCommit*, have been renamed to *cpu* in the model and the rate of the *cpu* action is calculated as follows, which gives the average rate of those actions:

$$c = \frac{2}{\sum \frac{1}{r_i}} \tag{1}$$

Where, $i \in \{3, 6\}$. The *cpu* action is used because PEPA does not allow us to directly limit the rates across multiple action types with a single bound. Hence we model a single action (*cpu*) and limit the total rate of this action, however it is in effect a combination of all the actions of which it is comprised. So, *cpu* cannot run faster than all the comprising actions (bounded capacity).

This PEPA model has been parametrised by the rates illustrated in Table 1. Note that these assumed rates were chosen as indicative based on measurements taken from a related system, but we have not measured implementations of 2PC or 3PC.

Table 1. The rates of 2PC PEPA model

Rate	Value	Rate	Value	Rate	Value
r_1	7.42	r_3	61.96657616	r_4	7.42
r_5	7.42	r_6	437.1975906	r_7	7.42
r_8	26.33486286	r_9	1330.250132	c	2/(r3+r6)

3   Decentralized Three Phase Commit Protocol

The Three Phase Commit Protocol (3PC) is a non-blocking protocol that extends the 2PC protocol. In the absence of failure the 2PC protocol works without any problems. However, if a server from a quorum fails for any reason, the other peers of the server will be blocked until it is recovered. This affects the performance of the 2PC protocol significantly. Consequently, a non-blocking commit protocol is presented by [15], where it allows operational servers to continue processing of a transaction even in the presence of a failure. Skeen [15] has presented two models of non-blocking commit protocol central and decentralized. Herein, we will only discuss and analyse the decentralized non-blocking protocol performance.

Skeen [15] has extended the 2PC protocol to be non-blocking by introducing a buffer state, which is called *"prepare to commit"*, between the wait state and the final state (commit) as depicted in Fig. 2. The *prepare to commit* state is only reachable if all other participants have voted to commit a transaction. Otherwise, the transaction will be aborted by other participants after a waiting time. So, as stated in [15] the protocol will be non-blocking if the concurrency set of a local state has no commit and abort at the same time; and also, the concurrency set of a non-committable state has no commit state. For the details about the fundamental non-blocking theorem we refer the interested reader to [15]. Accordingly, the non-blocking protocol is the canonical protocol. Nonetheless, the newly introduced state will introduce additional costs in both time and message interchanges communications.

3.1   Decentralized Three Phase Commit Protocol PEPA Model

The PEPA model of the non-blocking decentralized 3PC protocol is the same as the PEPA model presented in the Sect. 2.1 with the following changes: the action *prepareToCommit* has been added to the *Request* component. Additionally, the 3PC PEPA model has been parameterized by the same values illustrated in Table 1.

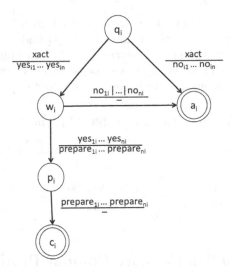

Fig. 2. Nonblocking decentralized three phase commit protocol [15].

Furthermore, two assumptions have been made: First (*Assumption1*), the *prepareToCommit* action has been defined as an independent action. In the second assumption (*Assumption2*) the action *prepareToCommit* is used as a centralized action. Whereas, this action is used as one of the *cpu* actions, and it has been renamed to *cpu*. Also, it has been assumed that it will present more contention for the available resources. The following is only showing some parts of the PEPA model of the 3PC protocol, where the changes have been made which differ from the model shown in Sect. 2.1.

– Assumption1
$$Request_{6a} \stackrel{def}{=} (ackServer3, r_5).Request_{6b}$$
$$Request_{6b} \stackrel{def}{=} (prepareToCommit, r_8).Request_7$$
$$Request_7 \stackrel{def}{=} (cpu, c).Request_8$$
$$Server \stackrel{def}{=} (cpu, c).Server$$

– Assumption2
$$Request_{6a} \stackrel{def}{=} (ackServer3, r_5).Request_{6b}$$
$$Request_{6b} \stackrel{def}{=} (cpu, c).Request_7$$
$$Request_7 \stackrel{def}{=} (cpu, c).Request_8$$
$$Server \stackrel{def}{=} (cpu, c).Server$$

4 Performance Metrics

Two performance benchmarks will be considered, latency and throughput, with the intention of inspecting the model behaviour.

Latency is considered as the total time for a data transmitted to a destination and then returned back to its source (measuring the round trip) on average. This calculation requires computing the cumulative average delay for all independent actions, plus the waiting time (queueing and service time) accrued at the shared *cpu* actions. The waiting time is calculated from the component populations using Little's law and the PASTA property. Based on the derived ODEs analysis results the latency of the model which has 2 *cpu* actions (i.e. 2PC and 3PC *Assumption1*) has been calculated as follows:

$$(1 + Pop(Req2 + Req7))\frac{2}{c} + \sum_{\forall a \in \sigma} \frac{1}{rate_a} \tag{2}$$

Also, the latency of the 3PC *Assumption2*, which has 3 *cpu* actions, has been calculated as:

$$(1 + Pop(Req2 + Req6b + Req7))\frac{3}{c} + \sum_{\forall a \in \sigma} \frac{1}{rate_a} \tag{3}$$

Where σ = {*sendRequest, receiveRequest, cpu, snd_xactToServer2, snd_xactTo Serve- r3, processAckServer2, processAckServer3, ackServer2, ackServer3, commit Server2, commitServer3, getRequest*}. Also, *a* is the type action for all actions in the set σ. It is clear that the model here is a closed queuing network. The reason of calculating the latency for the non-*cpu* actions $1/rate$ because they are not subject to queuing, just the time it takes to do that action. But for the shared *cpu* actions they have to queue (competitive actions) and they take serving time and the average waiting time. Hence, the population average queue length equals one (the current request) plus the queue length for a system with 1 less entity in it, i.e. the population for the system with $N - 1$ requests.

Throughput is one of the important performance metrics that is used to measure the number of times an action is performed per unit time. It has been calculated using the PEPA Eclipse Plug-in tool scalable analysis (Throughput).

5   Experiments and Results

The experiments which follow have been made based on the assumption that the processing is in the absence of failures. The performance benchmarks latency and throughput are used to evaluate the protocol behaviour. Whereas, it has been used the steady state analysis (CTMC) and the fluid flow analysis (ODEs) provided by PEPA Eclipse Plug-in tool, through the derived metrics (Throughput and Population). Additionally, latency and throughput have been measured from the server side. In the first set of experiments we have compared the performance of 2PC and 3PC (Assumption1 and Assumption2) using CTMC. Next, by using the ODEs the decentralized 2PC protocol is examined, and then the decentralized 3PC protocol has been evaluated.

5.1 CTMC Analysis

The CTMC steady state analysis can only be used with small scale systems due to the well-known problem of state space explosion; however, in this section we are going to explore the scale of clients that can be analysed using the steady state (CTMC) analysis. So, this will allow us to exactly identify what is the scaling limit of the CTMC analysis. The following figures illustrate the throughput and the population of only two most significant actions (*cpu* and *getRequest*). The throughput of the 2PC protocol and 3PC protocol (*Assumption1* & *Assumption2*) are shown in Figs. 3 and 4. The maximum number of clients that can be derived using the CTMC is 4, and also, the comparison shows that the behaviour of the 2PC and the 3PC *Assumption1* are extremely similar with a very slight difference as depicted in Figs. 3 and 4. However, the throughput of the 3PC *Assumption2* (i.e. *cpu* action) is much higher than the others. This is because we have used 3 *cpu* actions in *Assumption2*; replacing *prepareToCommit* action with a *cpu*. So, we have now an extra *cpu* action which is another part of processes which increases the overall throughput on the *cpu* action. Also, in these cases the load is very low, and the server has sufficient capacity. Nevertheless, in terms of the request completion (i.e. *getRequest* action) that is shown in Fig. 4, the performance of all protocols are consistent. For the reason that the number of clients is only varied from 1 to 4 (low load) due to the state space problem. Hence, in this case always there is enough resources and negligible queuing time.

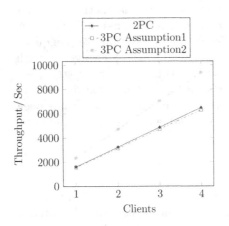
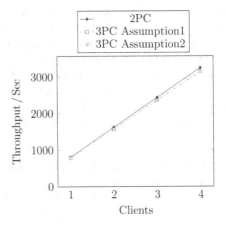

Fig. 3. Throughput of 2PC, 3PC (Assumption1 & Assumption2) using CTMC (*cpu* action).

Fig. 4. Throughput of 2PC, 3PC (Assumption1 & Assumption2) using CTMC (*getRequest* action).

The population shown in Figs. 5 and 6 show the system evolution and the maximum number of jobs at a specific time. Again, we can see the variation only in the Fig. 6, which is intuitive due to the number of *cpu* actions that has been used in *Assumption2*. The population is very small because there is a very

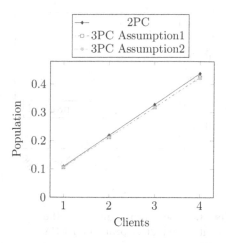

Fig. 5. Population of 2PC, 3PC (Assumption1 and Assumption2) using CTMC (*Request10*).

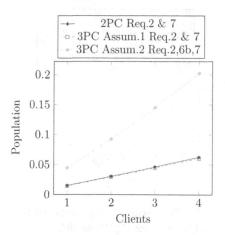

Fig. 6. Population of 2PC, 3PC (Assumption1 and Assumption2) using CTMC (*Request2,6b,7*).

small number of clients, which spend a very little time waiting for response from the server. Hence it is clear that the utilization is very low. As the CTMC only allows to analyse the system with a small scale (specifically 4 clients), therefore in the next section we will consider the fluid flow analysis (ODEs), which is used to overcome the state space problem of the CTMC and allows us to analyse very large scale systems.

5.2 Decentralized Two Phase Commit Protocol Using ODEs

The decentralized 2PC protocol PEPA model latency is shown in Fig. 7. The latency of the model is displayed as a flat line at the beginning, which means there is no waiting time and there are enough resources for manipulating the coming requests. Then it rises gradually as the load increases to reach its maximum latency (w = 3.55023) when the number of clients equals (200). The linear increase shown arises because the maximum capacity of the system is reached, and there are not enough resources for handling the incoming requests, thus the queuing time is increased for each additional client.

Figure 8 illustrates the throughput of the decentralized 2PC protocol. It is obvious that the saturation point of the 2PC protocol is given when the number of clients is 80, which gives the maximum throughput the model can have. After this point, the throughput is displayed as a flat line, which means that the system is saturated and throughput cannot increase further as all resources are taken. It is worth noting that the throughput of the *getRequest* action is used to represent the throughput of the protocol, whereas it has been used in the model to represent successful request completion.

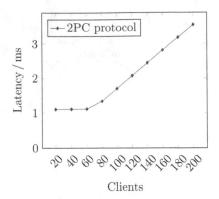

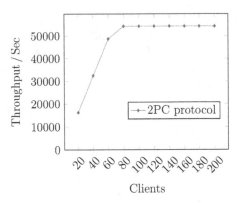

Fig. 7. The latency of the decentralized 2PC protocol PEPA model.

Fig. 8. The throughput of the decentralized 2PC protocol PEPA model.

5.3 Decentralized Three Phase Commit Protocol Using ODEs

In this section we are going to introduce the results of the decentralized 3PC protocol. The latency and the throughput of the decentralized 3PC protocol *Assumption1* that are shown in Figs. 9 and 10 are very similar to the results of the decentralized 2PC protocol which are illustrated in Figs. 7 and 8 respectively. The comparison shows that, there are very slight differences in both latency and throughput before reaching the saturation point (i.e. clients = 80), then the 3PC *Assumption1* and the 2PC are giving exactly the same performance with the use of 2 *cpu* actions. Unlike the *Assumption1* of the decentralized 3PC protocol, the use of *Assumption2* gives much higher latency and lower throughput as shown in Figs. 9 and 10 respectively. That is because in the *Assumption2* we have used

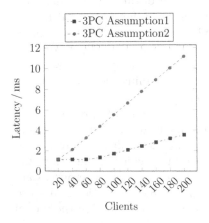

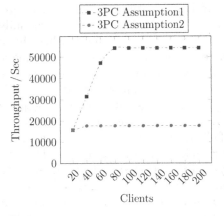

Fig. 9. The latency of the decentralized 3PC protocol PEPA model (*Assumption1* & *Assumption2*).

Fig. 10. The throughput of the decentralized 3PC protocol PEPA model (*Assumption1* & *Assumption2*).

3 *cpu* actions which has led to more contention for the resources, and also the rate of *cpu* action c in the *Assumption2* is slower than the same rate in the *Assumption1*. In particular, the rate of the *cpu* action in *Assumption2* is calculated as $c = 3/(1/r3 + 1/r6 + 1/r8)$, thus $c = 53.19384$ which is less than the c rate in the *Assumption1*, i.e. $c = 2/(1/r3 + 1/r6)$, thus $c = 108.548007$. So, the *Assumption2* shows more contention for the resources (i.e. *cpu* actions). Consequently, this limits the model throughput to the rate of *cpu* actions, reducing the performance significantly.

5.4   Comparing Decentralized 2PC and 3PC Protocols Using ODEs

Here, we are compare the performance of the decentralized protocols 2PC and 3PC (*Assumption1* and *Assumption2*), as shown in Figs. 11 and 12. Obviously, the use of 3PC *Assumption1* gives a performance that is extremely similar to the 2PC in both the latency and the throughput. That is because assuming that *prepareToCommit* is an independent action in *Assumption1* has a very slight effect on the model behaviour. Whereas, it is noticeable before reaching the saturation point in both cases (latency and throughput). So, although a new action (*prepareToCommit*) is added, the system is not overloaded. On the other hand, in the case of *Assumption2* the system is saturated rapidly (40 clients), which allows us to understand why the linear increase of the latency (i.e. $w = 11.144$) and the lower throughput where it is at most equals (17731.3).

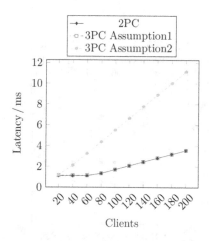

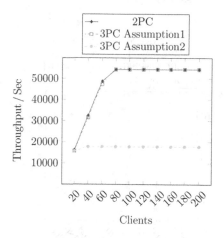

Fig. 11. The latency of the decentralized 2PC and 3PC protocols PEPA models (*Assumption1* & *Assumption2*).

Fig. 12. The throughput of the decentralized 2PC and 3PC protocol PEPA models (*Assumption1* & *Assumption2*).

5.5   Investigating the 3PC *Assumption2* bottleneck using ODEs

In this section, we have investigated further the problem of the 3PC *Assumption2* illustrated in Sect. 5.3. As the rate $r8$ was slow and it was limiting the performance of 3PC *Assumption2* this led to a bottleneck. Therefore, the performance implications of varying the rate $r8$ have been investigated. The main purpose of varying $r8$ is to consider the sensitivity of our results to deal with this rate. In the previous section the outcomes shown that there is one end where $r8$ is very slow (*Assumption2*) and the other end $r8$ is independent (*Assumption1*). So, by varying this rate we just consider the range of possibilities between those two extreme points.

As seen in Fig. 13, the latency of 3PC *Assumption2* is gradually decreased as the value of $r8$ becomes higher, hence, eventually the latency becomes very similar to the latency of *Assumption1*. Also, in Fig. 14 the model throughput is increased with the increase of the rate $r8$, and the saturation point has been shifted (from 40 to 60 clients). Obviously, the performance of 3PC *Assumption2* has improved significantly in comparison with the results that were shown in Sect. 5.3 by only increasing the rate $r8$ of the *prepareToCommit* action, which is used as a *cpu* action. So, $r8$ causes us to vary the behaviour and it is obvious that if we are wanting to speed up a real system (i.e. make the throughput better or latency less), it is that action which is the dominant feature. The *prepareToCommit* action has the biggest impact on the overall performance. Therefore, making that faster in some way will free the server up, whether that means giving more resources or faster servers.

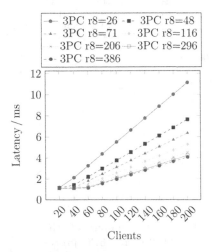

Fig. 13. The latency of the 3PC protocols (Assumption2) varying $r8$.

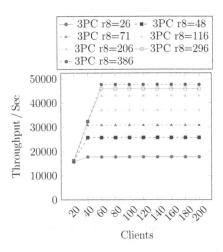

Fig. 14. The throughput of the 3PC protocols (Assumption2) varying $r8$.

6 Conclusion

The commit phase of the Two Phase Commit Protocol (2PC) and the Three Phase Commit Protocol (3PC) have been formally specified using the Markovian process algebra PEPA. Also, performance metrics of latency and throughput have been used to evaluate the protocol behaviours in the absence of failures. Initially, we have investigated scaling the demand (i.e. number of clients that can be derived) to show the limitations of CTMC analysis. By using a fluid (ODE) analysis we are able to consider much larger systems. The experimental ODEs results have shown that both the 2PC (blocking) protocol and the *Assumption1* of the 3PC (nonblocking) protocol, have consistent latency and throughput with an extremely slight variation before the saturation point is reached (Figs. 11 and 12). That is because the experiments are executed in the absence of failure, and the use of the *prepareToCommit* action in the *Assumption1* as independent action, has a very insignificant impact on the model behaviour. However, *Assumption2* of the 3PC has shown that the system is overloaded because of the new introduced *prepareToCommit* action as it is used as a centralized action *cpu* (see Sect. 5.3). Where, the system is saturated, hence, the latency increased linearly and the throughput is decreased in the comparison with the *Assumption1* (see Figs. 11 and 12).Therefore, we have varied the rate *r8*, which is the rate of the action *prepareToCommit*, that have been used as a *cpu* action (*Assumption2*). As shown in Figs. 13 and 14 increasing this rate has an important impact on the model behaviour, and the performance is increased significantly.

Although, the work presented here is at an abstract level and ignores many details, it does provide insight into the behaviour of the presented protocols and allows us to approximate the saturation point at various loads. The broader context in undertaking this work has been to better understand the performance of the ZooKeeper Atomic Broadcast protocol (ZAB). By modelling 2PC and 3PC we have seen how the different action sequences affect performance under different parameters. ZAB shares some properties with 3PC and so the comparison with 2PC also gives us some insight as to the expected ZAB performance which we will investigate further in our future work. The numerical figures in this paper are based on assumptions of rates and it would clearly be desirable to take measurements from actual implementations on different systems to achieve a more realistic parameterisation and to conduct validation.

Finally, the models specified here are very close to the class of PEPA model which has been defined as being amenable to mean value analysis (MVA) [18]. Unlike the ODE approximation used here, MVA can give exact results, so extending the class of PEPA model to include the form of model used here for 2PC and 3PC would clearly be of theoretical and practical interest.

Acknowledgements. The authors would like to acknowledge the contribution of Ibrahim El-Sanosi, a PhD student at Newcastle University, for providing measurements from another system which we have used to make assumptions for the rates in our models.

References

1. Al-Houmaily, Y.: Incompatibilty dimensions and integration of atomic commit protocols. Int. Arab J. Inf. Technol. **5**(4), 381–392 (2008)
2. Bernardo, M., Gorrieri, R.: A tutorial on empa: A theory of concurrent processes with nondeterminism, priorities, probabilities and time. Theoret. Comput. Sci. **202**, 1–54 (1998)
3. Bernstein, P.A., Hadzilacos, V., Goodman, N.: Concurrency control and recovery in database systems (1987)
4. Chrysanthis, P.K., Samaras, G., Al-Houmaily, Y.J.: Recovery and performance of atomic commit processing in distributed database systems. Recovery Mechanisms in Database Systems, pp. 370–416 (1998)
5. Donatelli, S.: Superposed generalized stochastic Petri Nets: definition and efficient solution. In: Valette, R. (ed.) ICATPN 1994. LNCS, vol. 815, pp. 258–277. Springer, Heidelberg (1994). doi:10.1007/3-540-58152-9_15
6. El Abbadi, A., Skeen, D., Cristian F.: An efficient, fault-tolerant protocol for replicated data management. In: Proceedings of the Fourth ACM SIGACT-SIGMOD Symposium on Principles of Database Systems. PODS 1985, pp. 215–229. ACM, New York (1985)
7. Gray, J.N.: Notes on data base operating systems. In: Bayer, R., Graham, R.M., Seegmüller, G. (eds.) Operating Systems: An Advanced Course. LNCS, vol. 60, pp. 393–481. Springer, Heidelberg (1978). doi:10.1007/3-540-08755-9_9
8. Haritsa, J.R., Ramamritham, K., Gupta, R.: The PROMPT real-time commit protocol. IEEE Trans. Parallel Distrib. Syst. **11**(2), 160–181 (2000)
9. Harrison, P.G., Strulo, B.: SPADES - a process algebra for discrete event simulation. J. Logic Comput. **10**(1), 3–42 (2000)
10. Hayden R.A., Bradley, J.T.: Fluid-flow solutions in PEPA to the state space explosion problem. In: 6th Workshop on Process Algebra and Stochastically Timed Activities (PASTA), p. 25 (2007)
11. Hillston, J.: Fluid flow approximation of PEPA models. In: Second International Conference on the Quantitative Evaluation of Systems (QEST 2005), pp. 33–42, September 2005
12. Hillston, J.: A Compositional Approach to Performance Modelling. Cambridge University Press, Cambridge (2008). New Ed edition (21 Aug. 2008)
13. Hillston, J., Gilmore, S.: Performance Evaluation Process Algebra (2011). http://www.dcs.ed.ac.uk/pepa/about/. Accessed 05 April 2016
14. Lampson, B.W.: Atomic transactions. In: Davies, D.W., Holler, E., Jensen, E.D., Kimbleton, S.R., Lampson, B.W., LeLann, G., Thurber, K.J., Watson, R.W. (eds.) Distributed Systems — Architecture and Implementation. LNCS, vol. 105, pp. 246–265. Springer, Heidelberg (1981). doi:10.1007/3-540-10571-9_11
15. Skeen, D.: Nonblocking Commit Protocols. In: Proceedings of the 1981 ACM SIGMOD International Conference on Management of Data. SIGMOD 1981, pp. 133–142. ACM, New York (1981)
16. Skeen D.: A Quorum-Based Commit Protocol. Technical report, Cornell University, Ithaca, New York (1982)
17. Skeen, D., Stonebraker, M.: A formal model of crash recovery in a distributed system. IEEE Trans. Softw. Eng. **SE–9**(3), 219–228 (1983)
18. Thomas, N., Zhao, Y.: Mean value analysis for a class of pepa models. Comput. J. **54**(5), 643–652 (2011)

19. Tribastone, M., Duguid, A., Gilmore, S.: The PEPA Eclipse Plugin. SIGMETRICS Perform. Eval. Rev. **36**(4), 28–33 (2009)
20. Weikum, G., Vossen, G.: Transactional Information Systems: Theory, Algorithms, and the Practice of Concurrency Control and Recovery. Elsevier, Burlington (2001)
21. Weihai, Y., Calton, P.: A dynamic two-phase commit protocol for adaptive composite services. Int. J. Web Serv. Res. **4**(1), 80–88 (2007)

Stochastic Models for Solar Power

Dimitra Politaki[1]([✉]) and Sara Alouf[2]([✉])

[1] Université Côte d'Azur, Inria, CNRS, I3S, Sophia Antipolis, France
`Dimitra.Politaki@inria.fr`
[2] Université Côte d'Azur, Inria, Sophia Antipolis, France
`Sara.Alouf@inria.fr`

Abstract. In this work we develop a stochastic model for the solar power at the surface of the earth. We combine a deterministic model of the clear sky irradiance with a stochastic model for the so-called clear sky index to obtain a stochastic model for the actual irradiance hitting the surface of the earth. Our clear sky index model is a 4-state semi-Markov process where state durations and clear sky index values in each state have phase-type distributions. We use per-minute solar irradiance data to tune the model, hence we are able to capture small time scales fluctuations. We compare our model with the on-off power source model developed by Miozzo et al. (2014) for the power generated by photovoltaic panels, and to a modified version that we propose. In our on-off model the output current is frequently resampled instead of being a constant during the duration of the "on" state. Computing the autocorrelation functions for all proposed models, we find that the irradiance model surpasses the on-off models and it is able to capture the multiscale correlations that are inherently present in the solar irradiance. The power spectrum density of generated trajectories matches closely that of measurements. We believe our irradiance model can be used not only in the mathematical analysis of energy harvesting systems but also in their simulation.

Keywords: Solar power · Semi-Markov process · Photovoltaic panel

1 Introduction

In the past decade, there has been an awareness rising concerning the energy cost and environmental footprint of the fastly growing Information and Communication Technology (ICT) sector. In [17] Van Heddeghem et al. assess how did the electricity consumption of the ICT sector evolve between 2007 and 2012. They report an increase in the relative share of ICT products and services (communication networks, personal computers and data centers, excluding TVs' set-top boxes and (smart)phones) in the total worldwide electricity consumption from about 3.9% in 2007 to 4.6% in 2012. Even though devices from new technologies are more energy efficient, this is outweighed by the fast growth in their numbers.

Among the most promising approaches recently pursued to reduce the environmental footprint of the ICT sector, we focus on the use of renewable energy

© Springer International Publishing AG 2017
P. Reinecke and A. Di Marco (Eds.): EPEW 2017, LNCS 10497, pp. 282–297, 2017.
DOI: 10.1007/978-3-319-66583-2_18

sources and in particular solar energy. As photovoltaic panels are being used worldwide to power multiple components of the ICT sector, there is an increasing effort in the literature to consider the solar energy production when modeling computer and communication systems. For illustration purposes, we mention two recent papers modeling ICT systems involving renewable energy sources.

In [4], Dimitriou, Alouf and Jean-Marie consider a base station that is powered by renewable energy sources and evaluate in particular the depletion probability. The base station is modeled as a multi-queue queueing system where energy queues model the batteries that store the harvested energy. The authors of [4] model the renewable energy production as a Poisson process whose rate is modulated by a Markov chain representing the random environment.

Neglia, Sereno and Bianchi consider in [13] the problem of geographical load balancing across data centers that have a dual power supply: grid and solar panels. They study the problem of scheduling jobs giving priority to data centers where renewable energy is available. The renewable energy source at each data center is modeled as an on-off process governed by a continuous time Markov chain. In the "on" state the data center can be fully powered by its renewable energy source; in the "off" state the data center is powered by the grid.

These examples among others illustrate the lack of a unified stochastic model for the solar energy to be used in the mathematical analysis of communication/computer systems. Our objective in this work is to develop such stochastic models for the solar power at the surface of the earth. Although there are a few models in the recent literature of the networking community [12], these rely on per-hour measurements. Therefore, such models do not capture the fluctuations in the solar irradiance at smaller time scales.

Our main contribution combines a deterministic model of the *clear sky* irradiance with a stochastic model of the so-called *clear sky index* to obtain a model of the actual irradiance hitting the surface of the earth. We will compare our model (after converting the actual irradiance to power generated by photovoltaic panels) to the night-day clustering model developed by Miozzo et al. in [12] for the generated power. We will propose for the latter a modified night-day clustering model. Our model for the harvested power is that of an on-off source in which the power generated in each state is frequently resampled from an appropriate distribution capturing the short-time scale fluctuations observed in practice.

To evaluate our models, we consider the autocorrelation functions and the periodograms of the generated trajectories. The autocorrelation function illustrates how well do our proposed models capture the multiscale correlations found in the data, whereas the spectral analysis allows to determine which characteristic time-scales are reproduced by the models.

In the following, we review the main notions used in the paper in Sect. 2 and discuss the related work in Sect. 3. Section 4 discusses the model of the clear sky index, and Sect. 5 is devoted to the model of the generated power. We assess our models in Sect. 6 before concluding the paper in Sect. 7.

2   Problem Definition

In this work, we are interested in two stochastic processes: the first one is the solar irradiance hitting a given surface on the earth, the second one is the power generated by a photovoltaic (PV) panel. We will define precisely each one of these processes in the following sections.

2.1   The Solar Irradiance

The amount of the solar energy that arrives per unit of time at a specific area of a surface is the solar irradiance and is expressed in W/m^2. In the following, the solar irradiance will refer to the *global* irradiance $I_G(t)$ accounting for all radiations arriving at a surface except for the ground-reflected ones. The reason for this is that we will rely on daily measurements of the global irradiance [1] to tune our models. No measurements of the ground-reflected radiations are available for download from [1]. However, their corresponding irradiance is usually insignificant compared to direct and diffuse irradiance.

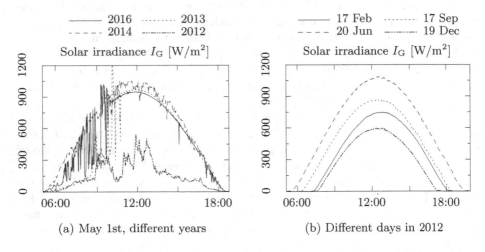

Fig. 1. Variations in the daily pattern of the solar irradiance are due to (a) the weather conditions and (b) the day of the year

The solar irradiance exhibits a night-day pattern that is affected by weather conditions which may induce burstiness at multiple time scales. Beside the obvious dependency on the geographic location, the solar irradiance depends also on the day of the year. Figure 1 illustrates these variations: per-minute measurements of the solar irradiance in Los Angeles [1] are depicted for the same day of different years (Fig. 1a) and for different days of the same year (Fig. 1b).

The solar irradiance $I_G(t)$ can be seen as the result of applying a multiplicative noise to the *clear sky* solar irradiance $I_{CS}(t)$. This multiplicative noise, denoted $\alpha(t)$ in this paper and called *clear sky index* in the literature, captures the perturbations seen in the solar irradiance with respect to the clear sky solar irradiance. We have $I_G(t) = \alpha(t)I_{CS}(t)$. Figure 2 illustrates $I_G(t)$, $I_{CS}(t)$ and $\alpha(t)$ for a sample day.

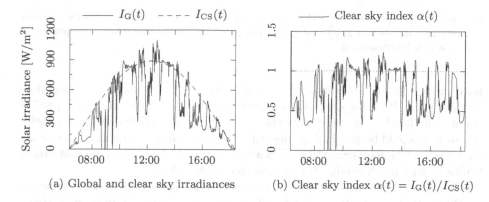

(a) Global and clear sky irradiances   (b) Clear sky index $\alpha(t) = I_G(t)/I_{CS}(t)$

Fig. 2. Illustrating the global irradiance $I_G(t)$, the clear sky irradiance model $I_{CS}(t)$ given in Eq. (2) and the resulting clear sky index $\alpha(t)$ on September 28th, 2010, in Phoenix, Arizona [16]

2.2   The Power Generated by a PV Panel

The solar irradiance can yield electricity through the use of a PV panel as shown in Fig. 3. The usable power is directly related to the solar irradiance arriving at the panel (that is I_G) as thoroughly explained in [12] and implemented in the tool SolarStat that is available online [6]. The general idea is the following:

1. The solar irradiance effectively used by the PV panel is the component of $I_G(t)$ that is perpendicular to its surface, that is $I_{eff}(t)$.
2. The PV panel translates the effective solar irradiance I_{eff} into electric power with current $i_{PV}(t)$ and voltage $v_{PV}(t)$.
3. A Schottky diode reduces slightly the voltage but preserves the current.
4. A power processor extracts the maximum power from the PV panel and the output power has current $i_{out}(t)$ and voltage v_{out}.

The fluctuations seen in the solar irradiance $I_G(t)$ are still present in the output current $i_{out}(t)$. There may be additional fluctuations due to the local temperature and humidity that affect the functioning of the PV cells.

3   Related Work

Studies on the solar irradiance are abundant in the literature. Given the paramount role of the solar energy in many biological ecosystems, it is crucial to

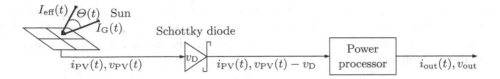

Fig. 3. Using a fraction I_{eff} of the solar irradiance I_G, the PV cells generate a power (current i_{PV} and voltage v_{PV}) that goes through a Schottky diode and a power processor before it can be consumed

have models for the solar irradiance as measurements are not always available. For instance, Piedallu and Gégout develop in [14] a model that can predict the accumulated solar energy anywhere, providing annual figures for an entire country, as would be required for predictive vegetation modeling at a large scale. However such biology-oriented models are not fit for ICT applications that evolve typically on a much smaller time scales than vegetation.

Targeting the design of a solar system, there is a large body of work focusing on the clear sky irradiance. To cite a few references, Dave, Halpern and Myers overview in [3] several clear sky irradiance models and compare the accumulated daily and annual energy. They consider a tilted surface and account for both sky radiations and ground-reflected radiations. They find in particular that the effective irradiance at a surface is proportional to the cosine of the angle between the sunlight direction and the normal to the surface. Bird and Hulstrom compare in [2] five models for the maximum clear sky solar irradiance and propose a sixth model based on algebraic expressions. All these models require many meteorological input parameters (e.g., the surface pressure, the total ozone, the precipitable water vapor).

Another important component when modeling the solar irradiance is the clear sky index. Jurado, Caridad and Ruiz characterize the clear sky index using 5-minute measurements of the solar irradiance [10]. They partition the data according to the solar angle, considering two one-hour intervals at a time (both intervals corresponding to the same range of solar angle). They find that the density of the clear sky index in each partition is bimodal and can be modeled as a mixture of Gaussian distributions. The parameters of the distributions and the mixing factor are obtained from measurements by least squares approximation. The authors observe that the standard deviations of the Gaussian distributions depend on the solar angle. Also the bimodal behavior observed over 5-minute intervals is no longer observed when the interval in the data is larger. This is an important outcome that indicates that a model tuned with data having a given frequency of measurements can not match data having a different measurements rate. This observation supports our intuition that if one wants to use a model of solar power at a given time scale, then the model must be tuned with data at the same time scale. The authors of [10] are not clear on how do they compute the clear sky index from the measurements of the solar irradiance. Surprisingly, the computed clear sky index is always below 1 suggesting that they consider a very large maximum clear sky irradiance.

Gu et al. consider in [7] a related metric which is the relative change of solar irradiance (this would be $100(\alpha - 1)$) under the impact of clouds. They analyze per-minute measurements of solar irradiance collected in Brazil over a period of two months during the wet season. They observe that broken cloud fields create a bimodal distribution for the relative change: shaded areas receive attenuated solar irradiance while sunlit areas may receive higher irradiance than under a clear sky. This effect is caused by radiations scattering and reflections from neighboring clouds. Conducting a spectral analysis on the time series of measured surface irradiance, they observe that clouds are responsible for two different regimes according to their types and density causing either large or small scale fluctuations. This study highlights the effect of clouds and have certainly impacted the development of subsequent models for the solar irradiance.

Miozzo et al. focus on the solar power generated by small embedded photovoltaic panels such as those used in sensor networks. They develop in [12] two stochastic models in which the dynamics of the power source is described by a semi-Markov process with $N \geq 2$ states. The first model is an on-off power source and the authors tune the sojourn time and power in each state by using a night-day clustering on hourly measurements of the solar irradiance. In the second model, the power source goes through a number of N states in a round-robin way and all sojourn times are equal and constant. A time slot based clustering enables the authors to estimate the power distribution in each state.

Ghiassi-Farrokhfal et al. consider also the solar power generated by photovoltaic panels but in the context of dimensioning an energy storage system. To near-optimally size a storage system, they develop in [5] a new *envelope* model for the generated power. In the general envelope model, the solar power is characterized by a statistical sample path lower envelope such that the probability of having the maximum of the distance envelope-solar energy exceed a given value is upper bound by a characteristic bounding function evaluated at the given value. Inspired by the findings of [7], the authors of [5] adapt the general envelope model to enable a separate characterization of the three underlying processes of solar power (diurnal, long-term, and short-term variations).

4   Modeling the Solar Irradiance I_G

In this section, we focus on the solar irradiance $I_\mathrm{G}(t)$. Our aim is to define a model able to capture the small time-scale fluctuations inherently present in the global irradiance. To that end, we model separately the clear sky irradiance $I_\mathrm{CS}(t)$ and the clear sky index $\alpha(t)$. By definition, we have

$$I_\mathrm{G}(t) = \alpha(t)I_\mathrm{CS}(t). \tag{1}$$

We discuss $I_\mathrm{CS}(t)$ in Sect. 4.1 and model $\alpha(t)$ in Sect. 4.2.

4.1 Modeling the Clear Sky Irradiance $I_{CS}(t)$

The solar irradiance arriving at a surface during a clear sky day without any perturbations due to a change in the meteorological conditions exhibits a predictable pattern as shown in Fig. 1b. The models discussed in [3] for the hourly clear sky irradiance and in [2] for the maximum clear sky irradiance are not easily applicable given the unavailability of many input parameters. Instead, we use the so-called "simple sky model" [9] which defines a simple sinusoidal form for each day, taking into account the times of sunrise and sunset and the maximum clear sky irradiance. The clear sky irradiance $I_{CS}(t)$ is given by the following equation:

$$I_{CS}(t) = \text{MaxClearSky} \cdot \sin\left(\frac{t - \text{sunrise}}{\text{sunset} - \text{sunrise}}\pi\right). \tag{2}$$

The values of "sunrise", "sunset" and "MaxClearSky" are astronomical data that can be easily obtained in practice for any date and many selected locations from the website [15] (the maximum clear sky irradiance is called there "maximal solar flux"). An illustration of Eq. (2) is in Fig. 2a.

4.2 Modeling the Clear Sky Index $\alpha(t)$

The clear sky index $\alpha(t)$ captures the fluctuations over time of the global irradiance with respect to clear sky conditions, as illustrated in Fig. 2b for a sample day and a sample location. Consequently, one thinks of defining a state for each macro weather condition. Based on our review of the literature, we define four states for $\alpha(t)$ that correspond to: heavy clouds between the sun and the surface (very low values of $\alpha(t)$), medium to light clouds between the sun and the surface (values of $\alpha(t)$ around 0.6), clear sky (values of $\alpha(t)$ around 1), and high reflection and diffusion in the atmosphere (values of $\alpha(t)$ larger than 1). We assume all transitions between different states to be possible.

We propose to capture the dynamics of $\alpha(t)$ by a discrete-time semi-Markov process.[1] Our model works as follows. When the process $\alpha(t)$ enters a state i, it will remain there for a duration τ_i governed by a probability density function f_i. While in state i, the clear sky index $\alpha(t)$ behaves like $\alpha_i(t)$, a stochastic process with probability density function g_i. When the sojourn time τ_i expires, the process *changes* its state. The distributions f_i and g_i, for $i \in \{1, 2, 3, 4\}$ will be fitted to empirical distributions of the sojourn times and values of $\alpha(t)$.

To tune our model of $\alpha(t)$ we use per-minute measurements of the solar irradiance $I_G(t)$. The data is for the region of Los Angeles from April 2010 until March 2015 [1]. We compute $\alpha(t) = I_G(t)/I_{CS}(t)$ using the data and Eq. (2) for each minute during the five years.[2] For illustration purposes, we compute the

---

[1] Using a discrete-time Markov process does not yield satisfactory results as correlations are not described well.

[2] We observe that we may well have in the real measurements $I_G(t) > 0$ around sunset and sunrise due to diffusion. As $I_{CS}(t) = 0$ at sunrise (and before) and sunset (and after), this implies that infinite values for the ratio $I_G(t)/I_{CS}(t)$ can occur. To discard such values when computing $\alpha(t)$, we enforce the (arbitrary) bound $\alpha(t) < 3$.

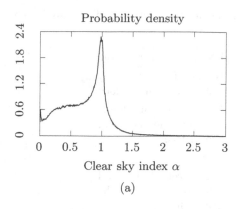

Probability density

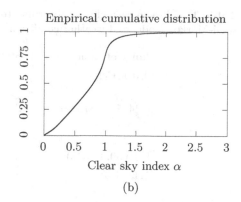

Empirical cumulative distribution

(a)   (b)

Fig. 4. Density and cumulative distribution curves of the clear sky index $\alpha(t)$ computed using Eq. (2) and per-minute solar irradiance data [1]

density and the cumulative distribution of the clear sky index and depict them in Fig. 4.

Remark 1. The density of the clear sky index depicted in Fig. 4a is not bimodal as found in [10]. The measurements used in [10] were made every 5 min and the densities were computed over two intervals of 1 h each corresponding to the same range of the solar angle. Instead, the density shown in Fig. 4a is for all 1-minute measurements over a period of 5 years.

Once that we have computed the values of $\alpha(t)$, we first aim to validate the number of states of our semi-Markov model. We apply the k-means clustering algorithm [11] and use the Davies-Bouldin index to define the optimal number of clusters. The Davies-Bouldin index is based on a ratio of within-cluster and between-cluster distances. The smaller its value the better the clustering.

We tested nine different clustering (for $k \in \{2, \ldots, 10\}$) and computed the Davies-Bouldin index for each clustering obtained. The values of the index were between 0.5017 and 0.5290. The smallest value was obtained for $k = 4$ implying that ideally the values of $\alpha(t)$ should be classified into four clusters. This analysis supports our choice of having four states in the model for the clear sky index and each state is mapped to one of the four clusters obtained. The details on the four clusters/states obtained when applying the k-means clustering algorithm are given in Table 1.

Now that we have clearly identified the four states of our semi-Markov model, our next step is to identify the transition probabilities among the states. We estimate them using the computed values of $\alpha(t)$ and the identified clusters. We first map each computed value of $\alpha(t)$ to its corresponding state, then we count the number of transitions between any pair of states. The transition probability from state i to state j is estimated as the ratio of the number of transitions from state i to state j to the total number of transitions out of state i. We find the following transition probability matrix for the four-state semi-Markov model:

Table 1. Values in each cluster according to k-means, their corresponding state in the semi-Markov model and weather condition

Range of values of $\alpha(t)$	State	Physical interpretation
$[0, 0.44152)$	1	Heavy clouds between the sun and the surface
$[0.44152, 0.81639)$	2	Medium to light clouds between the sun and the surface
$[0.81639, 1.4343)$	3	Clear sky
$[1.4343, 3)$	4	High reflection and diffusion in the atmosphere

$$P = \begin{bmatrix} 0 & 0.8361 & 0.0549 & 0.1090 \\ 0.3645 & 0 & 0.6296 & 0.0059 \\ 0.0274 & 0.9019 & 0 & 0.0707 \\ 0.0484 & 0.0536 & 0.8980 & 0 \end{bmatrix}. \tag{3}$$

The last step is to characterize the densities f_i and g_i for $i = 1, \ldots, 4$. We carry out a statistical analysis on the computed values of $\alpha(t)$ in order to determine the distributions of the sojourn times $\{\tau_i\}_{i=1..4}$ and the values $\{\alpha_i(t)\}_{i=1..4}$. Observe that the sojourn time τ_i in a given state i corresponds to the number of consecutive values of $\alpha(t)$ inside the corresponding cluster. Recall that $\alpha(t)$ is a discrete-time process and as the measurements used for tuning the model are minute-based, therefore the time unit in our model is the minute.

We opt to fit the data with phase-type (PH) distributions given their attractive analytical tractability and their high flexibility in fitting data. We use the PhFit tool [8] to find the phase-type distribution that best fits each one of the empirical distributions. In the PhFit tool, we choose the relative entropy as distance measure according to which the fitting is performed.

We repeatedly fit the data related to each variable changing the number of phases. We use probability plots to assess the quality of the fit and select the number of phases that yields the best fit. We report the chosen number of phases for each fitted variable in Table 2.

The probability plots of the selected phase-type distributions are displayed in Figs. 5 and 6. Each graph in Fig. 5 depicts on the y-axis the probabilities of the fitted distribution against the probabilities of the sojourn times in a given state on the x-axis. We observe that the phase-type distribution fits reasonably well the sojourn times for all states.

Regarding the values of $\alpha(t)$ in each state, we can see in Fig. 6 that the selected phase-type distributions fit very well the values of $\alpha(t)$. We observe that the quality of fit for $\alpha_1(t)$ and $\alpha_2(t)$ is obtained at the cost of having a significantly larger number of phases (that is 20; see Table 2) with respect to the other variables.

Table 2. Number of phases of the phase-type distribution fitting the (shifted) sojourn times and values in each state

Variable	Number of samples used in the fitting	Number of phases of the phase-type distribution fitting the variable
$\tau_1 - 1$	19678	5
$\tau_2 - 1$	8456	6
$\tau_3 - 1$	2094	6
$\tau_4 - 1$	15400	6
$\alpha_1(t)$	298141	20
$\alpha_2(t) - 0.44152$	345973	20
$\alpha_3(t) - 0.81639$	563411	6
$\alpha_4(t) - 1.4343$	34432	3

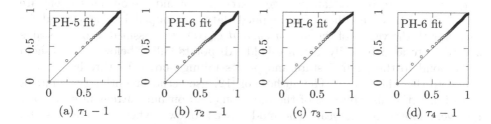

(a) $\tau_1 - 1$ (b) $\tau_2 - 1$ (c) $\tau_3 - 1$ (d) $\tau_4 - 1$

Fig. 5. Probability plots of the phase-type fitting for sojourn times in each state

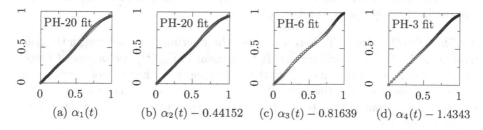

(a) $\alpha_1(t)$ (b) $\alpha_2(t) - 0.44152$ (c) $\alpha_3(t) - 0.81639$ (d) $\alpha_4(t) - 1.4343$

Fig. 6. Probability plots of the phase-type fitting for $\alpha(t)$ values in each state

5   Modeling the Harvested Power

To account for the power generated by PV panels when evaluating solar-powered systems, one has mainly two options. The first option is to use a model for the solar irradiance such as the one developed in Sect. 4 and then infer the power generated by the PV cells (or equivalently $i_{out}(t)$; see Fig. 3). This second step may be a simple linear model (i.e. the power generated by a panel of unit size is the solar irradiance effectively received multiplied by the efficiency of the panel) or a more detailed model such as the one implemented in the SolarStat tool [6]. The second option is to use directly a model for the power generated by a given PV panel (i.e. a model for i_{out}). Miozzo et al. have developed two such models in [12]. In this section, we propose a modification to their on-off model. We will compare our modified model to theirs in Sect. 6 and also to the model of Sect. 4 after we translate the solar irradiance to generated power using the SolarStat tool. We present briefly the on-off model in [12] before explaining our modification.

The dynamics of the harvested current $i_{out}(t)$ are captured by a two-state semi-Markov process. The distributions of the sojourn times and of $i_{out}(t)$ in each state are statistically defined using hourly measurements of the solar irradiance. In practice, Miozzo et al. apply the procedure summarized in Sect. 2.2 to map the solar irradiance data into the power generated by a PV panel of given size (number of solar cells connected in series/parallel) and characteristics (open circuit voltage, short circuit current, and reference temperature). Assuming the output voltage to be constant, the generated power and the output current are proportional to each other. The mapped data is grouped by month and for each month the values of the *output current* $i_{out}(t)$ are classified into two states according to an arbitrarily low threshold. All points falling below the threshold correspond to the "night" state and points falling above the threshold correspond to the "day" state. The authors of [12] use kernel-smoothing techniques to estimate the distributions of the durations and output current in each state for every month of the year. The model is as follows: when entering a state, a current and a duration are drawn from the corresponding distributions, then the source outputs the drawn current *constantly* for the drawn duration. At the end of the drawn duration, the source switches its state. In practice, the output current in the night state is set to 0.

Modified On-Off Model. To better capture the fluctuations observed in the solar irradiance $I_G(t)$ (which will inevitably be present in $i_{out}(t)$), we propose to modify the above-mentioned model in the following way: instead of keeping the current *constant* during the time the process remains in the "day" state, we frequently *resample* (every ten minutes) from the current distribution until a transition occurs.

6   Results

In this section we will evaluate the models presented in Sects. 4 and 5. We consider first the autocorrelation function (ACF) as a metric to test how well do generated synthetic data match the empirical data according to second order statistics. The empirical data is a 5-year long set of output current values sampled every minute. The current values are those matched by SolarStat (for a Panasonic solar panel of unit size) for the solar irradiance measurements [1]. We generate three synthetic data that are:

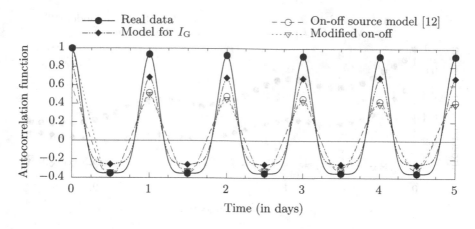

Fig. 7. ACF of the current harvested using Panasonic solar panels

1. a 5-year long set of output current values sampled every minute using the model of the solar irradiance presented in Sect. 4 and SolarStat to translate the irradiance into output current;
2. a 5-year long set of output current values sampled every 10 min using the on-off model in [12];
3. a 5-year long set of output current values sampled every 10 min using our modified on-off model (Sect. 5).

The autocorrelation functions of these four data sets are depicted in Fig. 7. Our solar irradiance model performs fairly well, capturing most of the correlations present in the empirical data. As already found by the authors of [12], the ACF of the on-off source model poorly resembles that of the empirical data. The ACF of our modified on-off model performs seemingly equally badly.

Strong correlations in the solar power exists over yearly lags due to the earth's annual circumnavigation of the sun. To assess how well does our solar irradiance model capture the correlations over very long periods, we sample the ACFs every 30 days and display the values in Fig. 8. We can make three observations: first, the ACF of the real data confirms the expected strong annual correlation; second, our solar irradiance model exhibits correlations that mimic those in the real data, even though to a lesser extent; third, the on-off models fail to track the ACF of the real data.

To complete this comparative analysis of the models, we compute the root mean square error (RMSE) between the ACF of the empirical data set and that of each of the synthetic data set. The RMSE metric is as follows: $\text{RMSE} = \sqrt{\frac{1}{n} \sum_i^n (y_i - \hat{y}_i)^2}$, where y_i and $\hat{y}_i$ are the ith samples of the empirical and synthetic data respectively, and n is the number of samples. The results reported in Table 3 confirm the superiority of the solar irradiance model over the on-off models.

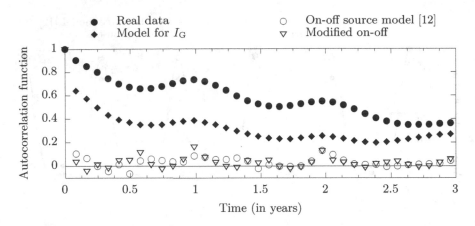

Fig. 8. Samples of the autocorrelation function of the output current (one sample per 30 days)

Table 3. Root mean square error (RMSE) between real and synthetic data

Model of solar irradiance I_G	Model of harvested power	
	On-off source model [12]	Modified on-off
0.1274	0.3231	0.2839

We can conclude from the comparison of the ACFs that our model of the solar irradiance outperforms the on-off models of the output current and captures well the multiscale correlations found in the real data.

We consider next the periodograms of the empirical data set and the synthetic data set generated by the solar irradiance model (see Sect. 4). The spectral analysis allows to determine which characteristic time-scales are reproduced by the model.

We compute the periodogram using the function with the same name in the Signal Processing Toolbox of Matlab. We adjust appropriately the x-axis in order to have frequencies (f, in Hertz) instead of the angular frequency ω. The power spectrum densities (PSD) are depicted in Figs. 9 and 10.

Observe that Gu et al. have analyzed in [7] the power spectrum of a 2-month set of 1-minute measurements of solar irradiance. The PSD had two clear peaks corresponding to 24 and 12 h but other than those the absence of other characteristic time-scale was striking. This is not the case of the PSD of the real data set displayed in Fig. 9. We can observe a series of peaks at larger frequencies that are the harmonics of $1.157\overline{407}\ 10^{-5}$ Hz (which corresponds to 24 h). The same observation applies to the PSD of the synthetic data set displayed in Fig. 10. The peak at the fundamental frequency corresponding to 1 day is clearly visible as well as those of its harmonics frequencies.

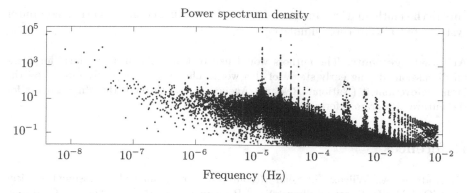

Fig. 9. Power spectrum of the 1-minute values of output current mapped from the real measurements [1] by the SolarStat tool

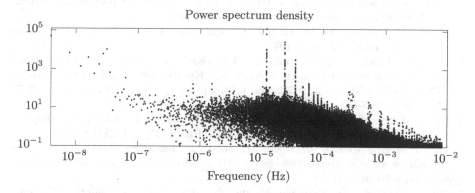

Fig. 10. Power spectrum of the 1-minute values of output current obtained after generating a 5-year trajectory from the model of Sect. 4 and translating it to current with the SolarStat tool

We conclude this section by stating that our solar irradiance model is able to generate synthetic data that exhibits all of the frequency peaks of real data, capturing its characteristic time-scales.

7 Conclusions

We have developed in this work a stochastic model for the solar irradiance. The model combines a deterministic model of the clear sky irradiance with a stochastic model for the so-called clear sky index to obtain a stochastic model for the actual irradiance hitting the surface of the earth. As per-minute solar irradiance data is used to tune our model, we are able to capture small time scales fluctuations as would be needed by ICT applications. Computing autocorrelation functions and periodograms of empirical and synthetic traces we found that our solar irradiance model performs very well. We believe our model can be used not only in the mathematical analysis of energy harvesting communication/computer systems but also in their simulation.

Acknowledgements. The authors would like to thank Alain Jean-Marie for fruitful discussions during early stages of this work. This work was partly funded by the French Government (National Research Agency, ANR) through the "Investments for the Future" Program reference #ANR-11-LABX-0031-01.

References

1. Andreas, A., Wilcox, S.: Solar Resource & Meteorological Assessment Project (SOLRMAP): Rotating Shadowband Radiometer (RSR); Los Angeles, California (Data). Report DA-5500-56502, NREL (2012). http://dx.doi.org/10.5439/1052230
2. Bird, R.E., Hulstrom, R.L.: A simplified clear sky model for direct and diffuse insolation on horizontal surfaces. Technical report Technical report SERI/TR-642-761, Solar Energy Research Institute, February 1981
3. Dave, J.V., Halpern, P., Myers, H.J.: Computation of incident solar energy. IBM J. Res. Dev. **19**(6), 539–549 (1975)
4. Dimitriou, I., Alouf, S., Jean-Marie, A.: A Markovian queueing system for modeling a smart green base station. In: Beltrán, M., Knottenbelt, W., Bradley, J. (eds.) EPEW 2015. LNCS, vol. 9272, pp. 3–18. Springer, Cham (2015). doi:10.1007/978-3-319-23267-6_1
5. Ghiassi-Farrokhfal, Y., Keshav, S., Rosenberg, C., Ciucu, F.: Solar power shaping: an analytical approach. IEEE Trans. Sustain. Energy **6**(1), 162–170 (2015)
6. Gianfreda, M., Miozzo, M., Rossi, M.: SolarStat: modeling photovoltaic sources through stochastic Markov processes. http://www.dei.unipd.it/~rossi/Software/Sensors/SolarStat.zip
7. Gu, L., Fuentes, J.D., Garstang, M., da Silva, J.T., Heitz, R., Sigler, J., Shugart, H.H.: Cloud modulation of surface solar irradiance at a pasture site in Southern Brazil. Agric. Forest Meteorol. **106**(2), 117–129 (2001)
8. Horváth, A., Telek, M.: PhFit: a general phase-type fitting tool. In: Field, T., Harrison, P.G., Bradley, J., Harder, U. (eds.) TOOLS 2002. LNCS, vol. 2324, pp. 82–91. Springer, Heidelberg (2002). doi:10.1007/3-540-46029-2_5
9. Iqbal, M.: An Introduction to Solar Radiation. Academic Press, New York (1983)
10. Jurado, M., Caridad, J., Ruiz, V.: Statistical distribution of the clearness index with radiation data integrated over five minute intervals. Sol. Energy **55**(6), 469–473 (1995)
11. Kanungo, T., Mount, D.M., Netanyahu, N.S., Piatko, C.D., Silverman, R., Wu, A.Y.: An efficient k-means clustering algorithm: analysis and implementation. IEEE Trans. Pattern Anal. Mach. Intell. **24**(7), 881–892 (2002)
12. Miozzo, M., Zordan, D., Dini, P., Rossi, M.: SolarStat: modeling photovoltaic sources through stochastic Markov processes. In: Proceeding of 2014 IEEE International Energy Conference, Dubrovnik, Croatia, pp. 688–695, May 2014
13. Neglia, G., Sereno, M., Bianchi, G.: Geographical load balancing across green datacenters. ACM SIGMETRICS Perform. Eval. Rev. **44**(2), 64–69 (2016)
14. Piedallu, C., Gégout, J.C.: Multiscale computation of solar radiation for predictive vegetation modelling. Ann. Forest Sci. **64**(8), 899–909 (2007)
15. ptaff.ca: Sunrise, sunset daylight in a graph. https://ptaff.ca/soleil/

16. Solar Resource & Meteorological Assessment Project (SOLRMAP), Southwest Solar Research Park (Formerly SolarCAT). http://midcdmz.nrel.gov/ssrp/
17. Van Heddeghem, W., Lambert, S., Lannoo, B., Colle, D., Pickavet, M., Demeester, P.: Trends in worldwide ICT electricity consumption from 2007 to 2012. Comput. Commun. **50**, 64–76 (2014)

Author Index

Printed in the United States
By Bookmasters